Agriculture and Climate Change

Chief Editors

Dr. R.K. Naresh
Sardar Vallabhbhai Patel University of Agriculture & Technology, Meerut, Uttar Pradesh, India

Dr. R.K. Gupta
Borlaug Institute for South Asia (Bisa), New Delhi, India

Dr. S.S. Dhaliwal
Punjab Agricultural University, Ludhiana, Punjab, India

Dr. Vivek
Sardar Vallabhbhai Patel University of Agriculture & Technology, Meerut, Uttar Pradesh, India

Editors

Dr. R.K. Singh
Banaras Hindu University, Varansi, Uttar Pradesh, India

Dr. S.S. Tomar
Rajmata Vijayaraje Scindia Krishi Vishwavidyalaya, Gwalior Zara-a, b. Road Morena, Madhya Pradesh, India

Dr. Vineet Kumar
Indian Institute of Farming System Research, Modipuram-Meerut, Uttar Pradesh, India

Dr. Saurabh Tyagi
Sardar Vallabhbhai Patel University of Agriculture & Technology, Meerut, Uttar Pradesh, India

Co-Authors

Dr. Sandeep Chaudhary
Sardar Vallabhbhai Patel University of Agriculture & Technology, Meerut, Uttar Pradesh, India

Miss. Lali Jat
Sardar Vallabhbhai Patel University of Agriculture & Technology, Meerut, Uttar Pradesh, India

Dr. Satyaveer Singh
Bihar Agricultural University, Sabour, Bihar, India

Sh. K.S. Krishna Prasad
Sardar Vallabhbhai Patel University of Agriculture & Technology, Meerut, Uttar Pradesh, India

AkiNik Publications
New Delhi

Published By: *AkiNik Publications*

AkiNik Publications
169, C-11, Sector - 3,
Rohini, Delhi-110085, India
Toll Free (India) – 18001234070

Chief Editors: *Dr. R.K. Naresh, Dr. R.K. Gupta, Dr. S.S. Dhaliwal and Dr. Vivek*

The author/publisher has attempted to trace and acknowledge the materials reproduced in this publication and apologize if permission and acknowledgements to publish in this form have not been given. If any material has not been acknowledged please write and let us know so that we may rectify it.

1st Edition: *2018*

Pages: *407*

ISBN: *978-93-5335-110-6*

Price: *₹ 795/-*

Contents

Chapter - 1
Climate Change and Agriculture: Impact and Mitigation

Authors

Dr. R.K.Gupta
Borlaug Institute for South Asia (BISA), New Delhi, India

Dr. R.K. Naresh
Sardar Vallabhbhai Patel University of Agriculture & Technology, Meerut, Uttar Pradesh, India

Chapter - 1

Climate Change and Agriculture: Impact and Mitigation

Dr. R.K. Gupta and Dr. R.K. Naresh

Climate change is no longer a distant scientific prognosis; rather, it is becoming a reality which has already started to make its impact felt on every sphere of life on this planet. The climate has been changing naturally at its own pace since the beginning of time, but recently it has gained momentum due to anthropogenic interventions. The most imminent climatic change factor in recent times is the increase in the atmospheric temperatures due to increased levels of greenhouse gases (GHGs) such as carbon dioxide (CO_2), methane (CH_4), nitrous oxide (N_2O), ozone (O_3) and chlorofluoro carbons (CFCs). The CO_2, CH_4 and N_2O concentrations in atmosphere were 280 ppm, 715 ppb and 270 ppb in 1750AD (pre-industrial era) which have now risen up to 381 ppm, 1774 ppb and 319 ppb, respectively (IPCC, 2007a). These increases in GHGs have resulted in warming of the climate system by 0.74°C between 1906 and 2005.Eleven of the twelve years spanning from 1995 to 2006 rank among the 12 warmest years in the instrumental record of global surface temperature (since 1850). The rate of warming has been much higher in recent decades. This has, in turn, resulted in increased average temperature of the global ocean, sea level rise, decline in glaciers and snow cover. There is also a global trend for increased frequency of droughts, as well as heavy precipitation events over most land areas. Cold days, cold nights and frost have become less frequent, while hot days, hot nights and heat waves have become more frequent.

Currently, atmospheric concentrations of CO_2, CH_4 and N_2O, which account for about 60%, 15% and 5% of global warming, are rising at the rate of around 0.5% (2 ppm per year), 1% and 0.22% per year respectively. The Intergovernmental Panel on Climate Change (IPCC) Scenario indicates that CO_2 concentrations will be at least 550 ppm by 2050, between 605 and 755ppm by 2070 and likely to be doubled by the end of the century. IPCC has further projected that temperature increase by the end of this century is likely to be in the range 2 to 4.5°C with the best estimate of about 3°C, and is very unlikely to be less than 1.5°C (Fig 1.1a & b). It is also likely that future tropical cyclones will become more intense, with larger peak wind speeds and heavier precipitation. Himalayan glaciers and snow cover are projected to contract. The projected sea level rise by the end of this century is likely to be 0.18 to 0.59 m. It is very likely that hot extremes, heat waves, and heavy precipitation events will continue to become more frequent. Increases in the amount of precipitation are very likely in high latitudes, while decreases are likely in most subtropical land regions, continuing observed patterns in recent trends.

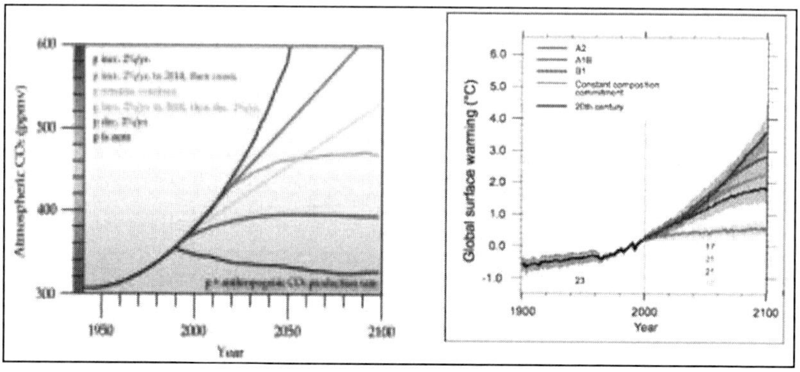

Fig 1.1: Projected rise in atmospheric CO_2 (a) and surface temperature (b) by different models (IPCC, 2007a)

Such global climatic changes will affect agriculture through direct and indirect effects on crops, soils, livestock and pests. Increase in atmospheric CO_2 has a fertilization effect on crops with C_3 photosynthetic pathway and thus promotes their growth and productivity. Increase in temperature, depending upon the current ambient temperature, on the other hand, can reduce crop duration, increase crop respiration rates, alter photosynthetic partitioning to economic products, affect the survival and distributions of pest populations thus developing new equilibrium between crops and pests, hasten nutrient mineralization in soils, decrease fertilizer use efficiencies, and increase evapotranspiration. Indirectly, there may be considerable effects on land use due to snowmelt, availability of irrigation, frequency and intensity of inter- and intra-seasonal droughts and floods, and availability of energy. All of these can have a tremendous impact on agricultural production and hence food security of any region.

According to the Intergovernmental Panel on Climate Change (IPCC), in order to avoid any major catastrophic impact of climate change on agriculture, atmospheric concentrations of CO_2 must be stabilized at around 450 ppm and rise in global temperature must be contained within 2°C over that of pre-industrial era. Since the CO_2 concentration has by now reached

390 ppm and the global temperature has already recorded a rise of around 1^0C over pre-industrial era, the world is left with the flexibility of emitting 60 ppm more CO_2 and pushing up 1^oC of temperature within which the high carbon lifestyle of developed countries like USA, Canada and Australia, the ambitious developmental needs of fast growing economy like India, China and Brazil, and the future growth prospects of underdeveloped countries in sub-Saharan Africa have to be accommodated which seems very unlikely. On top of these all, recurrent failures of world community to reach any legally binding agreement for curbing carbon emission lend further credibility to the projections of IPCC and ensure that these disastrous climatic projections do not stand wrong in future, signalling the world to be ready to face the catastrophic impacts of climate change on agriculture and hence food security.

Linkage between CO_2 and Agriculture: Past, Present and Future

Archaeological evidence indicates that a climatic factor not often considered, the CO_2 content, has throughout history had a great influence on agriculture; so much so that, it has been credited as the most important factor for the origin of agriculture at the beginning of Holocene. Agriculture originated independently in many distinct regions of the world at approximately the same time in human history. Humans in the Middle East domesticated lentils, barley, chick peas and wheat around 10,000 years ago. Rice, millets and the Brassica spp. were domesticated in the Far East 9,000 years ago. Beans and chili peppers were grown in Meso America 8,000 years ago. This synchrony in agricultural origins across the globe indicates that a common factor may have controlled the timing of the transition from foraging to food-producing societies.

According to Sage (1995), a global factor common to these widely diverse people was rise in atmospheric CO_2 from below 200 to near 270 ppm that occurred between 15 and 12 thousand years ago. Atmospheric CO_2 directly affects photosynthesis and plant productivity, with the largest proportional responses occurring below the current level of 390 ppm. In the late Pleistocene, CO_2 levels near 200 ppm were too low to support the level of productivity required for successful establishment of agriculture. Recent studies demonstrate that atmospheric CO_2 increase from 200 to 270 ppm stimulates photosynthesis and biomass productivity of C_3 plants by 25% to 50%, and greatly increases the performance of C_3 plants relative to weedy C_4 competitors. Rising CO_2 stimulates biological nitrogen fixation and enhances the capacity of plants to obtain limiting resources such as water and mineral nutrients. It also increases the ability of plants to remain productive during environmental stresses such as heat and drought. These results indicate that increases in productivity and reliability of food plants after 12000 years ago may have been substantial enough for plant husbandry to become a dependable mode of subsistence. CO_2 enrichment alone is unlikely to have caused the origin of agriculture; however, it could have removed a productivity barrier that inhibited plant cultivation, thereby allowing other causative agents to become important.

The 100 ppm rise in CO_2 from 270 to the present level during the last 120 years has coincided with another explosion in crop yields. The very successful plant breeding, irrigation, fertilization and better pest control would not have been as effective if the major plant nutrient, i.e., carbon had not been increasing during this time. Life on earth is dependent directly or indirectly on photosynthetic activities of the plants. Carbon dioxide, the first molecular link between atmosphere and biosphere, is the primary raw material for photosynthesis and thus become the nerve of the entire food chain. Plants with the C_4 photosynthetic pathway such as maize, sorghum, sugarcane etc. do not respond to elevated CO_2 because they have a biochemical CO_2 concentrating mechanism within the leaf. Consequently CO_2 fixation rate and growth are unaffected by increase in the atmospheric CO_2 concentration. In contrast, photosynthesis in C_3 plants like rice and wheat is affected by both short- and long- term changes in atmospheric CO_2 concentration because they do not have anyCO_2 concentrating mechanism. In C_3 plants carbon assimilation increases with a rise in CO_2 concentration: first, because the current level of atmospheric CO_2 (which is close to the Km. Michaelis-Menton constant, defined as the substrate concentration at which the rate of an enzymatic reaction is half of its maximum. of the enzyme for carbon fixation. Rubisco. to which CO_2 acts as a substrate), is sub-optimal to cause saturation in photosynthesis, any further increase in CO_2 concentration will increase the rate of photosynthesis; and second, because. Rubisco. has an affinity for O_2 as well as CO_2, and at ambient CO_2 concentrations competition between the photoreductive carbon cycle and photorespiration reduces net CO_2 assimilation rates by 30-50%, higher CO_2 concentration increase the ratio of CO_2 to O_2 at the site of fixation, thereby increasing the rate of CO_2 fixation and growth (Long *et al.*, 2004).

The increased rate of photosynthesis and reduced carbon loss by photorespiration under elevated CO_2 is expected to increase the biomass yield of the agriculturally important C_3 crops like wheat, rice, soybean etc. However, associated rise in temperature will hastens the ontogenetic development of the plants substantially reducing the crop duration. As the crops will get lesser time for biomass accumulation and grain filling, rise in temperature is expected to offset the beneficial carbon fertilization effects of elevated CO_2 on crop plants. Irrespective of the theoretical benefits of CO_2 on agriculture and bio-resources, the secondary influences of rising temperature such as extreme weather variability, changed precipitation pattern, reduced water availability, deterioration in soil health and nutrient dynamics, altered photosynthate partitioning to economic plant parts, effects on survival and distribution of pest population etc. will frequently be counterproductive. The way these secondary influences will interact among themselves and the extent to which they negate the positive direct influences of CO_2 fertilization is not at all clear which underlines the necessity of extensive research to be undertaken to establish which influence dominates yield outcomes and to further quantify the overall impact of climate change on agriculture.

Impacts on Crop Productivity and Food Production

Climate is a primary determinant of agricultural productivity. Rise in atmospheric carbon dioxide content, temperature and associated incidence of extreme weather events are the main climate change related drivers which impact crop productivity and food production across the globe. Although increase in CO_2 is likely to be beneficial to several crops such as rice, wheat and pulses, associated increase in temperatures, and increased variability of rainfall would considerably impact food production. According to the projections of IPCC (2007a), World agriculture will have to face a serious decline within this century due to global warming unless emissions of carbon dioxide and other greenhouse gases are substantially reduced from their rising path, and developing countries will suffer much steeper declines than the high income countries. Developing countries in the tropics and sub-tropics of the world where average temperatures are already near or above crop tolerance levels are predicted to suffer an average 10 to 25 percent decline in agricultural productivity by the 2080s assuming a so called business as usual.Scenario in which greenhouse gas emissions continue to increase. Individual developing countries face even larger declines. India, for example, could see a drop of 30 to 40 percent. Some smaller countries in Africa suffer what could only be described as an agricultural productivity collapse. Sudan, already wracked by civil war fuelled in part by failing rains, is projected to suffer as much as 56 percent reduction in agricultural productivity potential; Senegal, a 52 percent fall.

Some of the developed countries situated in the temperate and near the Polar Regions of the world where the present temperature regimes are suboptimal or not suited to agriculture could be benefitted by rising carbon dioxide as some additional land area is brought under cultivation by associated rise in surface temperature. IPCC (2007b) has projected that crop productivity is likely to increase slightly in temperate environments (e.g. in Northern Europe and North America) for local mean temperature increases of up to 1-3°C depending on the crop. This may decrease with further increase in temperature in some regions. China, further from equator than most developing countries, could escape major damage on average, although its south central region would be in jeopardy. The picture is similar in the United States, with projected reduction of 25 to 35 percent in the southeast and the south-western plains but significant increases in the northern states (Cline, 2007).

Table 1.1: Summary of the estimates for impact of global warming on world agricultural productivity potentialby 2080s (percent)

	Without Carbon Fertilization	With Carbon Fertilization
World	-16	-3
Rich countries	-6	-8
Developing countries	-21	-9
Median	-26	-15
Africa	-28	-17
Asia	-19	-7
Middle East- North Africa	-21	-9
Latin America	-24	-13

Based on Cline William, Global Warming and Agriculture: Impact Estimates by Country.

Some analysts have suggested that a small amount of global warming could actually increase global agricultural productivity. There might be some initial overall benefit to warming for a decade or two but - because future warming depends on greenhouse gas emission today- if timely and adequate action is not taken immediately with a sense of emergency, it would put global agriculture on an inexorable trajectory to serious damage. IPCC (2007b) and a few other global studies indicate considerable probability of loss in crop production in India with increases in temperature. Some of these projected loss estimates for the period 2080-2100 are 5 to 30% (Parry *et al.*, 2004; IPCC, 2007b). These long-time estimates generally assume business as usual scenario, and a limited or nil adaptation by all stakeholders. There have been growing studies on the impact of climate change in India as well.

As rice and wheat are the two most important cereals in the Indo-Gangetic Plains and are largely responsible for the food security of the country, most of the climate change related studies in India have focused on these two main crops. The recent trend of a decline or stagnation in the productivity of the rice-wheat cropping system in the Indo-Gangetic plain and north-western India has raised concerns. The wheat crop is sensitive to increases in the maximum temperature and rice is sensitive to increases in the minimum temperature in the tropical conditions of India. Although productivity declines may be partially offset by elevated CO_2 levels, possible water shortages and thermal stress would adversely affect the yield levels. Several crop growth simulation models show varying results of the impact on yields for wheat and rice in specific regions assuming specified higher temperature and CO_2 conditions.

Sinha and Swaminathan (1991) estimated that a 2°C increase in the mean temperature could decrease the rice yield by 0.75 tons/ha in high yielding areas, and 0.06 tons/ha in the low yield coastal region. A 0.5°C increase in the winter temperature could reduce wheat crop growth duration by 7 days and yield by 0.45 tons/hectare. A study by the Indian Agriculture Research Institute indicates irrigated wheat and rice yields will not be significantly affected due to the direct impact of the temperature increase and CO_2 concentration until 2050, but will show a large reduction in 2070 when the temperature increase is significant. Aggarwal *et al*., (2000) have shown that in northern India rice yields during last three decades are showing a declining trend and this is possibly related to increasing temperatures. Climate change is also projected to affect the agriculture and food production in northeast India. A detailed account of climatic changes and their likely impact on agriculture in this region can be found elsewhere (Manoj-Kumar, 2011a, b & c). More recent studies done at the Indian Agricultural Research Institute indicate the possibility of loss of 4-5 million tons in wheat production with every rise of 1°C temperature throughout the growing period even after considering benefits of carbon fertilization (Fig 1.2). This analysis assumes that irrigation would remain available in future at today's levels and there is no adaptation.

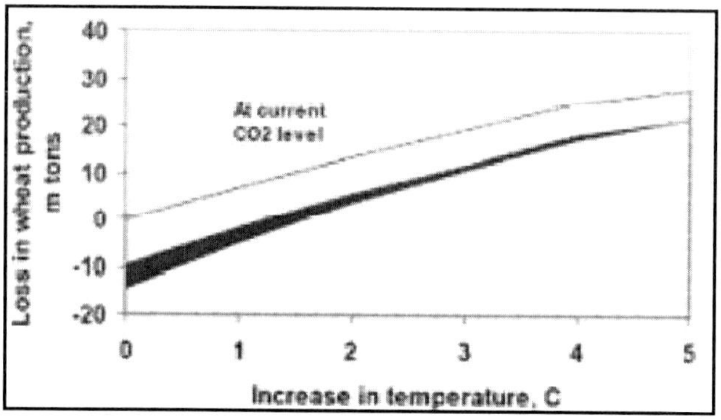

Fig 1.2: Projected loss in wheat production due to increasing temperature at current and 550 ppm CO_2 levels. The shaded areas of the curve indicate losses that can be offset by adaptation options such as change in planting dates and variety (Source: Aggarwal, 2007).

Keeping in view the current trend of rapidly declining groundwater table in wheat growing region of Indo-Gangetic plains, and future projections of erratic rainfall pattern, the assumption of irrigation availability in future at today's levels does not seem much convincing; further, as the wheat is currently grown mostly in irrigated areas and the irrigation availability at the critical growth stages of the crop is very crucial, actual loss of wheat production with rising temperature could be more severe than estimated by this study.

It is, however, possible for farmers and other stakeholders to adapt to a limited extent and reduce the losses (subsequent section discusses possible adaptation options). Simple adaptations such as change in planting dates and crop varieties could help in reducing impacts of climate change to some extent. For example, the study carried out at the Indian Agricultural Research Institute indicates that losses in wheat production at 1°C increase in temperature can be reduced from 4-5 million tons to 1-2 million tons if a large percentage of farmers could change to timely planting and changed to better adapted varieties (Fig 2). These adaptation benefits become smaller as temperature increases further.

The effect of rising temperature on grain yield of some winter crops namely wheat and barley- both cereals, mustard- an oilseed crop and gram- a pulse crop, in northern India was evaluated on the basis of historic datasets on meteorological sub-divisional scales, and through a dynamic crop growth model WTGROWS by Kalra *et al.*, (2008) (Fig 1.3). The reduction factors per degree rise in temperature, on average, were 4.26, 2.77, 0.32 and 1.32 q/ha for wheat, barley, gram and mustard respectively. Studies indicate that the direct impact of rising temperature could be smaller on the kharif (early winter harvested) crop, but the crop will be highly vulnerable due to increased incidence of extreme weather (rainfall duration/intensity, drought/flood), pest and diseases, and virulence. The Rabi crops (spring harvested) will be comparatively more vulnerable due to higher variation in temperature and uncertainties of rainfall and irrigation water availability. However, disappearance of the Himalayan glaciers and consequent drying up of the currently perennial rivers in the north and east may dry up the irrigation water supply to the rabi crop. In the long run, Indian agriculture may be seriously affected depending on the season, region and adaptation.

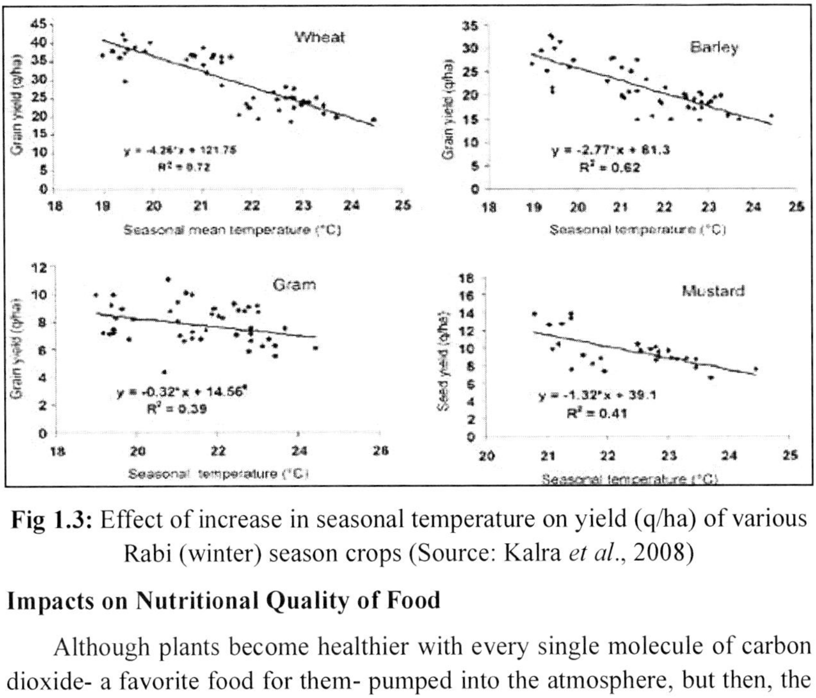

Fig 1.3: Effect of increase in seasonal temperature on yield (q/ha) of various Rabi (winter) season crops (Source: Kalra *et al*., 2008)

Impacts on Nutritional Quality of Food

Although plants become healthier with every single molecule of carbon dioxide- a favorite food for them- pumped into the atmosphere, but then, the food which plant produces for human being is nutritionally not so healthy and could trigger a pandemic of human malnutrition in future high CO_2 world. In a meta-analysis of the available literature on effect of CO_2 on protein content of several food grains, Taub *et al*., (2008), reported decline in the protein concentrations in the grains of several food crops under elevated atmospheric CO_2. Wheat, rice and potato provide 21%, 14% and 2%, respectively, of protein in the human diet (FAOSTAT, 2007), and therefore, any decline in protein concentrations of these crops would have a devastating impact on human nutrition world over. As the nitrogen concentration in different crops has been widely reported to decline almost invariably when grown under elevated CO_2, it is bound to decrease the protein concentration in the food grains because nitrogen constitutes 16% of the grain proteins. At elevated CO_2 and standard fertilizer levels, wheat produces 10 to 13 % less grain protein (Kimball *et al.*, 2001, Manoj-Kumar and Patra, 2010). Similarly, grain protein in rice (terao *et al.*, 2005) and tuber nitrogen in potato Fangmeier *et al.*, 2002) is reduced by about 10% at elevated CO_2 concentrations.

In a study of impact of increased temperature on quality of rice and wheat grains (Nagarajan *et al.*, 2009), amylose content of rice grains which is an important determinant of cooking qualities decreased linearly with increase in mean temperature. This decline varied from 0.65% per degree rise in temperature in IR 64 rice to 0.30% in super basmati rice. High temperature also affected grain elongation and aroma in basmati rice. Protein content in wheat increased with increasing temperature (0.3% and 0.15% per 0C respectively in bread wheat and durum wheat) due to increase in soluble (gliadin) protein as high temperature increases gliadin: glutenin ratio. However, according to IPCC (2001) the protein content of grain would decrease under both temperature and CO_2 increases. Since the rise in carbon dioxide and temperature affects protein content in mutually opposite direction as suggested by majority of the studies, the overall impact of these two factors, which will be more closer to reality, will depend on their interaction which has not been studied adequately, and calls for further research to reach a convincing conclusion. According to an analysis by Loladze (2002), crops that grow in high CO_2 are nutritionally barren, denuded of vital micronutrients such as iron, zinc, selenium and chromium. If this is the case, nutritional well-being of human being will be in jeopardy, but the impact will hit some people harder than others. Most of the developing countries are already burdened by "hidden hunger"- chronic mineral and vitamin deficiencies caused by eating green revolution crops bred in 1960s and 70s. These high- yield crops staved off starvation, but turned out to be low in essential nutrients, particularly iron, zinc and vitamin A. Now those people will face a second dietary whammy, while millions more will be pushed over the edge into malnutrition. Loladze (2002) reported 17% and 28% decline respectively in iron and zinc content of rice (the world.s most important crop) grown under elevated CO_2. Similar kind of discomforting trend of micronutrient reduction has been reported in wheat (the world's second most important cereal crop) and potato (most important vegetable crop) as well.

There are two mechanisms that potentially explain as to why does rising CO_2 concentration strip plants of some essential macro- and micronutrients.

The first is a "biomass dilution" effect. As plants absorb more airborne carbon, they produce higher-than-normal levels of carbohydrates but are unable to boost their relative intake of soil nutrients. The result of this dilution effect is increased yields of carbohydrate-rich fruits, vegetables, and grains that contain lower levels of macro- and micronutrients. To make matters worse, therc is another effect dragging element ratios down. Excess CO_2 stifles a plant's ability to absorb these nutrients in the first place. Normally, plants absorb chemicals through their roots in two ways. Nutrients can be sucked in along with the water absorbed by the plant, or they can just diffuse into the root down a concentration gradient. Increased CO_2 disrupts both the mechanisms. Higher levels of CO_2 put a squeeze on the rate at which plants absorb water - by making them "breathe" less deeply. Normally mean the plant loses water by evaporation. When the air contains more CO_2, plants can get away with narrowing the pores a little. That way, they get enough CO_2 while reducing their risk of drying out. But this has a profound effect on the water flowing through their tissues. If carbon-dioxide levels are doubled, transpiration decreases by about 23 percent. Roots suck in water using the pull of water evaporating through leaf pores. Reduced transpiration pull, even a little, under elevated CO_2 slows down the flow of water from the roots upward. With less water flowing through their system, plants suck in less of the micronutrients and it gets worse - reduced water flow makes the soil wetter, which dilutes its nutrient content so diffusion rates drop. Overall, the effect drastically reduces the availability of nutrients in root zone subsequently reducing the concentration of macro- and micronutrients in edible plant parts. Apart from reduction in micronutrients concentration (particularly Zn and Fe), their bioavailability is also expected to decline under the elevated CO_2 atmosphere (Manoj-Kumar, 2011d).Put simply, a mouthful of rice, has a lower concentration of micronutrients today than it did just a few generations ago, and a bite of bread in the CO_2-enriched atmosphere of the future will end up being less nutritious than the one in our current CO_2 atmosphere. Keeping in view the fact that over half of the world's population (~3.5 billion people) is already suffering from iron- and zinc deficiency induced health problems, any further decline in the concentration of these micronutrients will severely impact the nutritional security of human being in the high carbon dioxide world of future.

Impacts on Soil Health and its Suitability for Agricultural Production

Soil is nature's most precious gift to the mankind. The economic viability and environmental amenity of agricultural sector depends heavily on 'soil health'. Of the range of potential indicators used to infer soil health status, soil carbon is particularly important. Organic matter is vital because it supports many soil processes that are associated with fertility and physical stability of soil across the various ecosystem services. In particular organic matter provides an energy source for microbes structurally stabilizes soil particles, stores and supplies plant essential nutrients such as nitrogen, phosphorus and sulphur and provides cation/anion exchange for retention of ions and nutrients. Carbon within the terrestrial biosphere can also behave as either a source or sink for atmospheric CO_2 depending on land management, thus potentially mitigating or accelerating the greenhouse effect. Cycling of soil organic carbon is strongly influenced by moisture and temperature, two factors which are predicted to change under global warming. Overall, climate change will shift the equilibrium, both directly and indirectly of numerous soil processes. These include carbon and nitrogen cycling, nutrient dynamics, acidification, risk of erosion, salinisation, all of which will impact on soil health. The link between climate change and soil health impact is schematically presented in (Fig 1.4).

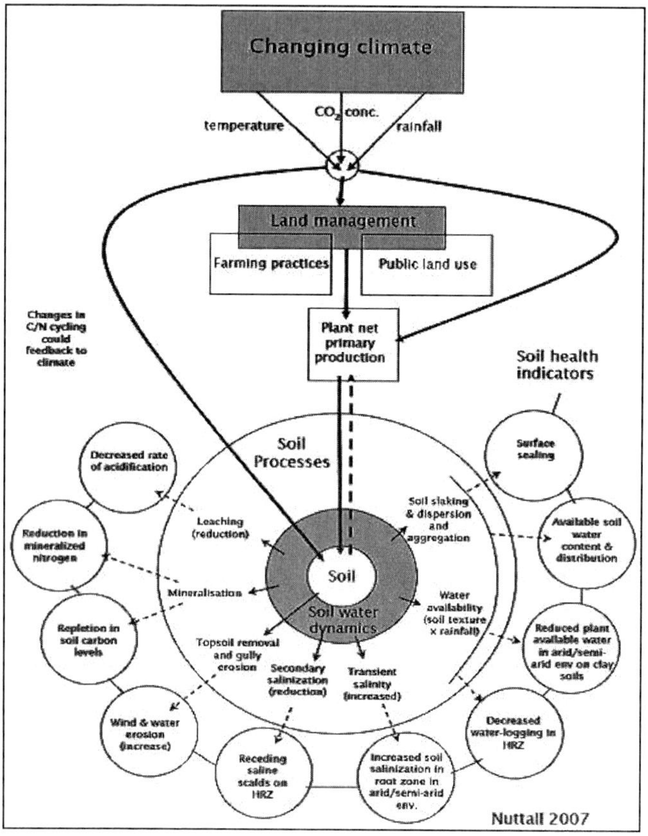

Fig 1.4: Potential links between climate change and soil health (Source: Nuttal, 2007)

Soil Organic Carbon and Nutrient Dynamics

Worldwide, Soil carbon levels are expected to decrease due to decreased net primary production as an effect of climate change. Any gains by increased plant water use efficiency, due to elevated CO_2 are likely to be outweighed by increased carbon mineralization after episodic rainfall and reduced annual and growing season rainfall. The quality of soil organic matter may also shift where the more inert components of the carbon pool prevail. The residues of crops under elevated CO_2 will have higher C: N ratio (Manoj-Kumar and Bhadraray, 2009, Manoj-Kumar, 2010), and this may reduce their rate of decomposition and nutrient supply. In contrast, increase of soil temperature is expected to increase nitrogen mineralization, but then increased gaseous losses of nitrogen through processes such as volatilization and denitrification may reduce its overall availability to plants. Increased microbial population in crop rhizosphere under elevated CO_2 may also compete with the crop plants for available nutrients in the soil. Soil biology and microbial populations are expected to change under conditions of elevated CO_2 and changed moisture and temperature regimes. As soil biology regulates nutrient dynamics and many disease risks, nutrient availability to crops could change as could the exposure to soil-borne diseases.

Soil Salinisation

Transient salinity may increase in soil under future climate change. Transient salinity increases as capillary rise dominates at higher temperature, bringing salts into the root zone in salt affected soils. Leaching during episodic rainfall events may be limited due to surface sealing. Increased subsoil drying increases concentration of salts in the soil solution. Conversely, the severity of saline scalds due to secondary salinisation may abate, as groundwater levels fall in line with reduced rainfall. Rise in sea level also may lead to salt-water ingression in the coastal lands turning them less suitable for conventional agriculture.

Soil Erosion and Nutrient Loss

An increased risk of soil erosion and nutrient loss due to reduced vegetation cover in combination with episodic rainfall and greater wind intensities is expected. Runoff and soil loss by erosion are most vulnerable to changing rainfall pattern under future scenario of climate change. In order to assess the impact of climate change on soil erosion and runoff, Tripathi *et al.*, (2009) selected seven watersheds in different agro-climatic regions across the country and used AVSWAT model to estimate the projected runoff and soil loss between 1961-1990 and 2071-2100. The frequency of occurrence of annual runoff of the magnitude of over 300 mm was projected to increase by 26.1% in Umiam (Meghalaya) to 200% in Antisar (Gujarat) thus increasing the frequency of flood. The frequency of annual runoff of less than 50 mm was expected to vary between 10% in Pogalure (Tamil Nadu) and 100% in Jonainala (Orissa) thus is increasing the frequency of drought. During the same period, soil loss is likely to increase by 3.7% in Udhagamandalam (Tamil Nadu) to about 757% at Pogalur (Tamil Nadu), with around 71% increase at Umiam (Meghalaya). These huge amounts of soil loss by erosion can potentially strip the top soil of their organic matter and nutrient content, thus rendering the soil unfit for crop cultivation in future.

Impacts on Insect Pests and Diseases

Crop-pest interactions will change significantly with climate change leading to impact on pest distribution and crop losses. Crop-weed competition will be affected by rising atmospheric CO_2 depending upon their photosynthetic pathway. C_3 crop growth would be favored over C_4 weeds affecting the need for weed control. However, the associated rise in temperature may further alter the competition depending upon the threshold ambient temperatures. Diseases and insect populations are strongly dependent upon temperature and humidity. Any increase in these parameters, depending upon their base value, can significantly alter their population, which ultimately results in yield loss. Even with small changes, the virulence of different pest's changes. For example, at 16°C, the length of latent period is small for yellow rust. Once the temperature goes beyond 18°C, this latent period increases, but that of yellow and stem rusts decreases (Nagarajan and Joshi, 1978). The appearance of black rust in north India in sixties and seventies was related to the temperature dependent movement of spores from south to north India (Nagarajan and Joshi, 1978).

Monocyclic diseases such as stem rot, sheath rot and false smut are less influenced by ambient weather conditions. Epidemics of monocyclic diseases are relatively rare in the sense of an explosive increase in their population. In contrast, polycyclic diseases such as blast, brown spot, bacterial leaf blight and rice tungro virus, which invade the aerial parts of the plants, are subjected to constant interaction with the weather, and are more likely to be affected by change in climate. They easily attain epidemic proportions to cause heavy losses (Abrol and Gadgil, 1999). For every insect species there is a range of temperatures within which it remains active from egg to adult stage. Lower values of this range are called 'threshold of development' or '**developmental zero**'. Within the favorable range, there is an optimum temperature at which most of the individuals of a species complete their development. Exposure to temperatures on either side of the range exerts an adverse impact on the insect by slowing down the speed of development. If ambient temperatures remain favorable for the pest after temperature increases, the pest incidence may be expected to rise due to increased rates of development, which may result in the completion of more pest generations. However, the pest population would be adversely affected once the ambient temperature exceeded the favorable range. Studies have shown that insects remain active within a temperature range from 15° to 32°C (Phadke and Ghai, 1994). In the case of red cotton bug at constant temperatures of 20, 25 and 30°C, the average duration of life-cycle was found to be 61.3, 38.3 and 37.6 days respectively, while at 12.5 and 35°C the pest did not show any development (Bhatia and Kaul, 1966). The most congenial temperatures for insect development have been suggested by Phadke and Ghai (1994). For the mustard aphid, Lipaphis erysimi, a maximum temperature ranging from 19 - 24°C is suggested, with a mean of 12-15°C; for rice stink bug, a maximum temperature between 26.9 and 28.2°C with a relative humidity of 80.6 - 82.1%; for rice green leafhopper, a temperature from 20 - 28°C ; for the brown plant hopper, a temperature from 24.8 - 28.6°; for aphids, thrips and leaf weevils, a mean temperature around 27.5 - 28.5°C; and a maximum temperature from 23 - 27.8°C is required for the gram pod borer.

Forecasting the appearance of aphids (Lipaphis erysine Kalt) on mustard crops grown during the winter season in the northern part of India based on the movement of western monsoon disturbances has been achieved (Ramana Rao *et al.*, 1994).Western disturbances bring in cold and humid air from the Mediterranean region, resulting in cloudy and favorable weather conditions for the occurrence of aphids on mustard crops. It was shown that there was a sharp increase in the population of aphids when the mean daily temperature ranged from 10° to 14°C, with a relative humidity of 67.85% and cloudiness greater than 5 octas. The swarms of locust produced in the Middle East usually fly eastwards into Pakistan and India during the summer season and lay eggs during the monsoon period. The swarms resulting from this breeding return during autumn to the area of winter rainfall, flying to all parts of India and influencing kharif crops (Rao and Rao, 1996). Changes in rainfall, temperature and wind speed may influence the migratory behaviour of locusts. Pests such as the armyworm, Mythimna separate, achieve higher population growth leading to outbreaks after heavy rains and floods. On the other hand, pests such as Pyrilla perpusilla become more damaging under drought conditions. Less frequent but intense rains in future will cause floods as well as droughts, which will thus influence the incidence of pests. Some pests, such as the cabbage white butterfly, Pieris brassicae, migrate to the plains in winter and back to the hills in summer. With milder winters in the hills and increasing temperatures, such migrations will also be affected. With shorter cold seasons, the onset of diapauses will be delayed in autumn, while its termination may be hastened in spring, thereby increasing the period of activity of pests. Pest population dynamics simulation models can be used to simulate population dynamics and assess the impact of climate change on pest incidence. Adynamic simulation model was developed for the rice stink bug, Leptocorisa acuta, using a thermal time concept (Fig 1.5). It was found that up to a 2°C rise in daily average temperature over 2001's weather would increase the pest population, while further increases in temperature would have an adverse effect on pest populations.

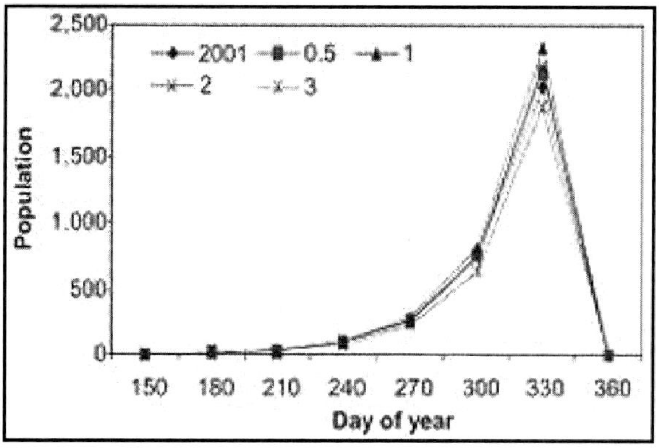

(Source: Reji *et al*., 2003)

Fig 1.5: Simulation of the effect of temperature rise on the rice stink bug population

Impacts on Forestry and Biodiversity

India is a country rich in biodiversity where forests account for about 20% (64 million ha) of the geographical area with nearly 200,000 villages classified as forest villages, there is obviously a great dependence on forest resources by communities. One tenth of the world's known species of higher altitude animals and plants occur in the Himalayas. In 1995, approximately 10 percent of the known species in the Himalayas were listed as 'threatened'. Climate is an important determinant of the geographical distribution, composition and productivity of forests. Rabindranath and Sukumar (1998) estimated the impacts of two climate change scenarios on tropical forests in India- one involving greenhouse gas forcing and the other incorporating the effects of sulphate aerosols. The first scenario, associated with increased temperature and rainfall, could result in increased productivity, migration of forest types to higher elevations and transformation of drier forest types to moister types. The second scenario involving a more modest increase in temperature and a decrease in precipitation in central and northern India could have adverse effects on forests and biodiversity therein.

Global climate change is also likely to lead to a shift of the temperate ecosystem to higher elevations and increased erosion and overland flows in the steep terrains of the Himalayan range. Increased temperature and rainfall may lower the productivity of forest resources. One study predicts that due to a decline in soil moisture availability caused by warmer temperatures, productivity of teakwood, India's most preferred wood species, could decline from 5.4m^3 per hectare to 5.07 m^3 per hectare, and a decline in productivity of deciduous forests may take place from 1.8m^3 per hectare to 1.5 m^3 per hectare. High temperature and moisture stress, coupled with extreme weather conditions could lead to higher natural degradation of forests.

Impacts on Coastal Agriculture and Fisheries

Densely populated and intensively cultivated low-lying coastal lands are vulnerable to coastal erosion and land loss, inundation, sea flooding and upstream movement of sea water into fresh water tributaries. India's 7,500 kilometres coastline will be particularly hard-hit by climate change. Large scale migration of people from coastal zones may occur due to submergence of coast lines after the rise in sea levels. Research shows that a one meter rise in the sea level will affect an area of 5,763 square kilometres and displace about 7.1 million people in India. Rice cultivation, commercial fishing and prawn farming practiced in coastal regions could be severely affected. Inundation of the coastal region and consequent problems of salinity in the adjoining tributaries may also adversely affect inland agriculture. Agriculture will be worst affected in the coastal regions of Gujarat and Maharashtra, where agriculturally fertile areas are vulnerable to inundation and salinisation.

Regional changes in the distribution and production of particular fish species are expected due to continued warming, with adverse effects projected for aquaculture and fisheries. Sea surface temperature in the Indian seas may increase by about 3°C by 2100. Increasing sea and river water temperature is likely to affect fish breeding, migration and harvests. Arise in temperature as small as 1°C could have important and rapid effects on the mortality of fish and their geographical distributions. Oil sardine fishery did not exist before 1976 in the northern latitude and along the east coast as the resource was not available, and sea surface temperatures were not congenial. With warming of sea surface, the oil sardine is able to find temperature to its preference especially in the northern latitudes and eastern longitudes, thereby extending the distributional boundaries and establishing fisheries in larger coastal areas. Corals in Indian Ocean will be soon exposed to summer temperatures that will exceed the thermal thresholds observed over the last 20 years. Annual bleaching of corals will become almost a certainty from 2050. Impacts on cost of cultivation, farm profitability and environmental quality Changes in climate are expected to affect the productivity and aggregate demand for inputs in agriculture such as water, labour, energy, equipments, fertilizer, and plant protection chemicals which will, in turn, increase the cost of cultivation and finally bring down farm profitability. Kumar and Parikh (1998) have showed that even with adaptation by farmers of their cropping patterns and inputs, the loss in farm-level net revenue is estimated to range between 9% and 25% for a temperature rise of 2°C-3.5°C. Increased incidences of pest infestation and disease outbreak in agricultural crops at higher temperature and humidity in future will increase the cost on plant protection chemicals. The rising levels of atmospheric CO_2 are likely to increase biomass production and yield of C_3 crops such as rice, wheat and soybean, but the absolute increase in productivity will occur only when nitrogen and phosphorus availability in soil is high (Manoj-Kumar *et al.*, 2011a, b & c, 2012 a & b). Manoj-Kumar *et al.*, (2009b) studied the impact of low (120 kg N ha-1), medium (180 kg Nha^{-1}) and high (240 kg N ha^{-1}) nitrogen applications on yield and quality of wheat grains in sub-tropical India under elevated atmospheric carbon dioxide (650 ppm) concentration. Although grain yield responded positively at all the three levels of nitrogen application but the response was maximum at higher doses. Relatively very small increase in yield at low N level was attributed to nitrogen deficiency faced by plants as indicated by appearance of N deficiency symptoms under elevated CO_2 concentration. Nitrogen concentration of wheat grains which is an important determinant of grain nutritional quality contributing 16% to the total grain protein, also decreased under elevated CO_2 at low level of N supply but medium and high doses of N caused an improvement in N content of wheat grains at high CO_2. The results underline the requirement of enhanced fertilizer-nitrogen application to wheat crop for sustaining higher productivity and maintaining nutritional quality of wheat under the futuristic ecosystem of elevated atmospheric carbon dioxide. However, the decrease in fertilizer nitrogen use efficiency at the higher doses, which is required under elevated CO_2, is a matter of great concerns both economically as well as environmentally (Table 1.2).

Table 1.2: Fertilizer-N use efficiency in wheat as influenced by interaction of atmospheric CO_2 and levels of fertilizer-N application

Fertilizer N Levels	Levels of Atmospheric CO_2	
	Ambient	Elevated (650 ppm)
100%	25.24^{ab}	27.49^{a}
150%	19.46^{cd}	22.33^{b}
200%	12.74^{c}	16.66^{d}

As the atmospheric CO_2 concentration rises, phosphorus requirements for plants growing in managed ecosystems will also need reassessing. Since most of the C_3 crop species fail to respond to highCO_2 when phosphorus is low- possibly because insufficient phosphorus is available to maintain maximum photosynthetic activity at high CO_2, higher rates of P fertilizer are likely to be needed to sustain maximum productivity of a number of crop species in future (Manoj-Kumar *et al.*, 2012a, b & c) (Table 1.3).

Table 1.3: Effect of phosphorus nutrition and CO_2 enrichment on wheat yield (Source: Conroy, 1992)

Phosphorus	CO_2 (ppm)	No. of Grains Per Head	Weight of Grains Per Head (g)
Low	340 660	13 12	0.5 0.5
High	340 660	22 33	0.8 1.3

Difference between a pair of values with any common letter as superscript is statistically non-significant ($p<0.05$). Values in percent (%) represent percent of the recommended dose (STCR recommendation) of fertilizer-N applied.

As nitrogen and phosphorus are the two most deficient nutrient elements in the soil world over, particularly in India, higher doses of these elements through external application of fertilizers will be required to sustain the higher productivity of plants under elevated CO_2. Fertilizer use efficiency in India is generally very low (30-50%). Increasing temperature in future is likely to further reduce fertilizer use efficiency. Higher frequency of heavy rainfall as projected in future will further aggravate the problem of nitrate pollution of ground water by facilitating more NO_3^{-1}leaching from the soil. As the fertilizer is most costly input in agriculture, and also its loss is associated with several environmental hazards (such as nitrate pollution of ground water, N_2O emission into atmosphere causing global warming and eutrophication of water bodies), application of higher doses of fertilizers with reduced use efficiency will increase the cost of cultivation as well as degrade the environmental quality in future scenarios of elevated CO_2 and temperature. Irrigation is another very costly and critical input in crop production which will be required more frequently at higher temperature due to accelerated drying of soil through higher evapo-transpirational loss of water in future. Higher cost will also be incurred on proper management of livestock to protect it from heat stresses induced by global warming. Put simply, reduced productivity of crops and livestock, and increased cost of agro-inputs will have a discomforting impact on farming by pulling down the overall profitability in agriculture under projected scenario of climate change in future.

Agriculture and Climate Change Mitigation

Changes in land use management sustaining the production, development of multiple stress-tolerant varieties, and evolvement of new agronomic management strategies are some of the adaptation strategies that are being talked about. Suitable resource-conservation technologies (RCTs) and proper contingency plans for temperature and rainfall-related risks need to be developed. Agroforestry systems and other biological carbon capture systems can also help in reducing the carbon emissions. Already large numbers of efficient water- and fertilizer-management practices that reduce emission of CH_4 and N_2O are available for paddy and wheat cultivation. Water harvesting and supplemental irrigation for drought-proofing areas, improved drainage and flood control measures in high rainfall regions, and techniques for conservation of resources are of immense importance for mitigating the adverse effect of climate change. Similarly, knowledge-based decision support systems for translating weather information into operational-management practices are required as adaptation mechanisms. These adaptations can be at the level of individual farmer, society, farm, village, watershed, or at national and regional level.

There are a large number of options in soil-, water-, and nutrient-management technologies, which contribute to both adaptation and mitigation.

Much of the research done in rain-fed agriculture relates to conservation of soil and rainwater (Venkateswarlu *et al.*, 2012). Important technologies include: CA (tillage, residue, and nutrient management), in situ moisture conservation, rainwater harvesting and recycling, efficient use of irrigation water, energy efficiency in agriculture, use of poor-quality waters, alternative land uses, and agroforestry on degraded lands. Agriculture systems will have to be more robust and resilient to extreme weather events, that is, drought and flooding. It is important that new agricultural practices will have to prevent further soil degradation (Sharma *et al.*, 2011 and Sharma and Jat, 2012), through increased soil organic matter, improved water-use efficiency (WUE) and nutrient-use efficiency and increased biodiversity (Hobbs and Govaerts, 2010). The improvement in rice-production technology and livestock/manure management can help in mitigation of 9% of the total anthropogenic CH_4 and better crop-land management can mitigate about 2% of the total N_2O emission from agriculture. The majority of the potential (70%) can be realized in developing countries (Smith *et al.*, 2007). The core challenge of climate change adaptation and mitigation in agriculture is to produce more with higher efficiency, under more volatile production conditions, and with net reductions in GHG emissions. Elimination or reduction of trace gas fluxes from soil would significantly impact carbon-equivalent impacts of agriculture (Gaunt and Lehmann, 2008). Recently, the concept of climate-smart agriculture has been proposed to overcome the adverse effects of climate change on agriculture. This concept includes many of the field based and farm-based sustainable agricultural land management practices, such as conservation tillage, agroforestry, residue management, and water management (FAO, 2011; World Bank, 2011). Most of the focus of climate-smart agriculture has been on the implementation of these field and farm practices, and the ways in which they can be improved in the context of a changing climate.

Promoting agricultural practices that mitigate climate change by reducing GHG emissions is important; but those same practices also have to improve farmer's production and income and buffer the production system against changes in climate Mitigation is a response strategy to climate change, and can be defined as measures that reduce the amount of emissions (abatement) or enhance the absorption capacity of greenhouse gases (sequestration). Regardless of the projected or actual impacts of climate change, agriculture is also likely to be directly or indirectly involved in climate change mitigation efforts. Agriculture acts as both, source as well sink of greenhouse gases. CO_2, CH_4 and N_2O are important greenhouse gases contributing 60%, 15% and 5% respectively to global warming. The IPCC (2007a) estimates that globally agriculture emits about 20%, 50% and 70% of the total anthropogenic emissions of these gases. Agriculture sector contributes 28% of the total GHG emissions from India (NATCOM, 2004, Fig 1.6).

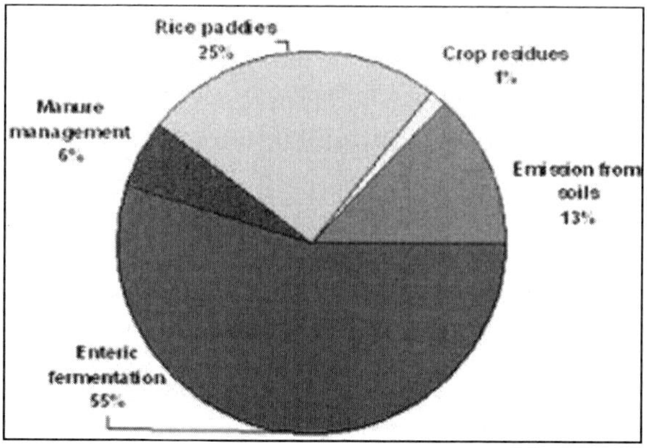

Fig 1.6: Relative contribution of different sectors in agriculture to GHG emissions (Source: NATCOM, 2004)

Carbon dioxide emissions arise from fossil fuel usage, soil tillage, deforestation, and biomass burning and land degradation. Biological generation of methane in anaerobic environment including enteric fermentation in ruminants, flooded rice fields, and anaerobic animal waste processing, are the principal sources of CH_4 from agriculture. Nitrous oxide emissions come from manure, legumes and inefficient fertilizer use. There are several approaches that can assist in reducing GHG emissions from agriculture. Mitigation of CO_2 emission from agriculture can be achieved by increasing carbon sequestration- a process of transferring carbon dioxide from the atmosphere into soil and plant in a form that is not immediately reemitted. This transfer of carbon helps off-set emission from fossil fuel combustion and other carbon emitting activities while enhancing soil quality and long term agronomic productivity. Soil management practices such as reduced tillage, manuring, residue incorporation, improving soil biodiversity, micro aggregation and mulching can play important roles in sequestering carbon in soil. Forests and stable grass lands are referred to as potential carbon sinks because they can store large amounts of carbon in their vegetation and root systems for long periods of time. Therefore, afforestation on marginal and degraded land is another carbon sequestration strategy.

Methods to reduce methane emissions from enteric fermentation include enhancing the efficiency of digestion with improved feeding practices and dietary additives. Feed composition should be carefully altered, either to reduce the percentage which is converted into methane or to improve the milk and meat yield. The efficacy of these methods depends on the quality of feed, breed and age of livestock. Strategies to reduce methane emission from rice cultivation could be altering water management, particularly promoting mid-season aeration by short-term drainage; improving organic matter management by promoting aerobic degradation through composting or incorporating it into soil during off-season drained period; use of rice cultivars with few unproductive tillers, high root oxidative capacity and high harvest index; and application of fermented manure like biogas slurry in place of unfermented farm yard manure. Site-specific nutrient management could be the most efficient management practice to reduce nitrous oxide emission from agricultural soil. Any measure which increases the use efficiency of applied nitrogenous fertilizer will reduce the N_2O emission from the soil. In a field experiment conducted at the farm of Indian Agricultural Research Institute, New Delhi, Pathak and Co-Workers studied the effectiveness of various nitrification inhibitors in mitigation of nitrous oxide and methane emission from rice-wheat system of Indo- Gangetic plain. They reported substantial reduction in N_2O-N emission on the application of nitrification inhibitors which ranged from 5% with hydroquinone to 31% with thiosulphate in rice and 7% with hydroquinone to 29% with DCD in wheat crop. Nitrification inhibitors also influenced the emission of CH_4. There are some plant-derived organics such as neem oil, neem cake and karanja seed extract which can also act as nitrification inhibitor.

Agriculture and Climate Change Adaptation

Adaptation to climate change can be defined as an adjustment made to a human, ecological or physical system in response to a perceived vulnerability of various sectors of agriculture, aiming to reduce the overall adverse impact of changing climate on agriculture. Potential adaptation strategies to deal with the impact of climate change include developing cultivars tolerant to heat and salinity stress and resistant to flood and drought, modifying crop management practices, improving soil and water management, improving fertilizer use efficiency, adopting new farm techniques such as resource conservation technologies, crop diversification, improving pest management, conserving biodiversity, agronomic and genetic interventions to produce nutrient-rich (protein, iron and zinc) crops, better weather forecast and crop insurance, and harnessing the indigenous technical knowledge of the farmers.

Changes in Land use-and Crop Management Practices

Small changes in climatic parameters can often be managed reasonably well by altering dates of sowing, spacing and input management. Adjustment of planting dates to minimize the effect of temperature increase induced spikelet sterility can be used to reduce yield instability, by avoiding the flowering period to coincide with the hottest period. Adaptation measures to reduce the negative effects of increased climatic variability, as normally experienced in arid and semi-arid tropics, may include changing the cropping calendar to take advantage of the wet period and to avoid extreme weather events during the growing season. Development of alternate cultivars and farming systems (such as mixed cropping, crop-livestock) that are more adaptable to changes in the environment can further ease the pressure.

Soil Water Management

A broad range of agricultural water management practices and technologies are available to spread and buffer production risks. Enhancing residual soil moisture through land conservation techniques assists significantly at the margin of dry periods while buffer strips, mulching and zero tillage help to mitigate soil erosion risk in areas where rainfall intensities increase. Water harvesting techniques and micro catchments are extremely beneficial in increasing biomass production in arid climates. Improvement in water use efficiency through measures such as advanced irrigation systems, viz. Drip irrigation technologies; centre-pivot irrigation systems, etc. coupled with reduction in operating hours, can significantly reduce the amount of water and nitrogen applied to the cropping system. This reduces emissions of nitrous oxide and water withdrawals.

Adoption of Resource Conserving Technologies

Resource- conserving technologies involving zero- or minimum tillage with direct seeding, permanent or semi-permanent residue cover, and crop rotations have potential to improve the use efficiency of natural resources, including water, air, fossil fuel and soil. Recent researches have shown that surface seeding or zero-tillage establishment of upland crops after rice gives similar yields to when planted under normal conventional tillage over a diverse set of soil conditions. In addition, such resource conserving technologies improve soil organic carbon status, and restrict the release of soil carbon thus mitigating increase of CO_2 in the atmosphere (Table 1.4).

Table 1.4: Effect of different tillage practices on soil organic carbon content (g kg^{-1}) after six years of soybean wheat cropping cycles in a Vertisol (Source: Anonymous, 2007)

Soil Depth (m)	Soil Organic Carbon Content (g kg^{-1})			LSD (P=0.05)
	No Tillage	Reduced Tillage	Convention Tillage	
0-0.05	13.08	12.47	11.01	1.23
0.05-0.15	8.01	8.70	7.20	0.92
0.15-0.30	6.69	6.42	5.57	1.04

Development of Climatic-Stress Resistant Crop

Most of the currently grown agricultural crops have not been developed to face the severity of the extreme climatic stress (such as temperature extremes, drought, flood, salinity, wind storms, nutrient deficiency in soil, frequent pest attack etc.) anticipated in future. Therefore, one important adaptive option includes developing climatic-stress resistant crop by intensive search for stress tolerance genes in wild genotypes, developing transgenic for biotic and abiotic stresses, transforming C_3 plants to C_4 photosynthetic pathway which is more efficient in utilizing atmospheric carbon dioxide.

Improved Risk Management Though Early Warning System and Crop Insurance

The increasing probability of floods and droughts and other uncertainties in climate may seriously increase the vulnerability of resource-poor farmers to global climate change. Early warning systems and contingency plans can provide support to regional and national administration, as well as to local bodies and farmers to adapt. Policies that encourage crop insurance can provide protection to the farmers in the event their farm production is reduced due to natural calamities. Harnessing the indigenous technical knowledge of the farmers: For centuries, Farmers in South Asia, often poor and marginal, have been facing, and are experimenting with the climatic variability. Although there is a large body of knowledge within local communities on coping with climatic variability and extreme weather events, rapidly changing climate conditions will require upgrading local knowledge with more scientific observations and establishing collaboration among neighbours and neighbouring countries to transfer knowledge from areas already experiencing these changes.

Conservation of Agro-Biodiversity

Biodiversity in all its components (e.g. genes, species, ecosystems) increases resilience to changing environmental conditions and stresses. Genetically-diverse populations and species-rich ecosystems have greater potential to adapt to climate change. Therefore, promoting use of indigenous and locally-adapted plants and animals as well as the selection and multiplication of crop varieties and autochthonous races adapted or resistant to adverse conditions is also an effective strategy for adaptation to climate change.

References

1. Abrol YP, Gadgil S. Rice ñ in a Variable Climate, APC Publishing House, New Delhi, 1999.
2. Aggarwal PK. Climate change: Implications for Indian agriculture. Jalvigyan Sameeksha. 2007; 22:37-46.
3. Aggarwal PK, Bandyopadhyay SK, Pathak H, Kalra N, *et al.*, Analysis of the yield trends of rice-wheat system in north-western India. Outlook on Agriculture. 2004; 29 (4):259-68.
4. Anonymous. Annual report (2006-07) of the Indian Institute of Soil Science, Bhopal. 2007.
5. Cline WR. Global Warming and Agriculture: Impact Estimates by Country. ISBN paper 978-0-88132- 403-7, 2007, 250.
6. Conroy JP. Influence of elevated atmospheric CO_2 concentrations on plant nutrition. Australian Journal of Botany. 2007; 40:445-456.
7. FAO. Strengthening Capacity for Climate Change Adaptation in the Agriculture Sector in Ethiopia. Proceedings from National Workshop held in Nazreth, Ethiopia, 2011.
8. FAOSTAT (Food and Agriculture Organization Statistics). Agricultural Data. Food and Agricultural Organization of the United Nations, 2007.
9. Gaunt JL, Lehmann J. Energy balance and emissions associated with biochar sequestration and pyrolysis bioenergy production. Environ Sci Tech. 2008; 42:4152–4158.
10. IPCC. Climate change: The scientific basis. Inter-government panel on climate change. Cambridge (UK): Cambridge Univ. Press, 2001.
11. IPCC. Climate Change: The Physical Science Basis. Summary for Policymakers. Inter-Governmental Panel on Climate Change, 2007, 2007a.
12. IPCC. Contribution of Working Group II to the Fourth Assessment Report of the Intergovernmental Panel on Climate Change-Summary for Policymakers, 2007(b).
13. Kalra N, Chakraborty D, Sharma A, Rai J *et al.*, Effect of Increasing Temperature on Yield of Some Winter Crops in Northwest India. Cur. Sci. 2008; 94(1):82-88.
14. Kimball BA, Morris CF, Pinter Jr, PJ Wall *et al.*, Elevated CO_2, drought and soil nitrogen effects on wheat grain quality. New Phytologist. 2001; 150:295-303.

15. Kumar KSK, Parikh J. Climate change impacts on Indian agriculture: The Ricardian Approach. In: Measuring the Impacts of Climate Change on Indian Agriculture. Eds: Dinar A, Mendelsohn R, Evenson R, Parikh J, Sanghi A, Kumar K, McKinsey J, Lonergon S. World Bank Technical Paper No. 402. Washington, DC. 1998.

16. Loladze I. Rising atmospheric CO_2 and human nutrition: toward globally imbalanced plant stoichiometry? Trends in Ecology & Evolution. 2002; 17(10):457-461.

17. Long SP, Ainsworth EA, Rogers A, Ort DR. Rising atmospheric carbon dioxide: Plants face the future. Annual Review of Plant Biology. 2004; 55:591-628.

18. Manoj-Kumar. Soil responds to climate change: is soil science in India responding to? Curr Sci. 2010; 99(7):892-89.

19. Manoj-Kumar, Patra AK. Nutritional security: a missing link in climate change debates. Curr Sci. 2010; 99(7):887-889.

20. Manoj-Kumar, Bhadraray S. Enhanced nitrogen demand by wheat: Achallenge under rising atmospheric CO_2, In: Abstracts of 4th World Congress on Conservation Agriculture-February 4-7, 2009, 415.

21. Manoj-Kumar. Impact of climate change on agriculture in North East India. Bhartiya Krishi Anusandhan Patrika. 2011a; 26(1-2):58-61.

22. Manoj-Kumar. North East India: soil and water management imperatives for food security in a changing climate? Curr Sci. 2011b; 101(9):11-19.

23. Manoj-Kumar. Evidences, projections and potential impacts of climate change on food production in Northeast India. Indian J Hill Farming. 2011c; 24(1-2):1-10.

24. Manoj-Kumar. Will rising CO_2 affect Zn and Fe bioavailability in food grains? Curr Sci. 2011d; 101(4):467.

25. Manoj-Kumar, Swarup A, Bhadraray S, Patra AK *et al.*, Yield, nutritional quality and phosphorus requirement of wheat (*Triticum aestivum* L.) grown under projected elevation of atmospheric CO_2 and temperature in subtropical India. Agrochimica. 2012a; 56(3):156-174.

26. Manoj-Kumar, Swarup A, Patra AK, Chandrakala JU, Manjaiah KM. Effect of elevated CO_2 and temperature on phosphorus efficiency of wheat grown in an Inceptisol of subtropical India. Plant Soil Environ. 2012b; 58(5):230-235.

27. Nagarajan S, Joshi LM. Epidemiology of brown and yellow rusts of wheat over northern India. II. Associated meteorological conditions. Plant Disease Reporter. 1978; 62:186-8.

28. Nagarajan S, Prasad ASH, Chaudhary HB, Anand A *et al.*, Impact of increased temperature on quality of rice and wheat grains. In: Global Climate Change and Indian Agriculture. (Ed.) Aggarwal PK, 2009, 54-59.

29. NATCOM. India's Initial National Communication to the United Nations Framework Convention on Climate Change, 268 Ministry of Environment and Forests, 2004.

30. Nuttal GJ. Climate change- identifying the impacts on soil and soil health, department of primary industries, Horsham, Victoria, Australia, 2007.

31. Phadke KG, Ghai S. Effect of global warming on insect populations and crop damage. Shashpa. 1994; 1(2):75-80.

32. Rabindranath NH, Sukumar R. Impact of climate change on forest cover in India. Commonwealth Forestry Review, 1998.

33. Ramana Rao BV, Katyal JC, Bhalra PC. Final Report on Indo-USAID-Sus-Project, CRIDA, Hyderabad, India, 1994.

34. Rao BVR, Rao MR. Weather effects on pest. In: Climate Variability and Agriculture. Abrol YP, Gadgil S, Pant GB. (Eds.), Narosa Publishing House, New Delhi, 1996, 281-296.

35. Rao GD, Sinha SK. Impact of climatic change on simulated wheat production in India. In: Implications of Climate Change for International Agriculture: Crop Modelling Study (Eds). C. Rosenzweig and I. Iglesias), EPA, USA, 1994, 1-10.

36. Reji G, Singh VS, Satish DG. Simulation of population dynamics of rice gundhi bug, Leptocorisa acutaí, Proceedings of the National Symposium on Frontier Areas of Entomological Research, 5-7 November, Entomology Division, IARI, New Delhi, 2003, 20-21.

37. Rosenzweig C, Parry ML. Potential impact of climate change on world food supply. Nature. 1994; 367:133-138.

38. Sage RF. Was low atmospheric CO_2 during the Pleistocene a limiting factor for the origin of agriculture? Glob Chang Biol. 1995; 1:93-106.

39. Sinha SK, Swaminathan MS. Deforestation, climate change and sustainable nutrition security. Clim. Change. 1991; 16:33-45.

40. Smith J, Smith P, Wattenbach M, Gottschalk P *et al.*, Projected changes in the organic carbon stocks of cropland mineral soils of European Russia and the Ukraine, 1990–2070. Glob Chang Biol. 2007; 13:342-356.

41. Tripathi KP, Sharda VN, Ali MS, Mudgal A. Impact of climate change on soil erosion and runoff. In: Global Climate Change and Indian Agriculture. (Ed.) Aggarwal PK. Indian Council of Agricultural Research, New Delhi-110012, India, 2004, 85-88.

Chapter - 2
Climate Change and Crop Production

Authors

Dr. R.K. Naresh

Sardar Vallabhbhai Patel University of Agriculture & Technology, Meerut, Uttar Pradesh, India

Dr. R.K. Singh

Division of Agronomy, Banaras Hindu University, Varansi, Uttar Pradesh, India

Chapter - 2

Climate Change and Crop Production

Dr. R.K. Naresh and Dr. R.K. Singh

The unimpeded growth of greenhouse gas emission is raising the earth's temperature. The consequences include melting glaciers, more precipitation, more and more extreme weather events, and shifting seasons. The accelerating pace of climate change, combined with global population and income growth, threatens food security everywhere. Agriculture is extremely vulnerable to climate change. Higher temperatures eventually reduce yields of desirable crops while encouraging weed and pest proliferation. Changes in precipitation pattern increase the likelihood of short-run crop failure and long-run production declines. Although there will be gains in some crops in some regions of the world, the overall impacts of climate change on agriculture are expected to be negative, threatening global food security. Populations in the developing world, which are already vulnerable and food insecure, are likely to be the most seriously affected. In 2005, nearly half of the economically active population in developing countries- 2.5 billion people-relied on agriculture for its livelihood. Today, 75 percent of the world's poor live in rural areas. Modeling of crop growth under climate change with insights from an extremely detailed global agriculture model, using two climate scenarios to simulate future climate-suggest that agriculture and human well-being will be negatively affected by climate change. In developing countries, climate change will cause yield decline for the most important crops. South Asia will be particularly hard hit. Climate change will have varying effects on irrigated yields across regions, but irrigated yields for all crops in South Asia will experience large decline. Climate change will result in additional price increase for the most important agricultural crops-rice, wheat, maize, and soybeans. Higher feed prices will result in higher meat prices. As a result, climate change will reduce the growth in meat consumption slightly and cause a more substantial fall in cereals consumption. Calorie availability in 2050 will not only be lower than that in the no-climate-change scenario but also it will decline relative to 2000 levels throughout the developing world. By 2050, the decline in calorie availability will increase child malnutrition by 20 percent relative to a world with no climate change. Climate change will eliminate much of the improvement in child malnourishment levels that would occur with no climate change. Thus, aggressive agricultural productivity investments of US$ 7.1.7.3 billion are needed to raise calorie consumption enough to offset the negative impacts of climate change on the health and well-being of children. The accelerating changes in the earth's environment are occurring due to the growth in human population through increasing level of resource consumption, changes in the technologies and socio-political organizations. Changes in global climatic scenarios, which affect agriculture, are: a) Changes in land use and land cover b) Worldwide decline of biodiversity c) Changes in atmospheric composition of gases, and d) Changes in climate. One of the major driving forces for global climate change is the rapid increase in the greenhouse gas content in the atmosphere. The per capita CO_2 emission from south Asia was 0.47 Tg (Tera gram) during 1995 whereas, it's projected value is 0.90 Tg for 2050. The total regional emission of GHG (in CO_2 equivalents) is 3 percent of the global total emission. Its annual growth rate is as high as 6 per cent for India, 8 per cent for Bangladesh and 10 per cent for Pakistan (IPCC, 2001).

These changes to the environment will most likely cause negative effects on society, such as poor health and decreasing economic development. However, some scientists argue that the global warming we are experiencing now is a natural phenomenon, and is a part of Earth's natural cycle. Presently, lot of debate is going on the theory, but one thing is certain, the world has been emitting greenhouse gases at extremely high rates and has shown only small signs of reducing emissions until the last few years. After the 1997 Kyoto Protocol, steps have finally been taken for reducing emissions.

What is Greenhouse Effect?

The "greenhouse effect" is the heating of the Earth due to the presence of greenhouse gases. It is named this way because of a similar effect produced by the glass panes of a greenhouse. Shorter-wavelength solar radiation from the sun passes through Earth's atmosphere and then is absorbed by the surface of the Earth, causing it to warm. Part of the absorbed energy is then radiated back to the atmosphere as long wave infrared radiation. Little of this long wave radiation escapes back into space; the radiation cannot pass through the greenhouse gases in the atmosphere. The greenhouse gases selectively transmit infrared waves, trapping some and allowing some to pass through into space. The greenhouse gases absorb these waves and re-emit the waves downward, causing the lower atmosphere warming.

Fourth assessment report of IPCC says that coastal belts of India are more prone to devastating impacts of global warming. The rising sea levels due to climate change may force many poor communities in low-lying coastal areas to move to higher ground. The livelihood of coastal people is mainly dependent on fishery, forestry and aquaculture. Hence, the displacement of people will result in competition between migrants and original inhabitants for both accesses to land and livelihood. Changing sea-ecosystem will further aggravate the misery of small fishing communities as the quantum of fish catch will diminish due to spread of algae and other diseases.

Greenhouse Gas Overview

In order, Earth's most abundant greenhouse gases are: water vapor, carbon dioxide, methane, nitrous oxide, ozone and chlorofluorocarbons.

a) **Carbon Dioxide (CO_2):** Carbon dioxide enters the atmosphere through burning of fossil fuels (oil, natural gas, and coal), solid waste, trees and wood products, and as a result of other chemical reactions (e.g., manufacture of cement). Carbon dioxide is removed from the atmosphere (or. sequestered.) when it is absorbed by plants as part of the biological carbon cycle.

b) **Methane (CH_4):** Methane is emitted during the production and transport of coal, natural gas, and oil. Methane emissions also result from livestock and other agricultural practices and by the decay of organic waste in municipal solid waste landfills.

c) **Nitrous Oxide (N_2O):** Nitrous oxide is emitted during agricultural and industrial activities, as well as during combustion of fossil fuels and solid waste.

d) **Fluorinated Gases:** Hydrofluorocarbons, perfluorocarbons, and sulfur hexafluoride are synthetic, powerful greenhouse gases that are emitted from a variety of industrial processes. Fluorinated gases are sometimes used as substitutes for ozone-depleting substances (i.e., CFCs, HCFCs, and halons). These gases are typically emitted in smaller quantities, but because they are potent greenhouse gases, they are sometimes referred to as High Global Warming Potential gases (High GWP gases).

Global warming potential: The global warming potential (GWP) depends on both the efficiency of the molecule as a greenhouse gas and its atmospheric lifetime. GWP is measured relative to the same mass of CO_2 and evaluated for a specific timescale. Thus, if a gas has a high GWP on a short time scale (say 20 years) but has only a short lifetime, it will have a large GWP on a 20 year scale but a small one on a 100 year scale. Conversely, if a molecule has a longer atmospheric lifetime than CO_2 its GWP will increase with the timescale considered. Examples of the atmospheric lifetime and GWP for several greenhouse gases include:

a) Carbon dioxide has a variable atmospheric lifetime, and cannot be specified precisely (Solomon *et al.*, 2007). Recent work indicates that recovery from a large input of atmospheric CO_2 from burning fossil fuels will result in an effective lifetime of tens of thousands of years (Archer and Ganopolski, 2005). Carbon dioxide is defined to have a GWP of 1 over all time periods.

b) Methane has an atmospheric lifetime of 12 ± 3 years and a GWP of 72 over 20 years, 25 over 100 years and 7.6 over 500 years. The decrease in GWP at longer times is because methane is degraded to water and CO_2 through chemical reactions in the atmosphere.

c) Nitrous oxide has an atmospheric lifetime of 114 years and a GWP of 289 over 20 years, 298 over 100 years and 153 over 500 years.

d) CFC-12 has an atmospheric lifetime of 100 years and a GWP of 11000 over 20 years, 10900 over 100 years and 5200 over 500 years.

e) HCFC-22 has an atmospheric lifetime of 12 years and a GWP of 5160 over 20 years, 1810 over 100 years and 549 over 500 years.

f) Tetrafluoromethane has an atmospheric lifetime of 50,000 years and a GWP of 5210 over 20 years, 7390 over 100 years and 11200 over 500 years.

g) Hexafluoroethane has an atmospheric lifetime of 10,000 years and a GWP of 8630 over 20 years, 12200 over 100 years and 18200 over 500 years.

h) Sulphur hexafluoride has an atmospheric lifetime of 3,200 years and a GWP of 16300 over 20 years, 22800 over 100 years and 32600 over 500 years.

i) Nitrogen trifluoride has an atmospheric lifetime of 740 years and a GWP of 12300 over 20 years, 17200 over 100 years and 20700 over 500 years.

Greenhouse Effects on Earth's Atmosphere

The contribution to the greenhouse effect by a gas is affected by both the characteristics of the gas and its abundance. For example, on a molecule-for-molecule basis, methane is about eight times stronger greenhouse gas than carbon dioxide (Houghton, 2005), but it is present in much smaller concentrations so that its total contribution is smaller.

Forth Assessment Report compiled by the IPCC -changes in atmospheric concentrations of greenhouse gases and aerosols, land cover and solar radiation alter the energy balance of the climate system., concluded that -increases in anthropogenic greenhouse gas concentrations is very likely to have caused most of the increases in global average temperature since the mid-20th century. (Kiehl and Trenberth, 1997), where 'most of' is defined as more than 50%.

Anthropogenic Greenhouse Gases

The main sources of greenhouse gases due to human activity are:

a) Burning of fossil fuels and deforestation leading to higher carbon dioxide concentrations. Land use change (mainly deforestation in the tropics) account for up to one third of total anthropogenic CO_2 emissions (Kiehl and Trenberth, 1997).

b) Livestock enteric fermentation and manure management (Steinfeld *et al.*, 2006) in paddy rice farming, land use and wetland changes, pipeline losses, and covered vented landfill emissions lead to higher atmospheric methane concentrations. Many of the newer style fully vented septic systems that enhance and target the fermentation process also are sources of atmospheric methane.

c) Use of chlorofluorocarbons (CFCs) in refrigeration systems, and use of CFCs and halons in fire suppression systems and manufacturing processes.

d) Agricultural activities, including the use of fertilizers, which lead to higher nitrous oxide (N_2O) concentrations.

The seven sources of CO_2 from fossil fuel combustion are (with percentage contributions for 2000-2004):

1. Solid fuels (e.g., coal): 35%
2. Liquid fuels (e.g., gasoline, fuel oil): 36%
3. Gaseous fuels (e.g., natural gas): 20%
4. Flaring gas industrially and at wells: <1%
5. Cement production: 3%
6. Non-fuel hydrocarbons: < 1%

7. The international bunkers. Of shipping and air transport not included in national inventories: 4% The US Environmental Protection Agency (EPA) ranks the major greenhouse gas contributing end-user sectors in the following order: industrial, transportation, residential, commercial and agricultural (Raupach *et al.*, 2007). Major sources of an individual's greenhouse gas include home heating and cooling, electricity consumption, and transportation.

Corresponding conservation measures are improving home building insulation, installing geothermal heat pumps and compact fluorescent lamps, and choosing energy-efficient vehicles. On December 7, 2009, the US Environmental Protection Agency released its final findings on greenhouse gases, declaring that greenhouse gases (GHGs) threatened the public health and welfare of the American people. The findings applied to the same. six key well-mixed greenhouse gases- named in the Kyoto Protocol: carbon dioxide, methane, nitrous oxide, hydrofluorocarbons, perfluorocarbons, and sulfur hexafluoride (Fig 2.1).

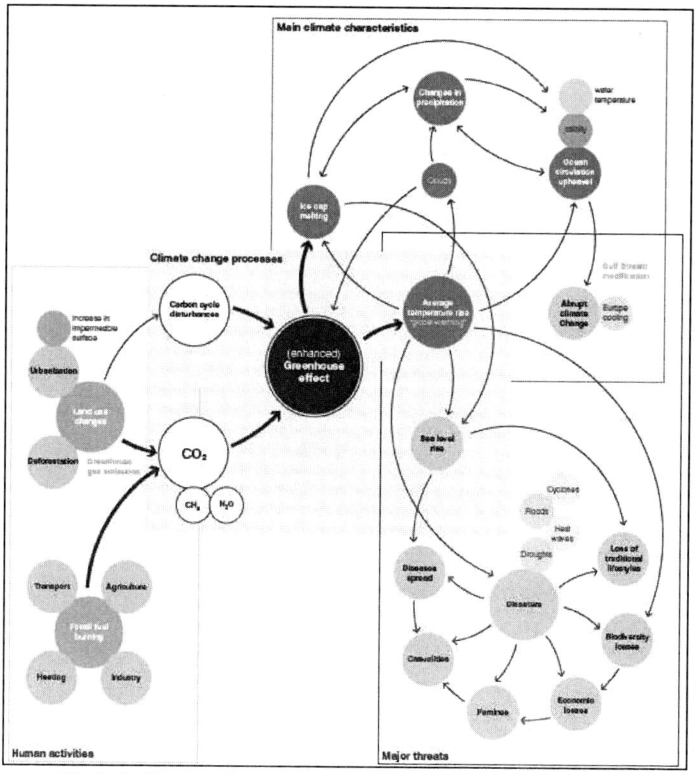

Fig 2.1: Climate change, processes, characteristic and threats

Role of Water Vapor

Water vapor accounts for the largest percentage of the greenhouse effect, between 36% and 66% from water vapor alone, and between 66% and 85% when factoring in clouds. However, the warming due to the greenhouse effect of cloud cover is, at least in part, mitigated by the change in the earth's albedo. According to NASA, the overall effect of all clouds together is that the Earth's surface is cooler than it would be if the atmosphere had no clouds.

Role of Water in Crop Production

Water is indispensable for human, animal and plant life. When plants are green, water constitute over 90 percent of plant body weight. Water availability influences various biochemical and physiological processes in plants. Water is one of the most important inputs essential for the production of crops. Plants need it continuously during their life and that too in huge quantities. It profoundly influences photosynthesis, respiration, absorption, translocation and utilization of mineral nutrients, and cell division besides some other processes.

Both its shortage and excess affect the growth and development of a plant directly and consequently, its yield and quality. In India, however rainfall is notoriously capricious, causing floods and droughts alternately. Its frequency distribution and amount are not in accordance with the needs of the crops. Artificial water supply through irrigation on one occasion, and the removal of excess water through drainage on another occasion, therefore, becomes imperative, if the crops are to be raised successfully. Water management in India, thus, comprises irrigation or drainage or both, depending considerably on the environmental conditions, soil, crops and climate. It is a situation oriented entity. Water affects the performance of crops not only directly but also indirectly by influencing the availability of other nutrients, the timing of cultural operations, etc. Water and other production inputs interact with one another. In proper combinations, the crop yields can be boosted manifold under irrigated conditions.

Water is Essential to Crops in the Following Ways

i. Water serves as solvent. Plants can absorb nutrients when these are present in solution form. Water itself acts as a nutrient also.
ii. Water is necessary for photosynthesis process and its end product is translocated to different parts of the plants in water solution.
iii. It helps in maintaining cell wall turgidity, cell enlargement and cell division which ultimately affects plant growth, when it is available

in plenty. Conversely, plants die when water supply is curtailed down.

iv. It is essential for seed germination, root growth and multiplication of microorganisms present in soil.

v. Due to its essentiality in the hydrolytic process, water helps in conversion of starch into sugar in the plants.

vi. Water is necessary in chemical, physical and biological reactions in soil.

vii. It protects the plant from adverse conditions such as high temperature, low temperature injury, drought and frost etc. Leaves get heated up with solar radiation.

viii. Plants dissipate heat by increased transpiration. Water acts as a buffer against high or low temperature injury as it has high heat of vaporization and high specific heat.

ix. The process of weathering and soil formation depends on water.

x. Good tilth can be secured at right stage of moisture content. Physical, chemical and biological activities in the soil are regulated by availability of water to a great extent.

xi. Transpiration is a vital process in plants and it occurs at a potential rate as long as water is available in adequate amount. If soil water deficit occurs, transpiration process is curtailed down seriously which affecting plant growth and yield.

Global GHG Emissions by Sector

Agriculture is particularly vulnerable to climate change. Projections to 2050 suggest both an increase in global mean temperature and increased weather variability, with implications for the type and distribution of agricultural production worldwide. Climate change will also worsen the living conditions for many who are already vulnerable, particularly in developing countries because of lack of assets and adequate insurance coverage. These impacts highlight key policy issues, including the need to produce more food for an increasing population. Projections of more than 9 billion people in 2050 suggest that food production will need to double from current levels.

Removal of Green House Gases

In order to limit future global warming to a 2°C temperature increase as recommended by IPCC, anthropogenic GHG emissions will have to decrease globally by at least 50% by 2050 from 1990 levels. Agriculture is not currently subject to emissions caps, although several OECD countries are already implementing mitigation action plans.

In addition to reducing its own emissions, carbon sequestration in agricultural soils can play an important role in offsetting emissions from other sectors. Some agricultural GHG mitigation options are cost competitive with a number of non-agricultural options in achieving long-term climate objectives.

Quantifying GHG emissions from agricultural activities is complex. First, the atomistic nature of production (many individual farmers) in a wide range of geographic and climatic conditions means that emissions are not only highly variable but also difficult and costly to measure precisely. Second, there continues to be a great deal of scientific uncertainty as GHG emissions from agriculture are subject to a complex interplay of many factors such as climate, soil type, slope, and production practices. Accounting for the indirect land use changes arising from agricultural production is another important challenge. The recent global surge in food prices highlighted the importance of agricultural policies for world food and energy markets. In particular, the links between production of biofuels from feedstock (in many cases subsidized), consequent land use changes, and food prices demonstrate the importance of foreseeing the range of consequences.

Greenhouse Gases can be removed from the Atmosphere by Various Processes

a) As a consequence of physical change (condensation and precipitation remove water vapor from the atmosphere).

b) As a consequence of chemical reactions within the atmosphere. For example, methane is oxidized by reaction with naturally occurring hydroxyl radical, OH^{-1} and degraded to CO_2 and water vapor (CO_2 from the oxidation of methane is not included in the methane global warming potential). Other chemical reactions include solution and solid phase chemistry occurring in atmospheric aerosols.

c) As a consequence of a physical exchange between the atmosphere and the other compartments of the planet. An example is the mixing of atmospheric gases into the oceans.

d) As a consequence of a chemical change at the interface between the atmosphere and the other compartments of the planet. This is the case for CO_2, which is reduced by photosynthesis of plants, and which, after dissolving in the oceans, reacts to form carbonic acid and bicarbonate and carbonate ions.

e) As a consequence of a photochemical change. Halocarbons are dissociated by UV light releasing Cl and F as free radicals in the stratosphere with harmful effects on ozone (halocarbons are generally too stable to disappear by chemical reaction in the atmosphere).

Climate Change Response Strategies for Agriculture

As a result of greenhouse gases already in the atmosphere from past and current emissions, our planet is already committed to at least as much warming over the 21st century as it has experienced over the 20th century (0.75°C). This implies that in addition to mitigation, adaptation to the anticipated warming is essential. Possible strategies for adapting food and forestry production to climate change have been identified (Schmidhuber and Tubiello, 2007).

Agriculture in the 21st century wills therefore, be undergoing significant challenges, arising largely from the need to increase the global food and timber supply for a world nearing a population of over 10 billion, while adjusting and contributing to respond to climate change. Success in meeting these challenges will require a steady stream of technical and institutional innovations, so that adaptation strategies to climate change are consistent with efforts to safeguard food security and maintain ecosystem services, including mitigation strategies that provide carbon sequestration, and offsets under sustainable land management (Easterling, 2007).

Physiological Changes and Agro-Ecological Impacts

Climate change will affect agriculture and forestry systems through a number of critical factors

a) Rising temperatures, can lead to negative impacts such as added heat stress, especially in areas at low-to-mid latitudes which are already at risk today. However, they can also lead to positive impacts, such as an extension of the growing season in high-latitude regions that are currently limited by low temperatures.

b) Elevated atmospheric CO_2 concentrations, which tend to increase plant growth and yield, and may improve water use efficiency, particularly in so-called C_3 carbon fixation plants such as wheat, rice, soybean, and potato. The impact on C_4 carbon fixation plants, such as maize, sugarcane, and many tropical pasture grasses, is not as pronounced due to different photosynthetic pathways (Easterling, 2007). How much agricultural plants in fields and trees in plantation forests benefit from elevated CO_2, given a number of limiting factors such as pests, soil and water quality, crop weed competition, remains an open question.

c) Changes in precipitation patterns, especially changes in frequency of the extremes, with both droughts and flooding events projected to increase in coming decades, leading to possible negative consequences for land-production systems. At the same time, a critical factor affecting plant productivity will be linked to simultaneous temperature and precipitation changes that influence soil water status and the ratio of evaporative demands to precipitation. All these factors, and their key interactions, must be considered together, across crops in different regions, in order to fully understand the impact that climate change will have on agriculture. Importantly, the experimental measurements of crop and pasture responses to changes in climate variables are still limited to small-scale plots, therefore, results are difficult to extrapolate to the field and farm level. As a consequence, current computer models of plant production, although quite advanced in their handling of soil-plant atmospheric dynamics as well as crop management, lack realistic descriptions of key limiting factors to real fields and farm operations. Therefore, the potential for negative impacts under climate change is not fully explored by current regional and global projections.

Key Interactions that are Currently Poorly Described by Crop and Pasture Models Include

a) No linearity and threshold effects in response to increases in the frequency of extreme events under climate change;

b) Modification of weed, pest, and disease incidence, including weed-crop competition;

c) Large-scale field response of crops to elevated CO_2 concentration; and

d) Interactions of climate and management variables, including effects of elevated CO_2 levels.

Regardless of these uncertainties, there is no doubt that plant development, growth, yield, and ultimately the production of crop and pasture species will be impacted by, and will respond to, increases in atmospheric CO_2 concentration, higher temperatures, altered precipitation and evapo-transpiration regimes, increased frequency of extreme temperature and precipitation events, as well as weed, pest and pathogen pressures (IPCC, 2001). Recent research has helped to better quantify the potential outcome of these key interactions.

Impacts

Higher Temperature

The Fourth Assessment Report of the IPCC (2007) provides a number of important considerations on the overall impacts of higher temperature on crop responses. The report suggests that at the plot level, and without considering changes in the frequency of extreme events, moderate warming (i.e., what may happen in the first half of this century) may benefit crop and pasture yields in temperate regions, while it would decrease yields in semiarid and tropical regions. Modeling studies indicate small beneficial effects on crop yields in temperate regions corresponding to local mean temperature increases of 1.3°C and associated CO_2 increase and rainfall changes. By contrast, in tropical regions, models indicate negative yield impacts for the major crops even with moderate temperature increases (1.2°C). Further warming projected for the end of the 21st century has increasingly negative impacts in all regions (Cline, 2007; Kimball *et al.*, 2002).

At the same time, farm-level adaptation responses may be effective at low to medium temperature increases, allowing coping with up to 1.2°C local temperature increases; an effect that may be considered as. Buying time. (Cline, 2007; Kimball *et al.*, 2002). Increased frequency of heat stress, droughts, and floods negatively affect crop yields and livestock beyond the impacts of mean climate change, creating the possibility for surprises, with impacts that are larger, and occurring earlier, than predicted using changes in mean variables alone.

Elevated Atmospheric CO_2 Levels

Studies conducted over the last 30 years have confirmed that plant biomass and yield tend to increase significantly as CO_2 concentrations increase above current levels. Such results are found to be robust across a variety of experimental settings-such as controlled environment closed chambers, greenhouses, open and closed field top chambers, as well as Free-Air Carbon dioxide Enrichment experiments (FACE). Elevated CO_2 concentrations stimulate photosynthesis, leading to increased plant productivity and modified water and nutrient cycles (Nowak *et al.*, 2004). Experiments under optimal conditions show that doubling the atmospheric CO_2 concentration increases leaf photosynthesis by 30.50 percent in C_3 plant species and by 10.25 percent in C_4 species, despite feedbacks that reduce the response of leaf photosynthesis by elevated atmospheric CO_2 concentrations (Ainsworth and Long, 2005). Increases in above ground biomass at 550 ppm CO_2 for trees are up to 30 percent. Plant physiologists and modelers recognize, however, that the effects of elevated CO_2, as measured in experimental settings and subsequently implemented in models, may over estimate actual field and farm-level responses, due to limiting factors such as pests, weeds, nutrients, competition for resources, and soil, water and air (Tubiello *et al.*, 2007).

Interactions of Elevated CO_2 with Temperature and Precipitation

High temperature during the critical flowering period of a crop may lower otherwise Positive CO_2 effects on yield by reducing grain number, size, and quality (Tubiello and Fischer, 2007). Increased temperature during the growing period may also reduce CO_2 effects indirectly, by increasing water demand. For example, yield of rainfed wheat grown at 450 ppm CO_2 increased up to 0.8°C warming, but declined beyond 1.5°C warming; additional irrigation was needed to counterbalance these negative effects (Centritto, 2005). Future CO_2 levels may favour C_3 plants over C_4; yet the opposite is expected under associated temperature increases. The net effects remain uncertain. Because of the key role of water in plant growth, climate impacts on crops significantly depend on the future precipitation scenario. Because more than 80 percent of total agricultural land -and close to 100 percent pastureland-is rainfed, Global Climate Model (GCM)-projected changes in precipitation will often shape both the direction and magnitude of the overall impacts (Olesen and Bindi, 2002). In general, changes in precipitation, and more specifically in evapo-transpiration to precipitation ratios, modify ecosystem productivity and function, particularly in marginal areas; higher water-use efficiency as a result of stomatal closure; greater root densities under elevated CO_2 may in some cases alleviate or even counterbalance drought pressures (Morgan *et al.*, 2004; Centritto, 2005).

Interactions of Elevated CO_2 with Soil Nutrients

Various FACE (Free air CO_2 enrichment) experiments confirm that high nitrogen content in the soil increases the relative response of crops to elevated atmospheric CO_2 concentrations (Nowak *et al.*, 2004). They demonstrated that the yield response of C_3 plant species to elevated atmospheric CO_2 concentrations is not significant under low nitrogen levels, but increases over 10 years with high levels of nitrogen rich fertilizer application. In fertile grasslands, legumes benefit more from elevated atmospheric CO_2 concentrations when compared to species that do not fix nitrogen (Ross *et al.*, 2004.). Therefore, to capitalize on the benefits of elevated CO_2 levels, declines in the availability of nitrogen may be prevented by biological N_2-fixation. However, other nutrients, such as phosphorus, an important nutrient for biological N-fixation, may act as a limiting factor and restrict legume growth response to higher atmospheric CO_2 concentrations.

Interactions with Air Pollutants

Tropospheric ozone has significant adverse effects on crop yields, pasture and forest growth, and species composition. Additionally, as global ozone exposures increase over this century, direct and indirect interactions with climate change and elevated CO_2 levels will further modify plant dynamics (Booker *et al.*, 2005). Although several studies confirm previous findings that elevated CO_2 concentrations may ameliorate otherwise negative impacts from ozone, it is important to note that increasing ozone concentrations in the future, with or without climate change, will negatively impact plant production and possibly increase exposure to pest damage. Ultraviolet (UV)-B exposure is in general harmful to plant growth, knowledge on the interactions between UV-B exposure and elevated CO_2 is still incomplete, with some experimental findings suggesting that elevated CO_2 levels mitigate the negative effects of UV-B on plant growth, while others show no effect.

Vulnerability of Carbon Pools

Impacts of climate change on the land that is under human management for food and livestock have the potential to significantly affect the global terrestrial carbon sink and to further perturb atmospheric CO_2 concentrations. Furthermore, the vulnerability of organic carbon pools to climate change has important repercussions for land sustainability and climate mitigation actions. Future changes in carbon stocks and net fluxes would critically depend on land use planning-policies, forestation /reforestation, and so on-and management practices such as nitrogen fertilization, irrigation, and tillage, in addition to plant response to elevated CO_2. Recent experimental research confirms that carbon storage in soil organic matter pools is often increased under elevated CO_2, at least in the short term (Allard *et al.*, 2005); yet the total soil carbon sink may become saturated at elevated CO_2 concentrations, especially when nutrient inputs are low (Gill, 2002). The effects of air pollution on plant function may indirectly affect carbon storage; recent research showed that tropospheric ozone resulted in significantly less carbon sequestration rates under elevated CO_2 (Rosenzweig, 2007) as a result of the negative effects of ozone on biomass productivity and changes to litter chemistry (Booker *et al.*, 2005). Finally, recent studies show the importance of identifying potential synergies between land-based adaptation and mitigation strategies, linking issues of carbon sequestration, emissions of greenhouse gases, land use change, and long-term sustainability of production systems within coherent climate policy frameworks (Antle, 2004). In addition, the predicted small global effects mask the fact that climate change is expected to disproportionately impact agricultural production in low-latitude, tropical developing countries, while some high-latitude, developed countries may benefit.

Impacts of Climate Change on Irrigation Water Requirement

A few new studies have quantified the impacts of climate change on regional and global irrigation requirements, irrespective of the positive effects of elevated CO_2 on crop water use efficiency. Considering the direct impacts of climate change on crop evaporative demand, in the absence of any CO_2 effects, an increase of net crop irrigation requirement is estimated, that is, net transpiration losses, of 5 to 8 percent globally by 2070, and larger regional signals, for example, 15 percent in southeast Asia (Doll, 2002). In another study, that included the positive CO_2 effects on crop water use efficiency, increases in global net irrigation requirement of 20 percent by 2080 were projected, with larger impacts in developed regions, due to increased evaporative demands and longer growing seasons under climate change.

Arnell (2004) has also projected increases in water stress- the ratio of irrigation withdrawals to renewable water resources -in the Middle East and Southeast Asia.

In developing countries like India, climate change could represent an additional stress on ecological and socioeconomic systems that are already facing tremendous pressure due to rapid urbanization, industrialization and economic development. With its huge and growing population, a 7500-km long densely populated and low-lying coastline, and an economy that is closely tied to its natural resource base, India is considerably vulnerable to the impacts of climate change. The various studies conducted in the country have shown that the surface air temperatures in India are going up at the rate of 0.4°C per hundred years, particularly during the post-monsoon and winter season. Using models, modelers predict that mean winter temperatures will increase by as much as 3.2°C in the 2050's and 4.5°C by 2080's, due to greenhouse gases. Summer temperatures will increase by 2.2°C in the 2050's and 3.2°C in the 2080's.

Extreme temperatures and heat spells have already become common over Northern India, often causing loss of human life. In 1998 alone, 650 deaths occurred in Odisha due to heat waves. Climate change has had an effect on the monsoons too. India is heavily dependent on the monsoon to meet its agricultural and water needs, and also for protecting and propagating its rich biodiversity. Scientists at IIT, Delhi, have already noted subtle changes in the monsoon rain patterns. They also warned that India will experience a decline in summer rainfall by the 2050s, summer rainfall accounts for almost 70% of the total annual rainfall over India and is crucial to Indian agriculture. Relatively small climatic changes can cause large water resource problems, particularly in arid and semi-arid regions such as northwest India. This will have an impact on agriculture, drinking water and on generation of hydroelectric power. Apart from monsoon rains, India uses perennial rivers, which originate and depend on glacial melt-water in the Hindukush and Himalayan ranges. Since the melting season coincides with the summer monsoon season, any intensification of the monsoon is likely to contribute to flood disasters in the Himalayan catchments. Rising temperatures will also contribute to the raising of snowline, reducing the capacity of this natural reservoir, and increasing the risk of flash floods during the wet season. Increased temperatures will impact agricultural production. Higher temperature reduces the total duration of a crop cycle by inducing early flowering, thus shortening the 'grain fill' period.The shorter the crop cycle, the lower the yield per unit area.

A trend of sea level rise of 1 cm per decade has been recorded along the Indian coast. Sea level rise due to thermal expansion of seawater in the Indian Ocean is expected to be about 25-040 cm by 2050. This could inundate low lying areas, down coastal marshes and wetlands, erode beaches, exacerbate flooding and increase the salinity of rivers, bays and aquifers. Flooding, erosion and salt intrusion will threaten deltas. Loss of coastal mangroves will have an impact on fisheries. The major delta area of the Ganga, Brahmaputra and Indus rivers, which have large populations reliant on reverie resources will be affected by changes in water regimes, salt water intrusions and land loss. Increase in temperatures will result in shifts of lower altitude tropical and subtropical forests to higher altitude temperate forest regions, resulting in the extinction of some temperate vegetation types. Decrease in rainfall and the resultant soil moisture stress could result in drier teak dominated forests replacing Sal trees in central India. Increased dry spells could also place dry and moist deciduous forests at increased risk from forest fires.

Medical Science suggests that the rise in temperature and change in humidity will adversely affect human health in India. Heat stress could result in heat cramps, heat exhaustion; heal stroke, and damage physiological functions, metabolic processes and immune systems. Increased temperatures can increase the range of vector borne diseases such as malaria, particularly in regions where minimum temperatures currently limited pathogen and vector development. Climate change will make monsoons unpredictable; as a result, rain-fed wheat cultivation in South Asia will suffer in a big way and the total cereal production will go down. The crop yield per hectare will be hit badly, causing food insecurity and loss of livelihood.

R. K. Pachauri, chairman of the IPCC estimated that a rise of 0.5 degree celsius in winter temperatures could cause a 0.45 tonne per hectare fall in India's wheat production. The average per hectare production in India is 2.6 tonnes. Worse still, Pachauri said, total agricultural land will shrink and the available land may not remain suitable for the present crops for too long. Farmers have to explore options of changing crops suitable to weather. He also pointed out that climatic changes could lead to major food security issues for a country like India. The report also predicts huge coastal erosion due to a rise in sea levels of about 40 cm resulting from faster melting of glaciers in the Himalayan and Hindukush ranges.

It can affect half-a-million people in India because of excessive flooding in coastal areas and also can increase the salinity of ground water in the Sunder bans and surface water in coastal areas. India needs to sustain an 8 to 10 per cent economic growth rate, over the next 25 years, if it is to eradicate poverty and meet its human development goals, according to a 2006 report on an integrated energy policy prepared by an expert committee of the Planning Commission. Consequently, the country needed at the very least to increase its primary energy supply three or four -fold over the 2003-04 level.

Impact of Climate Change on Agriculture - Factsheet on Asia

According to the study, the Asia-Pacific region will experience the worst effect on rice and wheat yields worldwide, and decreased yields could threaten the food security of 1.6 billion people in South Asia.

The crop model indicates that in South Asia, average yields in 2050 for crops will decline from 2000 levels by about 50 percent for wheat, 17 percent for rice, and about 6 percent for maize because of climate change.

In East Asia and the Pacific, yields in 2050 for crops will decline from 2000 levels by up to 20 percent for rice, 13 percent for soybean, 16 percent for wheat, and 4 percent for maize because of climate change.

With climate change, average calorie availability in Asia in 2050 is expected to be about 15 percent lower and cereal consumption is projected to decline by as much as 24 percent compared to a no climate change scenario.

In a no-climate change scenario, the number of malnourished children in South Asia would fall from 76 to 52 million between 2000 and 2050, and from 24 to 10 million in East Asia and the Pacific. Climate change will erase some of this progress, causing the number of malnourished children in 2050 to rise to 59 million in South Asia and to 14 million in East Asia and the Pacific, increasing the total number of malnourished children in Asia by about 11 million. To counteract the effects of climate change on nutrition, South Asia requires additional annual investments of 1.5 billion USD in rural development, and East Asia and the Pacific require almost 1 million USD more. Over half of these investments in both regions must be for irrigation expansion.

Additional Facts

The Asian countries most vulnerable to climate change are Afghanistan, Bangladesh, Cambodia, India, Lao PDR, Myanmar, and Nepal. Afghanistan, Bangladesh, India, and Nepal are particularly vulnerable to declining crop yields due to glacial melting, floods, droughts, and erratic rainfall, among other factors.

Asia is the most disaster-afflicted region in the world, accounting for about 89 percent of people affected by disasters worldwide.More than 60 percent of the economically active population and their dependents. 2.2 billion people- rely on agriculture for their livelihoods in developing parts of Asia.

In responding to the future challenges for agriculture of addressing climate change and increasing food demand, a coherent policy approach is needed that: Ensures a stable policy environment that sends clear signals to consumers and producers about the costs and benefits of GHG mitigating/sequestering activities.Provides a real or implicit price of carbon to create incentives for producers and consumers to invest in low-GHG products, technologies and processes. Fosters the application of existing technologies and invest in R&D for new technologies to reduce GHG emissions and increase productivity. Builds capacity to better understand and measure the GHG impact of agriculture for monitoring progress relative to national and international climate change goals.Facilitates adaptation by increasing producer resilience to climate change, and that compensate the most vulnerable groups.

References

1. Ainsworth EA, Long SP. What have we learned from 15 Years of Free-Air CO_2 Enrichment (FACE)? A Meta-Analytic review of the responses of photosynthesis, canopy. New Phytologist. 2005; 165(2):351-371.
2. Antle JM. Adaptation, spatial heterogeneity, and the vulnerability of agricultural systems to climate change and CO_2 fertilization: An integrated assessment approach. Climatic Change. 2004; 64(3):289-315
3. Archer D, Ganopolski A. A movable trigger: Fossil fuel CO_2 and the onset of the next glaciation. Geochemistry Geophysics Geosystems. 2005; 6(5):Q05003.
4. Antle JM, Capalbo SM, Elliott ET, Paustian KH. Climate change and global water resources: SRES emissions and socio-economic scenarios. Global Environmental Change-Human and Policy Dimensions. 2004; 14(1):31-52
5. Booker FL, Prior SA, Torbert, HA, Fiscus EL, Pursley WA, Hu S. Decomposition of soybean grown under elevated concentrations of CO_2 and O_3. Global Change Biology. 2005; 11(4):685-698
6. Centritto M. Photosynthetic limitations and carbon partitioning in cherry in response to water deficit and elevated CO_2. Agriculture Ecosystems & Environment. 2005; 106(2-3):233-242
7. Cline W. Global Warming and Agriculture: Impact Estimates by Country. Washington, DC: Peterson Institute for International Economics, 2007
8. Doll P. Impact of climate change and variability on irrigation requirements: A global perspective. Climatic Change. 2002; 54(3):269-293
9. Easterling W. Food, Fibre, and Forest Products. In: IPCC Working Group II Fourth Assessment Report. Cambridge: Cambridge University Press. 2007
10. Easterling WE. Climate change and the adequacy of food and timber in the 21st century. Proceedings of the National Academy of Sciences of the United States of America. 2007; 104 (50):19679.19679
11. Gill RA, Polley HW, Johnson HB, Anderson LJ, Maherali H, Jackson RB. Nonlinear grassland responses to past and future atmospheric CO_2. Nature. 2002; 417(6886):279-282
12. Houghton J. Global warming. Institute of Physics, 2005, 1362

13. IPCC. Climate Change 2001: Impacts, Adaptation, and Vulnerability: Contribution of Working Group II to the Third Assessment Report to the International Panel on Climate Change, J. J. McCarthy *et al*., (eds) Cambridge University Press: Cambridge, 2001

14. IPCC. Climate Change 2007: Mitigation of Climate Change, ed. B. Metz, *et al*., Cambridge and New York: Cambridge University Pres. 2007

15. Kiehl JT, Trenberth KE. Earth.s annual global mean energy budget. Bulletin of the American Meteorological Society. 1997; 78 (2):197-208

16. Kimball BA, Kobayashi K, Bindi M. Responses of agricultural crops to Free-Air CO_2 Enrichment. Advances in Agronomy. 2002; 77:293-368

17. Morgan JA, Pataki DE, Körner C, Clark H, Del Grosso, *et al*., Water relations in grassland and desert ecosystems exposed to elevated atmospheric CO_2. Oecologia. 2004; 140 (1):11-25

18. Nowak RS, Ellsworth DS, Smith SD. Functional responses of plants to elevated atmospheric CO_2.Do photosynthetic and productivity data from FACE experiments support early predictions? New Phytologist. 2004; 162(2):253-280

19. Olesen JE, Bindi M. Consequences of climate change for European agricultural productivity, land use and policy. European Journal of Agronomy. 2002.; 16(4):239-262

20. Raupach MR, Marland G, Ciais P, Le Quéré C, *et al.,* Global and regional drivers of accelerating CO_2 emissions. Proceeding of the National Academy of Science of the United State of America. 2007; 104 (24):10288-93

21. Rosenzweig C, Casassa G, Karoly DJ, Imeson A, *et al.,* Assessment of observed changes and responses in natural and managed systems. Climate Change 2007: Impacts, Adaptation and Vulnerability. Contribution ofWorking Group II to the Fourth Assessment Report of the Intergovernmental Panel on Climate Change, M.L. Parry, O.F. Canziani, J.P. Palutikof, P.J. van der Linden and C.E. Hanson, Eds., Cambridge University Press, Cambridge, UK, 2007, 79-131.

22. Ross DJ, Newton PCD, Tate KR. Elevated CO_2: Effects on herbage production and soil carbon and nitrogen pools and mineralization in a species-rich, grazed pasture on a seasonally dry sand. Plant and Soil. 2004; 260(1-2):183-196

23. Schmidhuber J, Tubiello FN. Global food security under climate change. Proceedings of the National Academy of Sciences of the United States of America. 2007; 104(50):19703-19708

24. Steinfeld H, Gerber P, Wassenaar T, Castel V, *et al.*, Lifestock's long shadow. Environmental Issues and Options. FAO, Rome. 2006

25. Tubiello FN, Amthor JS, Boote KJ, Donatelli M, *et al.*, Crop response to elevated CO_2 and world food supply.A comment on. Food for Thought. Science. 2006; 312:1918-1921

26. Tubiello FN, Fischer G. Reducing climate change impacts on agriculture: Global and regional effects of mitigation, 2000-2080. Technological Forecasting and Social Change. 2007; 74 (7):1030.1056

Chapter - 3
Climate Change and Resilient Agriculture

Authors

Dr. S.S. Dhaliwal
Department of Soil Science, Punjab Agricultural University, Ludhiana, Punjab, India

Dr. R.K. Naresh
Sardar Vallabhbhai Patel University of Agriculture & Technology, Meerut, Uttar Pradesh, India

Chapter - 3

Climate Change and Resilient Agriculture

Dr. S.S. Dhaliwal and Dr. R.K. Naresh

Green house gas (GHG) emissions in diversified agriculture on reclaimed sodic land: Estimated total GHG emission, i.e. global warming potential per hectare in terms of CO_2 equivalent (CO_2-eq) varied amongst different crop production systems. On an average, rice-wheat system emitted 1,823 kg CO_2-eq, whereas maize-wheat, fodder, vegetables and horticulture emitted 410, 245, 188 and 117 kg CO_2-eq from the respective areas of these cropping systems. The total emission from 1.8 ha area under diversified cropping system was 2,784 kg CO_2- eq as compared to 5,152 kg CO_2-eq from the rice– wheat system in the same area. On hectare basis, diversified agriculture system emitted 1,547 kg CO_2- eq/ha as compared to 2,862 kg CO_2-eq/ha in rice-wheat system. The global warming potential under diversified agriculture system was 46% (1,316 kg CO_2-eq/ha) less than that of rice-wheat.

Quantification of Green House Gas Emissions

Measurement of GHG fluxes (carbon dioxide and methane), moisture and heat in the soil-plant-atmosphere systems using the Eddy Covariance technique was carried out in rice-wheat rotation. The net ecosystem carbon dioxide exchange (NEE) and net ecosystem methane exchange (NEME) were monitored in rice and wheat crops and also during the fallow season (Fig 3.1). The gross primary productivity (GPP) and NEE was found to be highest between heading to ripening stage in rice, whereas it was maximum at flowering stage in wheat.The net ecosystem methane exchange during rice growth period was the highest between active tillering to maximum tillering stage in rice. The diurnal variations in mean net ecosystem exchange in submerged rice ecosystem in both dry and wet seasons varied between + 0.2 to - 1.2 and + 0.4 to - 0.8 mg $CO_2/m^2/s$. The cumulative seasonal methane emission was reduced by 75% in aerobic rice when compared with continuously flooded rice. The seasonal emissions were lower in slow release N fertilizer, especially, when applied on the basis of Customized Leaf Color Chart (CLCC).

Measurement of CH_4 and N_2O Fluxes in Rice

Fluxes of methane CH_4 and N_2O were measured under different establishment methods and source of nitrogen in irrigated rice in a Typic Ustochrept soil. Maximum methane flux was noticed under the treatment of transplanted puddle and normal urea application combination compared to all other treatment combinations at different stages of the rice crop. Closed chamber method was used to collect air samples from the experimental field and the air samples were analyzed by Gas Chromatography (Model SRI-8601C) equipped with a detector ECD and FID. The highest methane flux occurred in 72 days after sowing in all the treatment combinations. There is 54% reduction in CH_4 flux under the combination of aerobic rice and neem coated urea application followed by 49% reduction in aerobic rice and sulphur coated urea application compared to the combination of conventional puddling and normal application of urea. There is 33% reduction in N_2O flux under the combination of aerobic rice and neem coated urea application compared to conventional puddling and normal urea application.

Fig 3.1: Closed chambers used for collection of air samples at regular intervals from the paddy fields

Table 3.1: Predicted effects of climate change on agriculture over the next 50 years

Climatic Element	Expected Changes by 2050's	Confidence in Prediction	Effects on Agriculture
CO_2	Increase from 360 ppm to 450 - 600 ppm (2005 levels now at 379 ppm)	Very high	Good for crops: increased photosynthesis; reduced water use
Sea level rise	Rise by 10 -15 cm Increased in south and offset in north by natural subsistence/rebound	Very high	Loss of land, coastal erosion, flooding, salinisation of groundwater

Temperature	Rise by 1-2°C. Winters warming more than summers. Increased frequency of heat waves	High	Faster, shorter, earlier growing seasons, range moving north and to higher altitudes, heat stress risk, increased evapotranspiration
Precipitation	Seasonal changes by ± 10%	Low	Impacts on drought risk' soil workability, water logging irrigation supply, transpiration
Storminess	Increased wind speeds, especially in north. More intense rainfall events.	Very low	Lodging, soil erosion, reduced infiltration of rainfall
Variability	Increases across most climatic variables. Predictions uncertain	Very low	Changing risk of damaging events (heat waves, frost, droughts floods) which effect crops and timing of farm operations

Source: Climate change and Agriculture, MAFF (2000)

Current projections, from the 4th Assessment report by the Intergovernmental Panel on Climate Change (IPCC) published in 2007, suggest that global temperatures will rise between 1.8°C and 4.0°C (best estimate) by 2100 depending on emissions of greenhouse gases and that global sea level are likely to rise from anywhere between 180mm and 590mm.

Climate Resilient Agriculture can be defined as 'agriculture that reduces poverty and hunger in the face of climate change, improving the resources it depends on for future generations.' (Christian Aid, Time for Climate Justice 2015). Climate Resilient Agriculture wants to transform the current systems, and has a wider perspective than increased production only. It supports food production systems at local, regional and global level that are socially, economically and environmentally sustainable.

Strategies of Advancing Food Security while Mitigating Climate Change

Improving agricultural production is the basic strategy for both advancing food security and mitigating climate change. Indeed, agriculture must be the engine of economic development in SA, SSA, and elsewhere in the developing world. It is difficult to achieve poverty reduction without prior investment in agriculture (Lipton, 2005), especially in countries where majority of the population is rural and is dependent on agriculture. Developing countries in SA, SSA and elsewhere may not be able to advance economically and industrially without replacing extractive and subsistence farming (Katers, 2000; Funk and Brown, 2009) by science-based agriculture comprising of proven technologies and recommended management practices

(RMPs). In addition to advancing food security and alleviating poverty, raising low crop yields, and breaking the agronomic barriers in SA and SSA (especially in rain-fed/dry farming agro ecosystems), SCS may also be the judicious strategy to adapt to climate change (Table 3.2). The fact is that even the modest increase (20% to 30%) in production in rain-fed agriculture in SA and SSA through SCS can have a drastic positive impact on the standard of living and wellbeing of hundreds of millions of people now living below the poverty line.

Table 3.2: Per capita cereal production in 2007 and 2030 in different regions of Asia and the World (Modified from Funk and Brown, 2009)

Region	Per Capital Cereal Production (kg ha^{-1})		Per Capita Cropland Area (ha) 2007
	2007	2030	
World	354	306	0.11
North America	1374	1236	0.23
Asia			
Eastern Asia	314	276	0.06
Southern Asia	231	193	0.09
South- Eastern Asia	368	356	0.10
Western Asia	204	160	0.10
Central Asia	541	-	0.32
Sub-Saharan Africa			
Eastern Africa	131	84	0.09
Middle Africa	62	38	0.06
Northern Africa	190	180	0.11
Southern Africa	182	189	0.07
Western Africa	189	166	0.16

Adaptation to Climate Change

Strategies for adaptation to climate change, reducing risks to agriculture by adverse changes in growing season through alterations in precipitation and temperature regimes, are different at farm, national and global scales (Fig 3.2). At the farm level, the goal is to improve soil quality and adjust farming operations to buffer against the adverse effects of climatic disruption. Improvements in soil quality can be brought about by RMPs of soil management such as mulch farming (Acharya and Sharma, 1994), conservation agriculture (Acharya *et al.*, 1998), integrated nutrient management (INM), water harvesting and recycling through drip sub-irrigation (DSI) and judicious landscape management. Increasing fertilizer/N use efficiency, which in India decreased from 60 kg grain/kg of N applied in 1980 to 25 kg grain/kg of N applied in 2005 (IFA, 2007), is essential to increasing crop yields per unit consumption of energy-based inputs. Replacement of flooded rice by aerobic rice is important for saving water and sustaining crop yields (Kreye *et al.*, 2009; Bouman *et al.*, 2007; Ladha *et al.*, 2009). Application of bio-solids as soil amendments (e.g., compost, manure) is extremely important to improving productivity (Bejbaruha *et al.*, 2009), and creating a positive C budget and enhancing the ecosystem C pool. Growing improved varieties (including GM crops) and selecting alternate crops which create diverse cropping/farming systems are essential to enhancing soil resilience. Climate change will create numerous opportunities of growing alternate crops, and availing such an opportunity can improve household income, enhance production, and increase access to food.

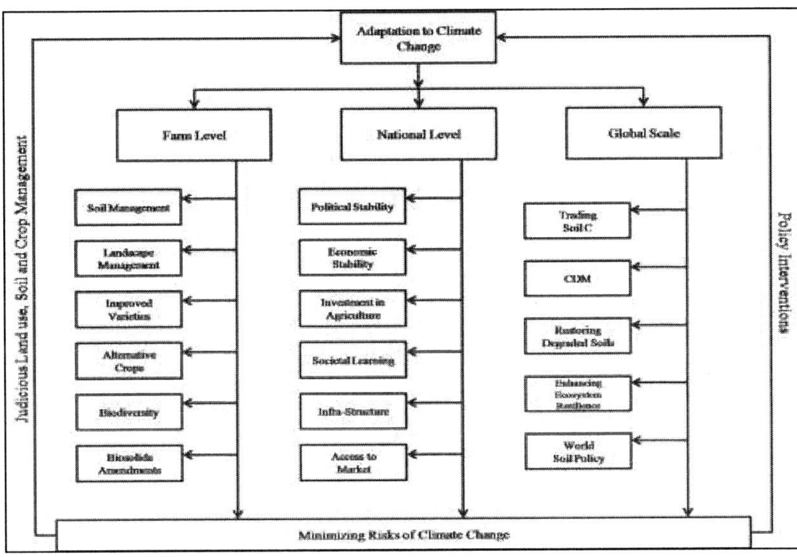

Fig 3.2: Strategies for adaptation to climate change

There are essential conditions of enhancing adaptation to climatic disruption at the national level. Important among these are political, economic and social stability, investments in agriculture, strengthening institutional support, and improving access to market and credit. At the global scale, the strategy is to restore degraded soils and desertified ecosystems, create opportunities for clean development mechanisms (CDM), formulate and implement World Soil Policy, and enhance ecosystem resilience (Fig 3.2). It is important to commoditize soil C, so that it can be traded as any other farm commodity. It is better to reward farmers for delivering ecosystem services (e.g., C sequestration, water quality improvement, biodiversity enhancement) than to provide subsidies, and give emergence aid as handouts.

Soil Carbon Sequestration to Mitigate Climate Change

World soils comprise the third largest global pool after the oceanic pool of 38000 Pg, the geologic/ fossil C pool of ~5000 Pg. The soil C pool is estimated at 2500 Pg, which has two distinct but related components: (i) the soil organic carbon (C) pool of about 1550 Pg, and (ii) the soil inorganic carbon (SIC) pool of about 950 Pg. Estimated to 1-m depth, the total soil C pool of about 2500 is about 3.2 times the atmospheric pool of 780 Pg, 40 times the biotic pool of 620 Pg, and about 1.8 times the terrestrial pool of 1400 Pg which is equal to combined atmospheric and biotic pools together. The atmospheric pool is increasing annually at the rate of 3.3 Pg C/yr (WMO, 2008; IPCC, 2007), while the terrestrial pool is decreasing because of deforestation, biomass burning, soil cultivation and drainage of wet/peat lands. There is a direct link between the soil and the atmospheric C pools, 1 Pg of soil C pool is equivalent to 0.47 ppm of CO_2 concentration in the atmosphere, and vice versa.

Conversion of natural and agricultural ecosystems leads to decline in the terrestrial (biotic and soil C) pools. The decline is caused by removal of the perennial vegetation cover (e.g., trees, shrubs, grasses) by seasonal/annuals which have low NPP. Thus, the amount of biomass-C returned to the soil is lower in agricultural than natural ecosystems. Furthermore, soils under agricultural ecosystems have higher decomposition rate than those under natural vegetation because of differences in soil temperature and moisture regimes. The rate of depletion of the SOC pool in agricultural ecosystems is exacerbated by soil degradation processes such as accelerated erosion by water and wind. Excessive tillage and drainage accentuate the rate of SOM decomposition. The rate and magnitude of depletion of SOC pool are also accentuated by removal of crop residues and uncontrolled grazing. Thus, most agricultural soils have lower SOC pool than their counterparts under natural ecosystems. The SOC pool in agricultural soils is lower than those under natural ecosystems by 25% to 50% in temperate climates and 50% - 75% in tropical regions (Lal, 2004). Most agricultural soils have lost 30 to 40 Mg C/ha, which can be restored. Therefore, conversion of degraded and depleted agricultural soils to restorative land use(s) and adoption of RMPs can restore the SOC pool. The process of restoration of SOC pool, through conversion of atmospheric CO_2 into humus via photosynthesis, is called soil C sequestration.

Similarly, the processes of increasing the terrestrial C pool (biota and soil) through photosynthesis and incremental addition of biomass C (above and belowground) is called terrestrial C sequestration. Assuming that the SOC pool to 1-m depth can be increased by 10% on a global scale over the next 50 to 100 years, difficult and challenging as the task may be, transfer of 250 Pg of atmospheric C 910% of 2500 Pg of soil C pool to 1-m depth) is equivalent to reducing atmospheric CO_2 concentration by 118 ppm (250 Pg x 0.47 ppm/Pg). If the process can be presumably accomplished today (which is not possible), increasing the soil C pool by 250 Pg would draw down the atmospheric CO_2 concentration from 383 ppm to 265 ppm, which is even lower than the pre-industrial concentration of 280 ppm. Indeed C sequestration in terrestrial ecosystems (soils and trees) is the only cost-efficient process of reducing the atmospheric abundance of CO_2 (McKinsey & Co., 2009). The best one can hope to achieve with the carbon capture and sequestration (CCS or geologic sequestration), is to stabilize the atmospheric concentration with 100% efficiency. In most cases, the efficiency of CCS with 15% additional cost of generating electricity is about 30%.

Ever since the dawn of settled agriculture about 10 millennia ago, the terrestrial ecosystems have lost as much as 478 Pg of C (Ruddiman, 2007; IPCC, 2007). Even if two-thirds of this can be restored (through afforestation, soil amelioration, and restoration of wetland/peat soils by inundation), this means increasing the terrestrial C pool by ~320 Pg or the equivalent to draw down the atmospheric CO_2 by 150 ppm. These estimates of 120 to 150 ppm of draw down of atmospheric CO_2 are indicative of the technical potential of the strategy of C sequestration in terrestrial ecosystems. The attainable and economic potentials are much lower and depend on the site specific and regional conditions. The rate of C sequestration in terrestrial ecosystems is 3-4 Pg C/yr until 2050 (Pacala and Socolow, 2006) through adoption of no-till farming and conservation agriculture, afforestation, and establishment of energy plantations.

Co-Benefits of Soil Carbon Sequestration

Sequestration of atmospheric CO_2 into the terrestrial ecosystems in general and soils in particular, is a win-win-win strategy. Among numerous co-benefits are: (i) advancing food security through improvements in soil quality, (ii) restoring degraded soils and desertified ecosystems, and (iii) mitigating climate change while improving the environment. It is extremely important to realize that enhancing SOC pool and maintaining it above the threshold (1.1% in the surface layer, Aune and Lal, 1998) is essential to advancing food security. For malnourished and resource-poor small landholders who are perpetually threatened by drought and crop failure, enhancing soil resilience and guaranteeing a minimum assured crop yield even during the worst seasons of below-normal rains is important to their survival. Thus, soil C sequestration is essential to their survival. Rather than being driven into the futile debate of who is responsible for gaseous emissions, food-insecure population would appreciate C-positive action towards elimination of hunger and malnutrition for which soil C sequestration is an important step and in the right direction. Commoditization of soil C can also create another income stream and provide the much-needed incentives for adoption of RMPs. Land managers can benefit by trading soil C in the emerging C-offset markets and by claiming the environmental benefits of the sustainable land management practices.

Similar to C, the farmers can also be paid for water management. Downstream farmers may be requiredto pay upstream land managers for specific practices which ensure availability of good quality water.

Payments for green water credits and charging users a fair price of the scarce water resource may be the only strategy to reverse the severe depletion of the ground water resources in northern India, as has been reported by Kerr (2009), Rodell *et al.*, (2009). Both water credits and C credits are important to adoption of RMPs, and sustaining the resource use. Sustainable crop management is important to alleviating poverty (Lipton, 2005). Furthermore, this cost-effective option of mitigating climate change is also a bridge to the future until C-neutral or C negative fuel sources take effect.

Diffusing the Population Bomb through Improvement in Agriculture

The world population has doubled numerous times since the primate started walking on two feet and descended from their habitat in trees. It was merely 5 million around 8,000 BC, 10 million in 3500 BC, 50 million by 1250, 800 million by 1750 and 1650 million by 1900. It has doubled twice during the 20th century to 6.7 billion in 2009. World population is expected to stabilize at ~10 billion around 2100, and will, thus, never double again. Evolution of agriculture about 10,000 years ago and its scientific advances especially during the 20th century have been responsible for increase in population. However, the future increases in population will also be moderated by improvements in agricultural production. Just as the quantum jump in agronomic production by The Green Revolution saved billions from hunger and malnutrition. Similarly, using improved agriculture as the engine of economic development in Africa and Asia would dampen the rate of population growth and stabilize world population lower than the 10 billion mark (6 or 7 billion by 2200 and beyond). Advancing food security, enhancing income, and improving standards of living are essential to reducing the rate of population growth. It is in this regard that the importance of improving soil quality and increasing agronomic production cannot be over-emphasized.

Managing Carbon-Climate Risks through Soil Sequestration

Two strategies of managing risks of climate change are: (i) reducing anthropogenic emissions, and (ii) sequestering emissions. Reducing anthropogenic emissions is the best strategy of which identifying low-C or no-C fuel sources and improving use efficiency of power are among numerous options. Sequestering emissions through the natural process of photosynthesis, and humification of biomass is the next best option and is a priority consideration. Carbon capture and storage (Chu, 2009; Normile, 2009) is of a lesser importance, especially when the first two options are judiciously and prudently implemented. While soil C sequestration is not a silver bullet for mitigating climate change, it is essential for numerous co-benefits which are important to human wellbeing and other ecosystem services. These co-benefits are illustrated in (Fig 3.3).

It is these co-benefits of soil C sequestration which are important consideration in choosing it as a strategy for managing carbon-related climate risks. Resource-poor farmers and food-insecure people care more about increase in agronomic production and minimum assured crop yields through improvements in soil quality than the potential of mitigating climate change. Improvements in the quantity and quality of renewable fresh water resources are another important co-benefit of C sequestration in soils and terrestrial ecosystems. Establishment of energy plantations, especially on degraded croplands and other lands of marginal soil quality can also enhance ecosystem services from land of otherwise low utility.

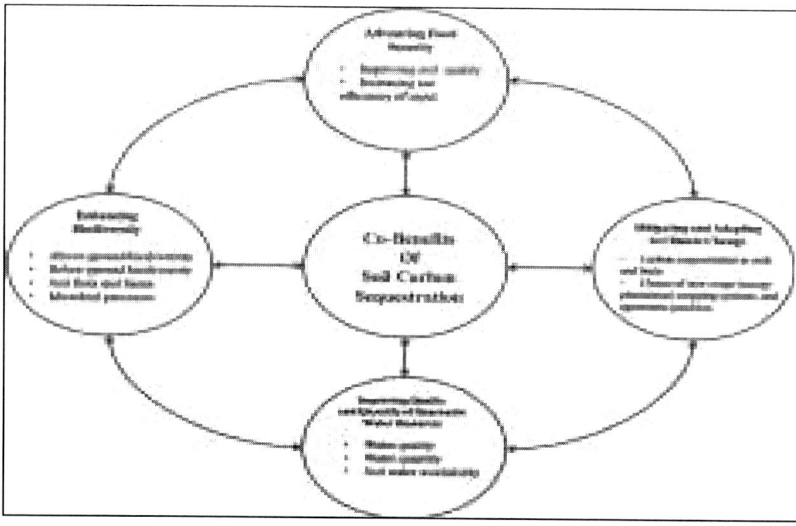

Fig 3.3: Co-benefits of carbon sequestration in soils and terrestrial ecosystems

Tenets of Soil Management

Soils must never be taken for granted. Soils must be used, improved and restored for generations to come. Sustainable management of soils is essential through observing 10 tenets of soil management (Lal, 2009a; b; Fig 3.4).

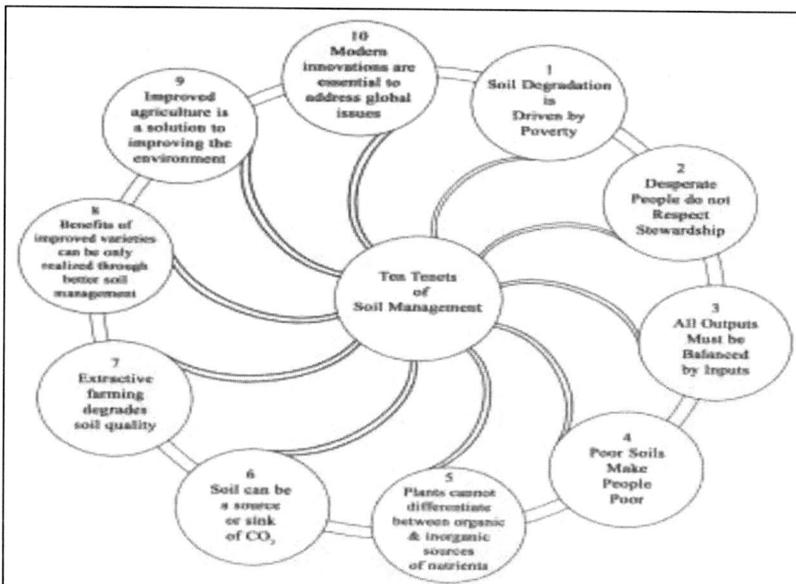

Fig 3.4: Ten tenets of soil management (Adapted from Lal, 2009b).

Ten tenets of soil management, constituting as spokes of a wheel representing the ecosystem services, include the following: (i) soil degradation by biophysical processes is driven by social, economic and political forces, (ii) desperate and helpless people do not care for the stewardship, (iii) to be sustainable, all outputs must be balanced by equivalent inputs in terms of nutrients and carbon, (iv) poor soils make people poor, and poor people make soils worst, (v) plants cannot differentiate between organic and inorganic sources of nutrients, and it is a matter of logistics and access to supply nutrients to crop at the most critical stages and for the amount required, (vi) soils can be a source or sink of atmospheric CO_2 depending on land use and management, (vii) a perpetual and widespread use of extractive farming practices can degrade the soils and the environment, (viii) benefits of improved varieties and GM crops can only be realized when grown under optimal soils and agronomic conditions, (ix) improved agriculture is a solution (rather than a cause) of the environmental problem, and (x) adoption of modern innovations of soil management, built upon the traditional knowledge, is essential to addressing the global issues.

Agriculture as a Solution for Climate Change

The agricultural sector holds significant climate change mitigation potential through reductions of GHG emissions and enhancement of agricultural sequestration. In addition, it also has significant role to adapt climate change. Adaptation alone is not enough to offset the effects of climate change, and thus still need to be supplemented by concerted mitigation efforts Vuren *et al.*, (2009). Mostly, when we implement adaptation measure, we enhance mitigation capacity of particular area such as practicing different land use managements (soil and water conservation measure, manure and fertilizer management) in the agricultural field will help us to sequester substantial amount of carbon in the field and reduce emission of methane and nitrous oxide which are the main GHG emission means. Therefore, the management activities are interrelated and help us to adapt and mitigate climate change. Agricultural activities are relatively affordable form of mitigation option, for which many technical options are already readily available FAO, (2010a).

Global adoption of organic agriculture (OA) has the potential to sequester up to the equivalent of 32% of all current man-made GHG emissions FAO, (2009). OA is a production system that sustains the health of soils, ecosystems and people. In OA, soil fertility is maintained mainly through farm internal inputs (organic manures, legume production, wide crop rotations etc); energy-demanding synthetic fertilizers and plant protection agents are rejected; and there is less or no use of fossil fuel DECD, (2015). In relation, improved cropland management (lower use of synthetic fertilizers, reduced tillage etc), Reducing industrial livestock production and improving feeding and grazing land management, Restoration of organic soils and degraded lands to increase soil carbon sinks, Improved water and rice management, Land-use change and agro-forestry, Increasing efficiency in fertilizer production and behavioural changes of food consumers (reducing the meat content) could also be main climate change mitigation measures in agriculture sector Robert *et al*., (2009).

As earlier mentioned, the agricultural sector has also a potential to adapt to climate change in many areas. Climate change adaptation is a continuous process requiring location-specific response. Adaptation should enable agricultural systems to be more resilient to the consequences of climate change Paul *et al*., (2009). Farming systems and farmers will differ enormously in their capacities to respond to climate change. Differentiated adaptation strategies and enhanced climate risk management support to agriculture and farming households are critical to counter the impacts of climate change IPCC, (2007b). These adaptation measures could include in particular the choice and change of species and varieties, the adaptation of the field works to the calendar (more flexibility), the adaptation of plant production practices (i.e. fertilization, plant protection, irrigation, etc.) or the adoption of plant production practices that increase the soil organic matter content or the soil coverage by plants, manure management and agroforestry practices. Some of them discussed below how these practices serve as adaptation means:

Change Crop Variety

It involves switching from one crop variety to another in response to climatic stresses and changes. Study done by Komba and Muchapondwa FAO, (2011), in Tanzania explained that Tanzania's farmers try to adapt climate change by using drought resistance crops. Introducing Avena species (Ingedo) species in Ethiopia as fodder crop and through time it replace the dominant stable crop i.e., barley in the highland and serve as one means to adapt CC Denman *et al*., (2007).

Change in Cropping Pattern

Application of changes in how crops are cycled within a season. Farmers in the drought-prone semi-arid areas of Brazil have realized that several varieties of a single crop species can occupy a common land area, incorporating several bean varieties, maize and sorghum, among others, to increase harvest potential arid climate stresses Komba and Muchapondwa, (2015). Likewise, In Ethiopia farmers try to adapt CC by planting diversifies crops (homestead maize and other crops).

Change in Cropping Calendar

It is another common adaptation to climate change at the farm level, which largely involves altering the timing of farm activities to suit climatic variations or changes. In Philippines, farmers adapt to the early onset of rainy season through early cultivation of upland farms, which results in high agricultural production for the season and higher household income from farm activities Action Aid, (2008). Tanzania's farmers also used planting date changing practice to adapt CC Lasco and Pulhin, (2009). In addition, according to Rhodes *et al.*, (2014) most West Africa counties such as Burkina Faso, Niger and Senegal already develop and implement a mathematical model for different crops to plant under changed climate by shifting planting date to adapt CC. Changing planting dates of crops helps greatly for the farmers live in East Gojam (Choke mountain in Ethiopia) and East Hararghe for CC adaptation respectively.

Farm Management Practices

Change in current farm management practices such as OA practice focus on maintaining diverse farming systems (i.e. planting different crop species) also helps diversify potential sources of income for farmers, making the farming household more resilient to adverse impacts of climate on agricultural production Lasco and Pulhin, (2009). According to Rhodes *et al.*, (2014) also crop residue management practice is considered one of the best climate smart actions. In addition, Smallholder farmers in sub-humid Southwestern Cameroon have been adapting to variations in rainfall through different soil and water conservation practices. Ditch and check dam constructions are the major soil and water conservation (SWC) practice that helps to adapt CC in Ethiopia. Similarly, study in Ethiopia revealed that terracing and different water harvesting practices widely used to adapt the changed climate.

Barriers to Adapt and Mitigate Climate Change in Agriculture Sector

According to Smith *et al.*, (2007b) study, the following condition considered as the main barrier to mitigate CC; it need long time (15-60 years) to obtain maximum storage that is carbon sequestration in soils or terrestrial biomass, depending on management practice, management history and the system being modified. Costs that are related to measurement and monitoring and transaction cost (incentive-based system) also another barrier: When there is lack of clear ownership about the land results in less incentive to implement different land management activities become another problem. On the other hand, According to IPCC Fourth Assessment Report, There are often technological and financial limits that prevent the scale of adaptation that we would need. The other barriers mentioned by FAO 2011, are funding and implementation barriers. Similarly, Okolo *et al.*, (2015) study finding revealed that policy barrier such as lack of local context and little attention to local situation, limited capacity to implement the policy, lack of coordination etc. hinder farmers' to practice different adaptation activities.

References

1. Acharya CL, Sharma PD. Tillage and mulch effects on soil physical environment, root growth, nutrient uptake and yield of maize and wheat on an Alfisol in northwest India. Soil till Res. 1994; 32:291-302.
2. Acharya CL, Kapur OC, Dixit SP. Moisture conservation for rainfed wheat production with alternative mulches and conservation tillage in the hills of northwest India. Soil till Res. 1998; 46:153-163.
3. Action Aid. He Time is NOW: Lessons from Farmers Adapting to Climate Change, Action Aid. 2008
4. Agarwal S, Sethi V, Gupta P, Jha M, *et al.*, Experiential household food insecurity in an urban underserved slum of North India. Food Security. 2009; 1:239-250.
5. Anonymous. "L'Aquile" joint statement on global food security: L'Aquile Food Security Initiative (AFSI). Food Security. 2009; 1:235-237.
6. Aune J, Lal R. Agricultural productivity in the tropics and critical limits of properties of Oxisols, Ultisols and Alfisols. Tropical Agriculture (Trinidad). 1998; 74:96-103.
7. Banik, D. Starvation and India's Democracy. Routledge, London. 2007
8. Bejbaruha R, Sharma RC, Banik P. Direct and residual effect of organic and inorganic sources of nutrients on rice-based cropping systems in the sub-humid tropics of India. Journal of Sustainable Agriculture. 2009; 33:674-689.
9. Bouman BAM, Lampayan RM, Toung TP. Water Management in Irrigated Rice: Coping with Water Scarcity. IRRI, Los Banos, Philippines, 2007, 54
10. Brown ME, Funk CC. Food security under climate change. Science. 2008; 319:580-581.

11. Chu S. Carbon capture and sequestration. Science. 2009; 325:1599
12. Denman KL, Brasseur G, Chidthaisong A, Ciais P, Cox PM, *et al.*, Couplings between Changes in the Climate System and Biogeochemistry. In: Boonpragob K, Heimann M, Molina M (Eds.). Climate Change 2007: He Physical Science Basis. Contribution of Working Group I to the Fourth Assessment Report of the Intergovernmental Panel on Climate Change. Cambridge University Press, Cambridge, United Kingdom and New York, NY, USA, 2007
13. OECD. Agriculture and Climate Change. Trade and Agriculture Directorate. 2015
14. FAO. Rome Declaration on World Food Security. World Food Summit. Rome, Italy. 1996
15. FAO. 1.02 Billion People Hungry. FAO New Room, Rome, Italy. 2009
16. FAO. Agriculture, food security and climate change in post Copenhagen processes: An FAO Information note. 2010a
17. FAO. Strengthening Capacity for Climate Change Adaptation in the Agriculture Sector in Ethiopia. Proceedings from National Workshop held in Nazreth, Ethiopia. 2011
18. Funk C, Dettinger MD, Brown ME, Michaelson JC, *et al.*, Warming of Indian Ocean threatens eastern and southern Africa, but could be mitigated by agricultural development. Proceedings of National Academy of Science. 2008; 105:11081-11086.
19. Funk CC, ME Brown. Declining per capita agricultural production and warming oceans threaten food security. Food Security. 2009; 1:271-289.
20. Hazzeldine R.S. Carbon capture and storage: How green can black be? Sci. 2009; 325:1647-1652.
21. IFA. Nutrient use efficiency in Indian agriculture. India Fertilizer Assoc., New Delhi, India. 2007
22. IPCC. The Climate Change 2007: The Physical Science Basis. Cambridge Univ. Press, Cambridge, U.K. 2007
23. IPCC. Contribution of Working Group II to the Fourth Assessment Report of the Intergovernmental Panel on Climate ChangeSummary for Policymakers. 2007b
24. ISRIC. On Solid Ground. World Soil Information. Wageningen. The Netherlands, 2009, 26

25. Katers RW. Cautionary tales: Adaptation and the global poor. Climatic Change. 2000; 45:5-17.

26. Kerr RA. Northern India's ground water is going, going, going. Science. 2009; 325:798

27. Komba C, Muchapondwa E. Adaptation to Climate Change by Smallholder Farmers in Tanzania. Environment for Development. Discussion Paper Series. 2015

28. Koning N, Mol APJ. Wanted: Institutions for balancing global food and energy markets. Food Security 1, 2009, 291-303.

29. Kreye C, Bouman BAM, Castañeda AR, Lampayan RM, *et al.*, Possible causes of yield failure in tropical aerobic rice. Field Crops Res. 2009; 111:197-206.

30. Kumar P, Pant M, Negi GCS. Soil physic-chemical properties and crop yield improvement following Lantana mulching and reduced tillage in rainfed croplands in the Indian Himalayan Mountains. Journal of Sustainable Agriculture. 2009; 33:636-657.

31. Ladha JK, Yadvinder Singh, Erenstein O, Hardy B(Eds). Integrated Crop and Resource Management in the Rice-Wheat System of South Asia. IRRI/ADB/RWC, Los Banos, Philippines, 2009, 395

32. Lal R. Soil carbon sequestration impacts on global climate change and food security. Sci. 2004; 304:1623-1627.

33. Lal R. Laws of sustainable soil management. Agron. Sustainable Soil Manag. 2009a; 29:7-9.

34. Lal R. Ten tenets of soil management. J Soil Water Con. 2009b; 64:20A-21A.

35. Lasco R, Pulhin F. Agroforestry for Climate Change Adaptation and Mitigation. An academic presentation for the College of Forestry and Natural Resources (CFNR), University of the Philippines Los Baños, Philippines. 2009

36. Lipton M. The family farm in a globalizing world: the role of crop science in alleviating poverty. IFPRI, Washington, D.C. 2005

37. McKinsey and Co. Pathways to a low-carbon economy. Version 2 of the global greenhouse gas abatement cost curve. McKinsey & Co., New York. 2009

38. Normile D. Round and round: a guide to the carbon cycle. Science. 2009; 325:1642-1643.

39. Okolo W, Twyman J, Ampaire E, Acosta M. Barriers to successful climate change policy implementation in Uganda: Findings from a qualitative policy study in Nwoya and Rakai Districts, Uganda. CCAFS. 2015

40. Info Note. Copenhagen, Denmark: CGIAR Research Program on Climate

41. Change, Agriculture and Food Security (CCAFS).

42. Orr Jr, FM. Offshore geologic storage of CO2. Science. 2009; 325:1656-1658.

43. Pacala S, Socolow R. Stabilization Wedges: Solving the Climate Problem for the Next 50 Years with Current Technologies. Science. 2006; 305:968-972.

44. Paul H, Ernsting A, Semino S, Gura S, Lorch A. Agriculture and Climate Change: Real Problems, False Solutions. Report published for the 15th Conference of the Parties of UNFCCC in Copenhagen, December. 2009

45. Robert J, Adrian M, Anne O. High Sequestration, Low Emission, Food Secure Farming. Organic Agriculture - a Guide to Climate Change & Food Security, IFOAM. 2009

46. Rochelle GT. Amine scrubbing for CO2 capture. Science. 2009; 325:1652-1654.

47. Rodelle M, Velicogn I, Famiglietti JS. Satellite- based estimates of ground water depletion in India. Nature (doi: 10.138/nature 08238), 2009, 1-4.

48. Ruddiman WF. Plows, Plagues, and Petroleum: How Humans Took Control of Climate. New York, Princeton UP. 2007

49. Schrag DP. Storage of carbon dioxide in offshore sediments. Science. 2009; 325:1658-1659.

50. Smith P, Martino D, Cai Z, Gwary D, Janzen H, *et al.*, Policy and Technological Constraints to Implementation of Greenhouse Gas Mitigation Options in Agriculture. Agri, Ecosyst Environ. 2007b; 118: 6-28

51. Verhagen J, Put M, Van Keulen H. Climate change and drought risks for agriculture. In: The Impact of Climate Change on Drylands: With a Focus on West Africa (A.J. Dietz, R. Ruben, A. Verhagen, Eds), Kluwer Academic Publishers, Dordrecht, 2004, 49-59.

52. Vuren DPV, Ochola WO, Riha S, Gampietro M, Ginze H, *et al.*, Outlook on agricultural change and its drivers. In: McIntyre BD, Herren HR, Wakhungu J, Watson RT (ed.). Agriculture at a Crossroads, Island Press, Washington, DC. 2009

53. Walker B, Barrett S, Polasky S. *et al.*, Looming global scale failures and missing institutions. Science. 2009; 325:1345-1346.

54. WMO. Greenhouse Gas Bulletin. The State of Greenhouse Gases in the Atmosphere Using Global ObservationsThrough 2007. World Meteorological Organization, Geneva, Switzerland. 2008

Chapter - 4
Climate Change Impact on Indian Agriculture and Adaptation Strategies

Authors

Dr. Vivek
Sardar Vallabhbhai Patel University of Agriculture & Technology, Meerut, Uttar Pradesh, India

Lali Jat
Sardar Vallabhbhai Patel University of Agriculture & Technology, Meerut, Uttar Pradesh, India

Chapter - 4

Climate Change Impact on Indian Agriculture and Adaptation Strategies

Dr. Vivek and Lali Jat

Climate change has become the talk of the day everywhere, be it international, regional and national arena and this phenomenon has got so much attention and importance that noble prize was awarded to Ale Gore and Intergovernmental Panel on Climate Change (IPCC) for their extensive work on climate change in 2007. IPCC in its recently released report has confirmed that global atmospheric concentrations of carbon dioxide (CO_2), methane (CH_4) and nitrous oxide (N_2O) and other green house gases (GHGs) have increased markedly as a result of human activities since 1750. The CO_2, CH_4 and N_2O concentrations in atmosphere were 280 ppm, 715 ppb and 270 ppb, respectively in 1750 AD. In 2005, these values have become 379 ppm, 1774 ppb and 319 ppb, respectively. The increase of GHGs was 70% between 1970 and 2004. The major cause for the increase of CO_2 is the fossil fuel use and land use change, while those of $CH_4$and N_2O are primarily due to agriculture (IPCC, 2007a) (Fig 4.1 and Fig 4.2). Per capita GHGs emissions of some countries are given in (Table 4.1).

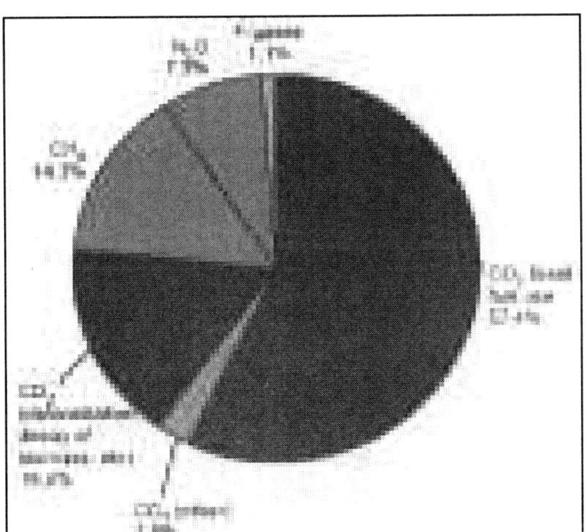

Fig 4.1: Emissions from different sources

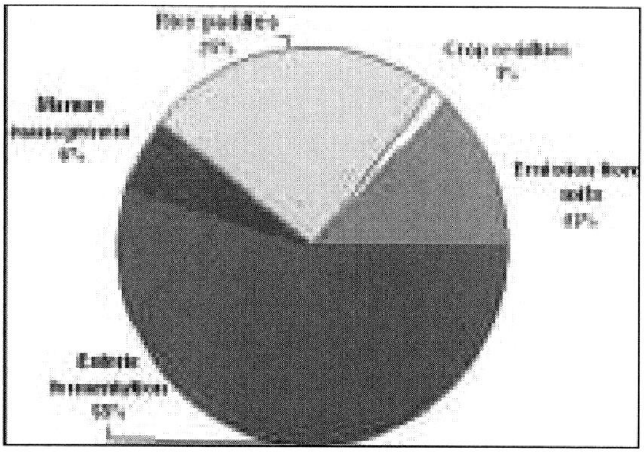

Fig 4.2: Relative contribution of different sectors in agriculture to GHG emissions (Source: NATCOM, 2004)

These increases in GHGs have resulted in warming of the climate system by 0.74°C between 1906 and 2005. Eleven of the twelve years (1995-2006) rank among the 12 warmest years in the instrumental record of global surface since 1850. This has in turn resulted in the increased average temperature of global ocean, rise in sea level, decline in glaciers and snow cover. It has also resulted in increase in the frequency of droughts, as well as heavy precipitation events over most of land areas in the recent time. IPCC has projected that temperature increase by the end of this century is likely to be in the range of 2.0 to 4.5°C. They also projected the sea level rise by the end of the century is likely to be 0.18 to 0.59 m (IPCC, 2007a). Climate change, in many parts of the world, adversely affected socio-economic sectors, including water resources, agriculture, forestry, fisheries and human settlements, ecological systems and human health, especially in developing countries due to their vulnerability.

Table 4.1: A comparison of India's per capita GHG emission with some other countries

Country	**Per-Capita CO_2 Emissions (Metric Tons)**
USA	20.01
EU	9.40
Japan	9.87
China	3.60
Russia	11.71
India	1.02
World Average	4.25

Indian Agriculture and Climate Change

India is a large developing country with nearly 700 million rural population that directly depends on climate sensitive sectors (agriculture, forests and fisheries) and natural resources (such as water, biodiversity, mangroves, coastal zones, grasslands) for their subsistence and livelihoods. Further, the adaptive capacity of dry land farmers, forest dwellers, fisher folk and nomadic shepherds is very low. Climate change is likely to impact all the natural ecosystems as well as socio-economic systems as per the National Communications Report of India to the UNFCCC (Dwivedi, 2011). In India, several studies have shown that temperature marked an increasing trend over the previous century; rainfall had no such significant change. However, at regional levels, the rainfall showed increasing or decreasing trends in the previous century (NATCOM, 2004). Variability of rainfall is found in the different parts of India in the last few decades. Surface air temperature for the period 1901-2000 indicates that a warming of 0.4°C over these 100 years. The spatial distribution indicates that there is significant warming trend over west coast, central India, interior peninsula and north east India. However, cooling trend was observed over northwest India and southern India. There are evidences that glaciers in the Himalayas are receding at a rapid pace (Kulkarni and Bahuguna, 2002; IPCC, 2007b). It is also projected that rainfall over India by the end of the century will increase by 15- 40% and mean annual temperature by 3-6°C (NATCOM, 2004). The warming will be more pronounced over land area and relatively greater over north India (Fig 4.3). The warming is also likely to be greater in winter and post monsoon season trend was observed over northwest India and southern India. There are evidences that glaciers in the Himalayas are receding at a rapid pace (Kulkarni and Bahuguna, 2002; IPCC, 2007b). It is also projected that rainfall over India by the end of the century will increase by 15- 40% and mean annual temperature by 3-6°C (NATCOM, 2004) (Fig 4.3). The warming will be more pronounced over land area and relatively greater over north India. The warming is also likely to be greater in winter and post monsoon season.

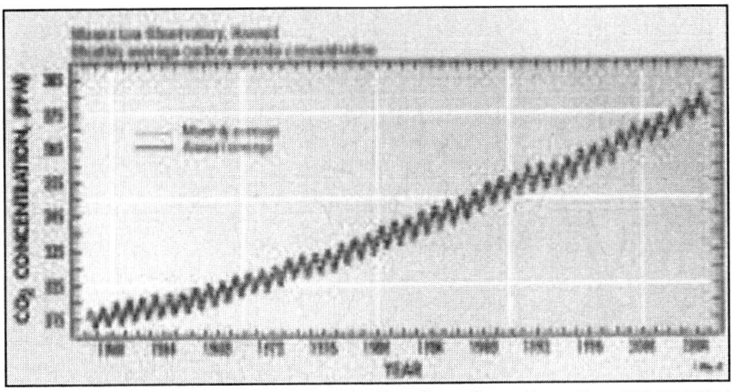

Fig 4.3: Increasing CO_2 concentration during last four decades (NATCOM, 2004)

Impact of Climate Change on Indian Agriculture

Climate change will have an economic impact on agriculture, including changes in farm profitability, prices, supply, demand and trade. The magnitude and geographical distribution of such climate-induced changes may affect our ability to expand the food production as required to feed the populace (Fig 4.4). Climate change may thus have far reaching effects on the patterns of trade among nations, development and food security.

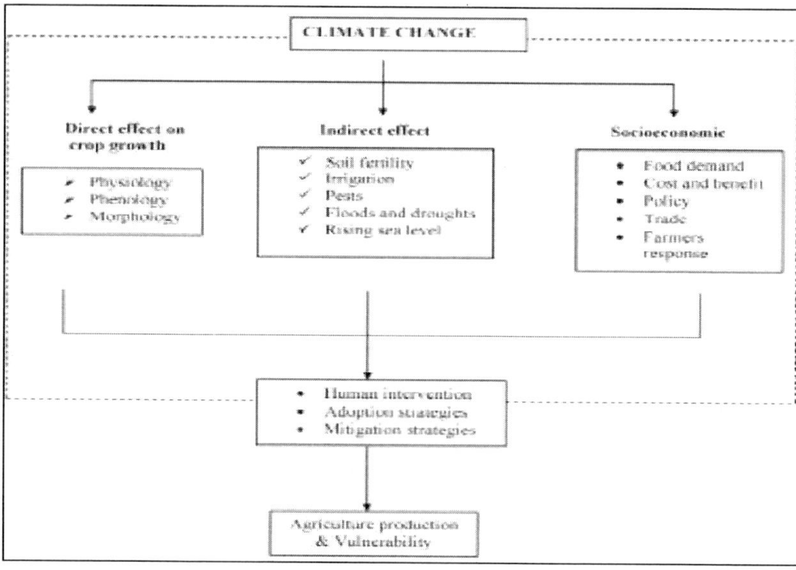

Fig 4.4: Impact of climate change

In the Indian context, Kumar and Parikh (2001) have estimated the macro level impacts of climate change using such an approach (Fig 4.5). They showed that under doubled carbon dioxide concentration levels in the later half of twenty-first century, the gross domestic product would decline by 1.4 to 3% points due to climate change. More significantly, they also estimated an increase in the proportion of population in the bottom income groups of the society in both rural and urban India as a result of climate change.The immediate problems that our farmers face relate to intra-seasonal variability of rainfall, extreme events and unseasonal rains. These aberrations cause heavy losses to our crops every year. There is therefore an urgent necessity for us to speed up our efforts to evolve climate-resilient crop varieties, cropping patterns and management practices. Agriculture is sensitive to short-term changes in weather and to seasonal, annual and long-term variations in climate. Crop yield is the culmination of a diversified range of factors.

Parameters like soil, seed, insect pest and diseases, fertilizers and agronomic practices exert significant influence on crop yield. The burgeoning population, along with human-induced climate change and environmental problems is increasingly proving to be limiting factors for enhancing farm productivity and ensuring food security for the rural poor. Very high temperatures will make it difficult, if not impossible, to store onion and potatoes. Preservation of vegetables and fruits would also become difficult. It would not be an exaggeration to say that the entire food processing industry would be in doldrums if the present cold storage equipment fails under the new climatic conditions. The paradox is, more is the refrigeration activities, the more will be the greenhouse emission gases and the greater would be the global warming.

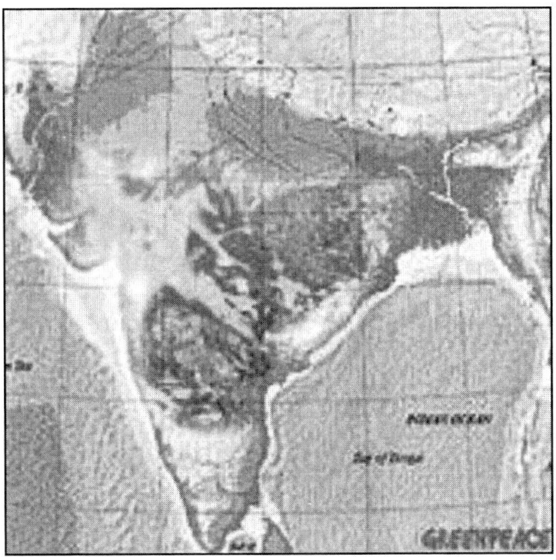

Fig 4.5: Different parts of India prone to regular drought and floods due to climate change

Soil

The soil system responds to the short-term events such as episodic infiltration of rainfall and also undergoes long-term changes such as physical and chemical weathering due to climatic change. The potential changes in the soil forming factors directly resulting from global climate change would be in the organic matter supply, temperature regimes, hydrology and changes in the potential evapotranspiration. Both the organic matter and carbon to nitrogen ratio (C: N ratio) will diminish in a warmer soil temperature regime. Drier soil conditions will suppress both root growth and decomposition of organic matter and will increase vulnerability to erosion. Increased evaporation from the soil and accelerated transpiration from the plants themselves will cause soil moisture stress.

Insect Pests, Diseases and Weeds

Incidence of insect pests and diseases would be most severe in tropical regions due to favorable climate/weather conditions, multiple cropping and availability of alternate hosts throughout the year. Climate change is likely to cause a spread of tropical and subtropical weed species into temperate areas and to increase the numbers of many temperate weed species currently limited by the low temperature at high latitudes.

Crop Production and Productivity

Estimates of impact of climate change on crop production could be biased depending upon the uncertainties in climate change scenarios, region of study, crop models used for impact assessment and the level of management. This study reports the results of a study where the impact of various climate change scenarios has been assessed on grain yields of irrigated rice with two popular crop simulation models- Ceres-Rice and ORYZA1N at different levels of N management. The results showed that the direct effect of climate change on rice crops in different agro-climatic regions in India would always be positive irrespective of the various uncertainties. Rice yields increased between 1.0 and 16.8% in pessimistic scenarios of climate change depending upon the level of management and model used. These increases were between 3.5 and 33.8% in optimistic scenarios. At current as well as improved level of management, southern and western parts of India which currently have relatively lower temperatures compared to northern and eastern regions are likely to show greater sensitivity in rice yields under climate change (Fig 4.6).

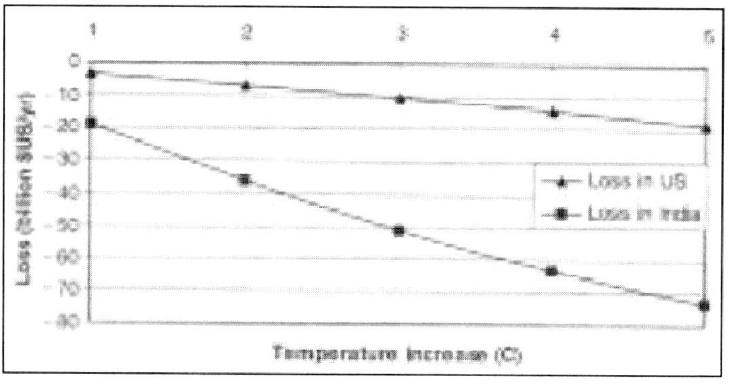

Fig 4.6: Warming impact in India: effect of temperature on farm value (source: Dinar *et al.*, 2002)

Overall, temperature increases are predicted to reduce rice yields. An increase of 2- 4°C is predicted to result in reduction in yields. Eastern regions are predicted to be most impacted by increased temperatures and decreased radiation, resulting in relatively fewer grains and shorter grain filling durations. By contrast, potential reductions in yields due to increased temperatures in Northern India are predicted to be offset by higher radiation, lessening the impacts of climate change (Fig 4.7). Although additional CO_2 can benefit crops, this effect may get nullified by an increase of temperature.

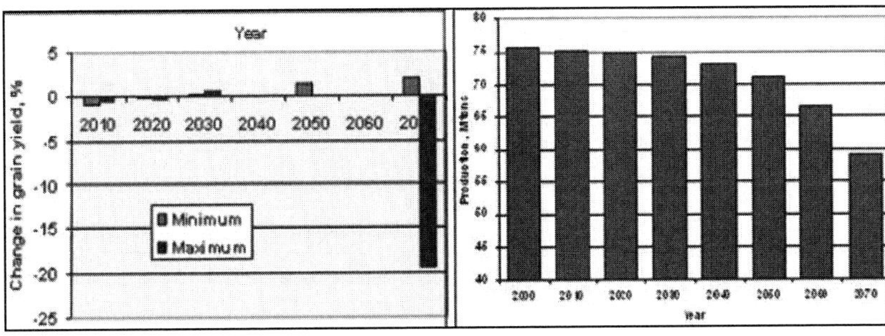

Fig 4.7: Impact of climate change on yield ability of rice and wheat (Aggarwal, 2003)

More recent studies done at the Indian Agricultural Research Institute, New Delhi indicated the possibility of loss of 4-5 million tons in wheat production with every degree rise of 1°C temperature throughout the growing period even after considering benefits of carbon fertilization (Fig 4.8). This analysis assumes that irrigation would remain available in future at today's levels and there is no adaptation.

Fertilizer use Efficiency

Fertilizer use efficiency in India is generally very low (15-30 %). Increasing temperature in future is likely to further reduce fertilizer use efficiency. This will lead to increased fertilizer requirement for meeting future food production demands. At the same time, greater fertilizer use leads to higher emissions of greenhouse gases. A large number of resource-poor farmers in tropics are not able to apply desired levels of fertilizers, irrigation and pest control. Simulation studies done at different levels of N management indicated that the crop response could vary depending upon the N management and the climate change scenario (Aggarwal, 2003).

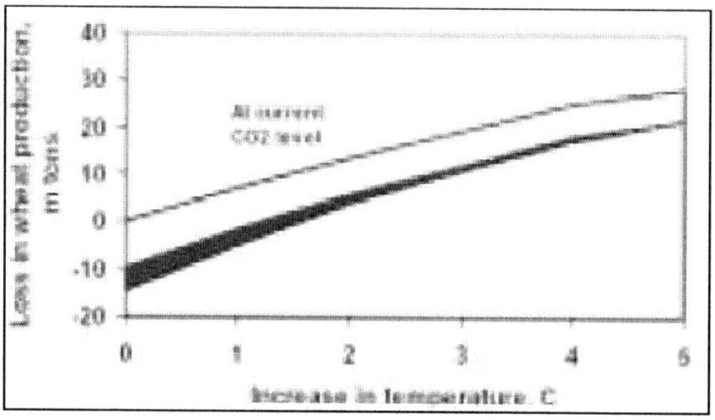

Fig 4.8: Projected loss in wheat production due to increasing temperature at current and 550 ppm CO_2 levels. The shaded area of the curve indicates losses that can be offset by adaptation options such as change in planting dates and variety

Crop Physiology

The greenhouse gases CH_4, N_2O and chlorofluorocarbons (CFCs) have no known direct effects on plant physiological processes. Increase in the atmospheric air and soil temperature result in the variation of cytokinin. ABA (Abscissic acid) affects many physiological processes like water uptake, photosynthesis, respiration; assimilate partitioning and ultimately poor source sink relationship, which is evident by low yield. C_3 plants (Rice, wheat) will always have advantages of elevated CO_2 compared to C_4 (sugarcane, millets and maize) and CAM (pineapple) plants. Many scientific studies revealed that the yield of C_3 plant increases with increase of CO_2 concentration in the atmosphere compared to C_4 plants. In the long run, increase of CO_2 concentration leads to increase in temperature, which again affects the yield of C_3 and other plants (Fig 4.9). Elevated CO_2 favors more vegetative growth through increased photosynthesis. Due to more CO_2, often more starch accumulates in the chloroplast and rupture of plastids which is visible through electron microscope observations. Leafy vegetable will have an advantage in the context of CO_2 in the production of more leaves.

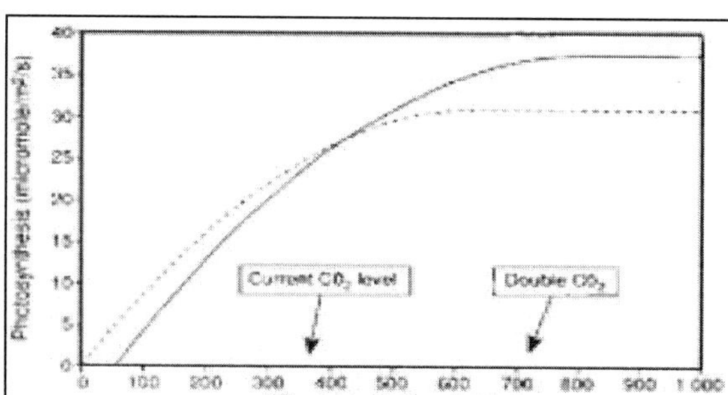

Fig 4.9: Effects of CO_2 concentrations onC_3 andC_4 plants (Wolfe and Erickson, 1993)

Vulnerability to Climate Change and Adaptative Capacity in India

O. Brien *et al.*, (2004) tried to find out the vulnerability profile of India on the basis of adaptive capacity, sensitivity and exposure to climate change. Adaptive capacity was measured on the basis of biophysical, socioeconomic, and technological factors that influence agricultural production in 466 districts of India. This map shows higher degrees of adaptive capacity in districts located along the Indo-Gangetic Plains (except Bihar) and lower adaptive capacity in the interior portions of the country, particularly in the states of Bihar, Rajasthan, Madhya Pradesh, Maharashtra, Andhra Pradesh, and Karnataka (Fig 4.10).

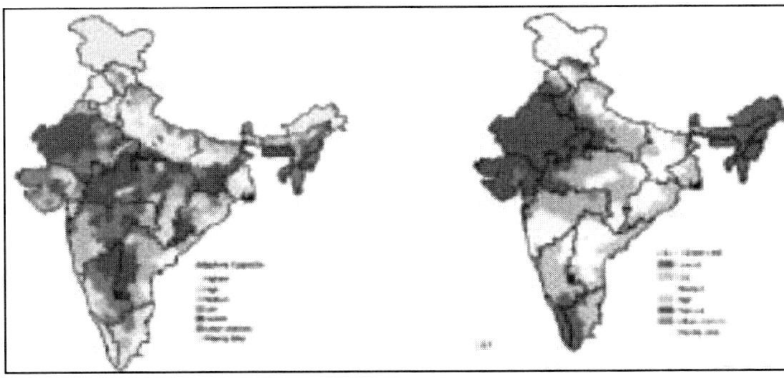

Fig 4.10: District-level mapping of (a) adaptive capacity and. (b) climate sensitivity in India, districts are ranked and presented as quintiles (O. Brien *et al.*, 2004)

To measure sensitivity under exposure to climate change, climate sensitivity index (CSI) was constructed that measures dryness and monsoon dependence, based on a girded data set for 1961-1990. The areas with high to very high climate sensitivity for agriculture are located in the semiarid regions of the country, including major parts of the states of Rajasthan, Gujarat, Punjab, Haryana, Madhya Pradesh and Uttar Pradesh (Fig 4.10). Finally, to depict climate change vulnerability in India, they summed the district-level index of adaptive capacity with the index of climate sensitivity underexposure. The resulting climate vulnerability map (Fig 4.11) represents current vulnerability to future climate change across the districts. The map depicts the range of relative climatic vulnerability at the district level in India. It is important to note that the districts with the highest (or lowest) climate sensitivity under the scenario of climate change used here are not necessarily the most (or least) vulnerable. For example, most districts in southern Bihar have only medium sensitivity to climate change, yet are still highly vulnerable to climate change as a the result of low adaptive capacity. By contrast, most districts in northern Punjab have very high sensitivity to climate change, yet are found to be only moderately vulnerable as the result of high adaptive capacity. Assessment of both adaptive capacity in combination with climate change sensitivity and exposure is thus crucial for differentiating relative vulnerability to climate change.

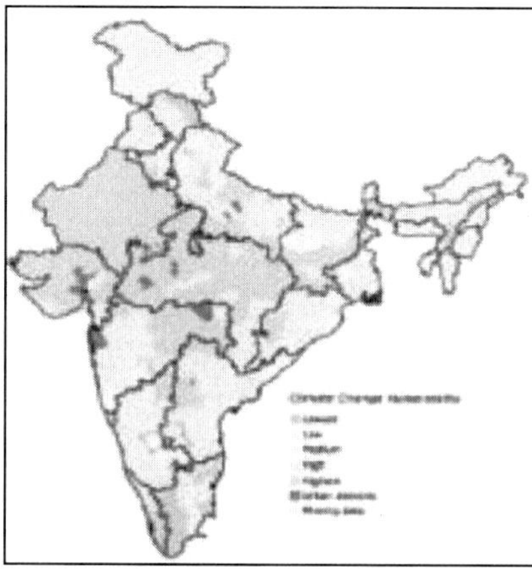

Fig 4.11: District-level mapping of climate change vulnerability (O. Brien *et al.*, 2004)

Adaptation Strategies in Agriculture

Any perturbation in agriculture can considerably affect the food security and thus increase the vulnerability of a large fraction of the resource poor population. We need to understand the possible coping strategies by different sections and different categories of producers to global climatic change. Such adaptation strategies would need to simultaneously consider the background of changing demand due to globalization and population increase and income growth, as well as the socio-economic and environmental consequences of possible adaptation options (Aggarwal *et al.*, 2004; Easterling *et al.*, 2004). Developing adaptation strategies exclusively for minimizing the negative impacts of climatic change may be risky in view of large uncertainties associated with its spatial and temporal magnitude (Table 4.2). We need to identify 'no-regrets' adaptation strategies that may anyway be needed for sustainable development of agriculture. These adaptations can be at the level of individual farmer, society, farm, village, watershed, or at the national level. Some of the possible adaptation options are discussed below:

Research for C_4 Strain

Scientists are working on a new strain of C_4 rice that could boost rice production by as much as 50 per cent, while potentially reducing the need for excessive water and fertilizers. The unique C_4 rice is expected to behave like corn and other plants that perform photosynthesis much more efficiently involving four carbon atoms' unlike the conventional rice strains, which use three carbon atoms. But significantly, the new strain is a genetically modified crop for which global acceptance is still extremely limited. And even if it finds acceptance in the years ahead, it has long way to go from experimental laboratories to successive field trials before being tried in real-world conditions.

Table 4.2: Crop water requirements to increase (wheat)

S. No	State	Area (ha)	Crop Water Requirement (mm)			Water Requirement (Million Cubic Meter)			% Deviation (2020-1990)	% Deviation (2050-1990)
			1990	2020	2050	1990	2020	2050		
1	J&K	253023	217.4	224.3	229.7	823.9	851.7	874.5	3.1	5.3
2	HP	367770	281.8	291.8	299.1	1051.1	1089.7	1119.2	3.7	6.3
3	Punjab	3468000	359.1	371.8	380.8	12553.6	13002.5	13317.5	3.6	6.1
4	Haryana	2316674	452.1	467.1	480.9	10475.4	10825.4	11158.0	3.4	6.4
5	UP	9443104	4 423.7	434.6	447.9	39717.8	40750.1	41990.4	2.6	5.8
6	Bihar	2076727	438.6	449.3	465.0	9046.2	9271.1	9593.9	2.5	6.1
7	West Bengal	366729	399.8	407.2	425.2	1449.5	1479.8	1543.1	1.9	6.4
8	Rajasthan	2010241	498.6	511.6	9923.6	10208.4	10208.4	10479.5	2.7	5.4
9	Gujarat	727400	604.7	614.7	630.9	4603.3	4683.2	4807.1	1.7	4.3
10	MP	4188248	502.0	513.3	526.2	21176.9	21655.3	22200.1	2.3	4.9
11	Maharashtra	932800	606.417	617.82	633.5	5613.8	5718.3	5859.4	1.9	4.5

AICRPAM - CRIDA

Changes in Land use and Management

Effect of small changes in climatic parameters can often be managed reasonably well by altering dates of planting, spacing and input management. Development of alternate cultivars and farming systems (such as mixed cropping, crop-livestock) that are more adaptable to changes in the environment can further ease the pressure.

Development of Resource Conserving Technologies

Recent researches have shown that surface seeding or zero-tillage establishment of upland crops after rice gives similar yields to the crop planted under normal conventional tillage over a diverse set of soil conditions. In addition, such resource conserving technologies restrict release of soil carbon into atmosphere thus mitigating increase of CO_2 in the atmosphere. Greater emphasis on water harvesting and improving regional as well as farm water use efficiency could help to face uncertain rainfall.

Improved Land use and Natural Resource Management Policies and Institutions

Adaptation to environmental change could be in the form of crop insurance, subsidies, pricing policies, and change in land use. Necessary provisions need to be included in the development plans to address these issues of attaining twin objectives of containing environmental changes and improving resource use productivity. Policies are needed that would encourage farmers to conserve water, energy and soil resources. For example, financial compensation/incentive for enriching soil carbon and increasing the efficiency of irrigation water through drip and sprinkler methods could encourage farmers to improve soil health, manage with less water, and assist in overall sustainable development. The National Mission for Sustainable Agriculture, which is one of the eight Missions under our National Action Plan on Climate change, also seeks to devise appropriate adaptation and mitigation strategies for ensuring food security, enhancing livelihood opportunities and contributing to economic stability at the national level (NAPCC, 2008).

Improved Risk Management Though Early Warning System and Crop Insurance

The increasing probability of floods and droughts and other uncertainties in climate may seriously increase the vulnerability of resource-poor farmers to global climate change. Early warning systems and contingency plans can provide support to regional and national administration, as well as to local bodies and farmers to adapt. Policies that encourage crop insurance can provide protection to the farmers in the event their farm production is reduced due to natural calamities.

Reducing Dependence on Agriculture

Although the share of agriculture in gross domestic product in India has declined to 16% but 58% population continues to remain dependent on this sector. Such trends have resulted in fragmentation and decline in the size of land holdings leading to inefficiency in agriculture and rise in unemployment, underemployment, and low volume of marketable surplus and, therefore, increased vulnerability to global climate change. Institutional arrangements, such as cooperatives and contract farming that can bring small and marginal farmers together for increasing production and marketing efficiencies are needed (Aggarwal, 2007).

Global climate change has considerable implications on Indian agriculture and hence on our food security and farmers' livelihood. We need to take steps urgently to increase our adaptive capacity. This would require increased support to adaptation research, developing regionally differentiated contingency plans for temperature and rainfall related risks, enhanced research on seasonal weather forecasts and their applications for reducing production risks, and evolving new land use systems, including heat and drought tolerant varieties, adapted to climatic variability and changes and yet meeting food demand. Strengthening current institutions and policy can also improve adaptive capacity. There is an urgent need to strengthen our surveillance mechanisms for various pests. We also need to support community partnerships in developing food and forage banks to manage scarcity during projected increased periods of drought and floods (Fig 4.12). Mechanisms for integrated management of rainwater, surface, and ground water need to be developed. Once a crop is planted, farmers need insurance cover to manage risks associated with extremes of temperature and precipitation events. Weather derivatives should be provided to farmers at an early date. For mitigation of GHGs from agriculture, there is a need to renew focus on nitrogen fertilizer use efficiency with added dimension of nitrous oxides mitigation. Widespread testing of neem cake, and neem coated urea, known to inhibit N_2O emissions, would be rewarding. Studies are also needed to determine optimal size of livestock population in different agro-ecological regions considering national milk requirement, GHGs emissions and social issues. Financial incentives for improved land management including resource conservation/ enhancement (water, carbon, energy), and fertilizer use efficiency should be considered. These could also assist in sustainable development.

Fig 4.12: Infrastructure Development

Crop Diversification

Crop diversification methods such as crop rotation, mixed cropping and double cropping, reduce the vulnerability of crop yields. Crop diversification has also been found to result in reduced erosion, improved soil fertility, improved crop yield, reduced risk of crop failure and enhanced water savings.

Food grain production in India has increased spectacularly due to the

Green Revolution from 50 Mt in 1951 to 212 Mt in 2002, and the mean cereal productivity has increased from 500 kg/ha to almost 1800 kg/ha (Government of India, 2004). These increases were largely the result of area expansion, large-scale cultivation of new high yielding semi-dwarf varieties since the early 1960s, and the increased application of irrigation, fertilizers and biocides, supported by progressive government policies. Currently, Indian agriculture has 190 million ha gross sown area (142 million ha net sown area), and 40% of this is irrigated. There have been similar revolutions in the production of milk, fish, eggs, sugar, and a few other crops. India is now the largest producer of milk, fruits, cashew nuts, coconuts and tea in the world, the second largest producer of wheat, vegetables, sugar and fish, and the third largest producer of rice. This growth in agricultural production has also led to considerable surplus food stocks purchased by the government. The droughts of 1987, 1999~2000, and of 2002~2003 could generally be managed and did not lead to severe problems of food security because of these buffer stocks. Despite this progress, food production in India, on an aggregated scale, is still considerably dependent on the rainfall quantity and its distribution. The summer monsoon (June through September) contributes to 78% of India's annual rainfall and is a major water resource. It is important to recognize that the Green Revolution was largely confined to the irrigated areas. In the past 50 years, there have been around 15 major droughts, due to which the productivity of rain fed crops in those years was affected. Limited options for other income and widespread poverty continue to threaten the livelihoods of millions of small and marginal farmers in this region. The food security of India may be at risk once again in the future due to continued population growth. By 2050, India's population is projected to grow to 1.6 billion. This rapid and continuing increase in the population implies a greater demand for food. The demand for rice and wheat, the predominant staple foods, is expected to increase to 122MT and 103Mt, respectively, by 2020, assuming medium income growth. The demand for pulses, fruits, vegetables, milk, meat, eggs and marine products is also expected to increase very sharply. This additional food will have to be produced from the same or possibly shrinking land resource base, because there is no additional land available for cultivation. It is estimated that the average yields of rice, wheat, coarse grains, and pulses need to increase by 56, 62, 36 and 116%, respectively, by 2020. Although there is pressure to increase production in order to meet higher demands, there has lately been a significant slow down of the growth rate in production and yield from the cultivated areas. The annual rate of growth in food production and yield peaked during the early years of the Green Revolution, but since the 1980s, it has declined since the 1980s.The perceived gradual increase in environmental degradation, the early signs of which are becoming visible in areas that benefited largely from the Green Revolution technologies, is further compounding the problem. There is now great concern about decline in soil fertility, change in water table depth, rising salinity, resistance of harmful organisms to many pesticides, and degradation of irrigation water quality as, for example, in north-western India. Nutrient removal by crops over time has exceeded its application and consequently, farmers now have to apply more fertilizers to realize the same yield as achieved 20~30 years ago. The introduction of canal irrigation in Haryana has resulted in almost 0.5 Mha being affected by soil salinity. The rapid increase in the number of tubewells during the last three decades has resulted in an over-exploitation of groundwater in many blocks, leading to declining water tables. In some canal irrigated districts, on the other hand, the water table has risen, resulting in increased problems of salinity. Several pathogens and insect pests have also shown a tendency to increase under the intensive farming systems such as the rice wheat system. In the 21st century, one of the great challenges for Indian agriculture will be, therefore, to ensure that food production is in line with both poverty reduction and environmental preservation. The roadmap of sustainable agricultural development may also have to consider two additional important global drivers of change in agriculture in the coming decades: globalization and climate change. The on-going globalization process and multilateral trade liberalization associated with the World Trade Organization (WTO) is forcing India to make structural adjustments in the agricultural sector to increase its competitiveness and efficiency.

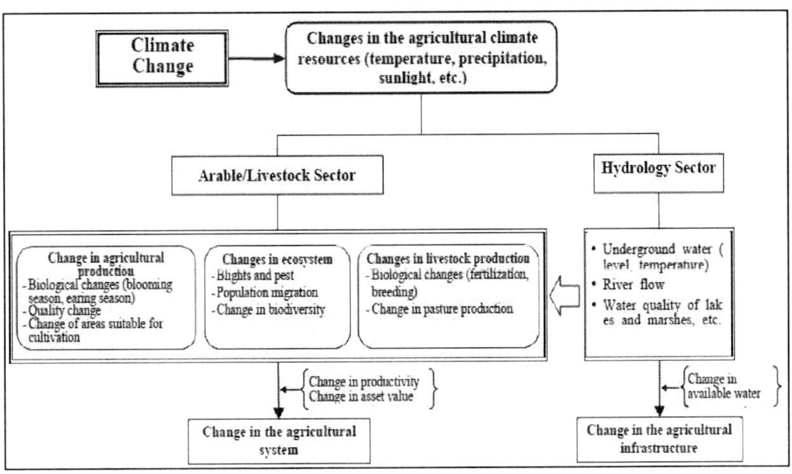

(a)

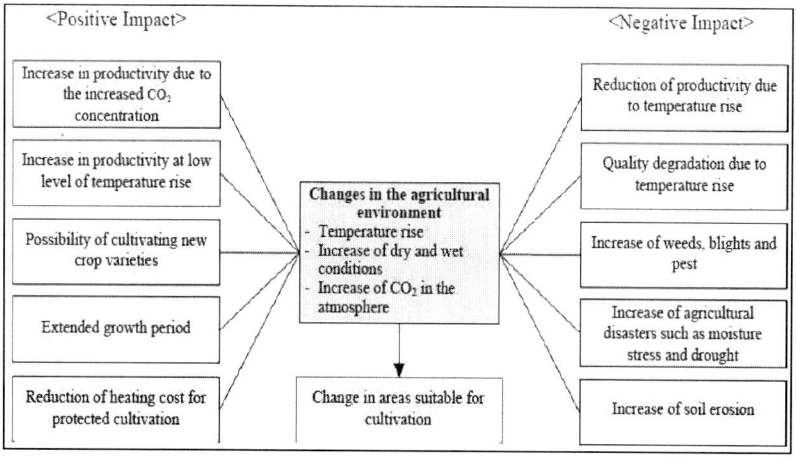

Source: Kim, Chang-Gil and *et al.*, (2009)

(b)

Fig 4.13(a): Flow of the climate change impact on the agricultural sector, **(b):** Potential impacts of global warming on the agricultural sector

Agricultural production is carried out through the selection of crops suitable for the climate of a specific region and application of proper farming methods. Therefore, agriculture is a climatedependent bio-industry with notable regional characteristics. Regional characteristics refer to the ecosystem characteristics determined by the climate of the region. Climate change disturbs the agricultural ecosystem, resulting in the change in agricultural climatic elements such as temperature, precipitation, and sunlight, while further influencing the arable, livestock, and hydrology sectors. The flow of the impacts of climate change on the agricultural sector can be illustrated as shown in <Fig 4.13a>. First of all, the impacts of climate change on the arable and livestock sector are made known by biological changes including the change of flowering and harvesting seasons, quality change, and shift of areas suitable for cultivation Young-Eun *et al.*, (2007). Climate change affects the agricultural ecosystem, giving rise to blights and pests and causing population movement and change in biodiversity. In the livestock sector, climate change brings about biological changes in areas such as fertilization and breeding and also affects the growing pattern of pastures.

Climate change affects the hydrology including underground water level, water temperature, river flow, and water quality of lakes and marshes, by impacting precipitation, evaporation, and soil moisture content. In particular, the increase of precipitation by climate change leads to an increase of outflow while the temperature rise increases evaporation, resulting in the reduction of outflow. In order to understand the quantitative impacts of climate change on water resources, a deterministic hydrology model, based on the general circulation model, is used.

To analyze how and to what extent the change in temperature and precipitation following global warming affects the agricultural sector various experiments, simulations and research methods are carried out both in laboratories and in the field. As the impacts of climate change on the agricultural sector vary with the related variables, it is difficult to generalize certain analytical results. Therefore, what is attempted here is to classify the impacts of climate change into positive and negative ones based on the results that researches have gathered thus far in related fields (Fig 4.13b). The positive impacts of global warming include the increase in crop productivity due to fertilization effect caused by the increase in carbon dioxide concentration in the atmosphere, expansion of the areas available for production of tropical and/or subtropical crops, expansion of two-crop farming due to the increased cultivation period, reduction of damages of winter crops by low temperature, and reduction of heating cost for agricultural crops grown in the protected cultivation facilities.

Low Carbon Green Growth Strategy

The concept of green growth was created in order to increase the possibility for policy enforcement in sustainable development, whose concept is abstract and broad-based covering the three aspects of economic feasibility, environmental protection and social equity (UN ESCAP, 2006). In other words, green growth is a qualitative growth that enhances the standard of living through the attainment of ecological and economic soundness. The relevant framework act defines green growth as "growth achieved by saving and using energy and resources efficiently to reduce climate change and damage to the environment, securing new growth engines through research and development of green technology, creating new job opportunities, and achieving harmony between the economy and the environment". As green growth is variably defined, it can be understood as a complex and open-ended concept that accepts a new civilization and an order of change.

It has been known that climate change has resulted from the disruption of the energy balance for the global climatic system caused by the increase of greenhouse gases and aerosol in the atmosphere and the changes in land covering and solar radiation. Especially, it has been suggested by the results of scientific analyses that global warming is very likely to have resulted from human activities (IPCC, 2007). Countermeasures that the agricultural sector can consider in the face of risks and challenges of climate change, represented by global warming, are largely divide into the mitigation method that reduces the scale and rate of climate change by mitigating and absorbing the greenhouse gas emission and the adaptation method that admits the inevitability of global warming, understands the impacts of climate change, and minimizes the damages it could cause (Fig 4.14a).

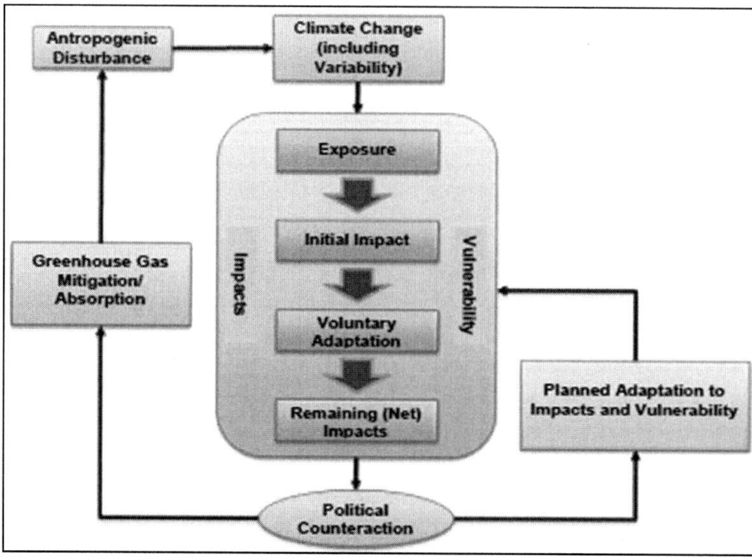

(a)

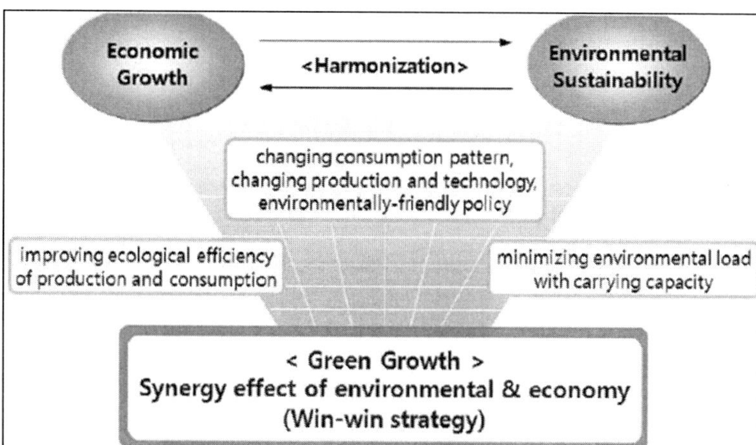

(b)

Fig 4.14(a): Approaches to the countermeasures against climate change, **(b):** Mechanism to Achieve the Goals of Green Growth

The operating principle of the green growth policy is to turn the vicious cycle between the environment and economic growth into a virtuous cycle through a transformation into a newly established growth pattern and economic structure. Under the new paradigm of qualitative growth, the essential factors of production are new ideas, transformational innovations, and state-of-the-art technology. The economic growth based on these drivers is expected to generate substantially intensive, qualitative growth unlike the extensive quantitative growth of the past. Therefore, green growth continuously enhances productivity by putting green capitals (i.e., green technology and green knowledge) into the production process, thereby reducing pollution and expanding natural capital (energy and environmental resources). As a result, green growth brings about changes in production, technology and consumption patterns and ecological efficiency by taking into account environmental capacity and economic and environmental aspects of production and consumption (Fig 4.14b).

Green growth involves policy regime and social value as it aims to achieve economic growth through a low-carbon society and green industrialization. The realization of green growth is accompanied by a significant amount of economic costs and efforts, and above all, it requires us to move away from our existing patterns of living. Thus, for green growth to be materialized, economic incentives, development and the distribution of green technology, and understanding and cooperation of relevant bodies, are of the utmost importance and urgently require a paradigm shift and swift response.

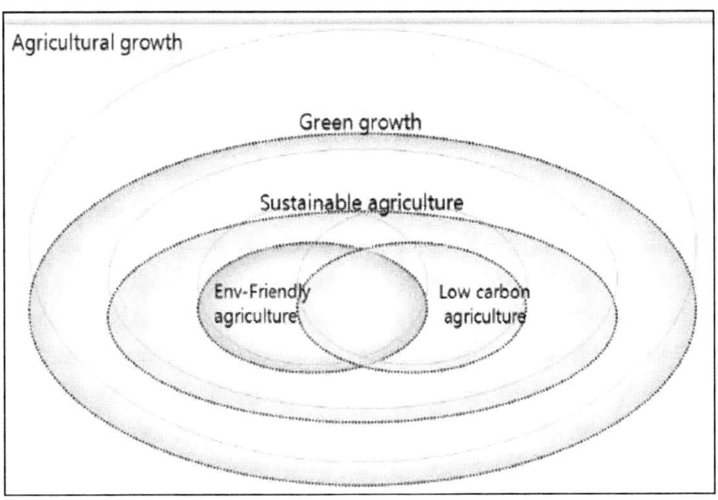

(a)

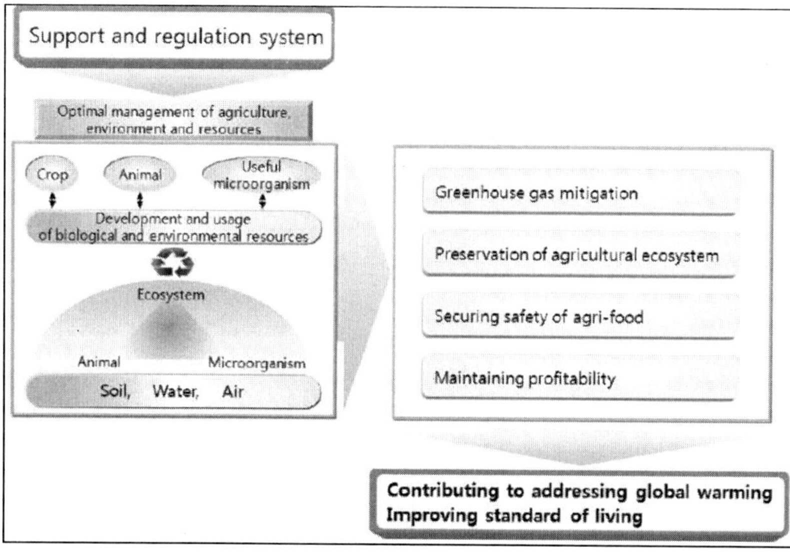

(b)

Fig 4.15(a): Conceptual Position of Green Growth in Agriculture, **(b):** Vision and Goals of Low Carbon Green Growth

Green growth in the agricultural sector is more comprehensive than sustainable agriculture, and it means a growth that guarantees environmentally sound and economically viable growth that takes into consideration environmental capacity in the agricultural ecosystem. Green growth in the agriculture sector is achieved through a shift in agricultural practice that takes into account the environmental capacity of each different region and the water system, low-carbon agriculture via greenhouse gas reduction and higher absorption capacity, and energy efficiency and savings. Green growth can be realized by shifting to a sustainable agricultural system including environmentally-friendly, low carbon agriculture. Thus, green growth in the agricultural sector can be regarded as more comprehensive than sustainable agriculture (Fig 4.15a). In the meantime, green growth in the agri-food sector means a growth through a shift towards an environmentally sound, low-carbon life cycle in the agricultural aspects of not only production but also distribution, processing and consumption.

Shifting from pre-existing conventional agricultural growth to low-carbon green growth (LCGG) requires the formulation of a vision and specific goals. An appropriate vision should reconcile agriculture with the environment, mitigate greenhouse gases and improve the agricultural ecosystem, which contributes to the lessening of negative effects of global warming and improves the living standards for present and future generations (Fig 4.15b). As a basic direction to develop an LCGG action plan, a resource-cycling agricultural system should take its root based on the 3R cycle of "Reduced ↔ Recycled ↔ Reuse". It should be followed by the transition of the agricultural production goal from maximization to optimization. In other words, the agricultural production goal should be shifted from maximizing production to optimizing production by taking into account the local agricultural environment and greenhouse gas emitting and absorbing conditions. For the reduction of greenhouse gases and improvement of living standards, a policy combination of support, regulation and compensation should be formulated so that the right policy mix entails consolidation and coordination for environmental policies. To minimize the inconvenience and economic costs that come along with green growth, green technology should be actively developed and disseminated

References

1. Aggarwal PK, Mall RK. Climate change and rice yields in diverse agro-environments of India. II. Effect of uncertainties in scenarios and crop models on impact assessment. Climatic change. 2002; 52(3):331-343.
2. Aggarwal PK. Impact of climate change on Indian agriculture. Journal of Plant Biology. 2003; 30:189-198.
3. Aggarwal PK. Climate change: implications for Indian agriculture. Jalvigyan Sameeksha. 2003; 22:37-46.
4. Aggarwal PK, Joshi PK, Ingram JS, Gupta RK. Adapting food systems of the Indo-Gangetic plains to global environmental change: Key information needs to improve policy formulation. Environmental Science and Policy. 2004; 7:487-498.
5. Dinar Ariel, Aaron Ratner, Dan Yaron. Evaluating Cooperative Game Theory inWater Resources. In: Economics ofWater Resources: The Contribution of Dan Yaron, (Eds. Dinar, Ariel and David Zilberman Kluwer) Academic Publishers, Boston, MA, 2002, 165-181.
6. Dwivedi J. Climate change and agriculture. Ground report India, 2011.
7. Easterling WE, Hurd B, Smith J. Coping with Global Climate Change: The Panel on Climate Change. Role of Adaptation in the United States. Pew Center on Global Climate Change, 2004.
8. IPCC. The physical science basis. Summary for Policymakers. Intergovernmental Panel on Climate Change, 2007a.

9. IPCC. Climate change Impacts, adaptation and vulnerability. Summary for Policymakers.Intergovernmental Panel on Climate Change, 2007, 2007b.
10. Kim Chang-gil. Strategies for Implementing Green Growth in Agricultural Sector. In Proceedings in Green Korea-Green Growth and Cooperation. National Research Council for Economics, Humanities and Social Science, 2009.
11. Briena K, Robin Leichenkob, Ulka Kelkarc, Henry Venemad *et al.* Mapping vulnerability to multiple stressors: climate change and globalization in India. Global Environmental Change. 2004; 14:303-313
12. Kulkarni A.V, Bahuguna IM. Glacial retreat in the Baspa basin, Himalayas. Journal of Glaciology. 2002; 48 (160):171-172.
13. Kumar KSK, Parikh J. Socio-economic impacts of climate change on Indian agriculture. International Review for Environmental Strategies. 2001; 2(2):275-293.
14. NATCOM. India.s Initial National Communication to the United Nations Framework Convention on Climate Change. National Communication Project, Ministry of Environment and Forests, Govt. of India, 2004, 266.
15. NAPCC (National Action Plan on Climate Change). Adaptation to climate change in agriculture, forestry and fisheries: Perspective, framework and priorities. Food and Agriculture Organization, Rome, Italy, 2008.
16. UN ESCAP. Kim Chang-gil and Jeong Hak-kyun, eds., The term "green growth" was newly coined by the Economist magazine (January 27, 2000). It is more widely used in journals than in academic circles. In UN ESCAP's Ministerial Conference on Environment and Development had an in-depth discussion on this new concept before it started to be widely used after Davos Forum, 2005, 2006, 2008, 31.
17. Wolfe DW, Erickson JD. Carbon dioxide effects on plants: uncertainties and implications for modeling crop response to climate change. In: Agricultural Dimensions of Global Climate Change. H.M. Kaiser and T.E. Drennen (eds.). St. Lucie press, Delray Beach, Florida, 1993, 153-178.
18. Young-Eun Na. The impacts of climate change on agricultural production are divided into primary impacts and secondary impacts. The primary impacts refer to the changes in the composition of the

atmosphere due to increased greenhouse gases, which include the change in crop growth response and the change in energy and moisture balance in the farmland. The secondary impact caused by the change in agricultural climate resources affected by the primary impacts include the shift in suitable places for cultivation and physical and chemical changes in agricultural soil, 2007, 94.

Chapter - 5
Climate Change and Sustainable Food Security

Authors

Dr. R.K.Gupta
Borlaug Institute for South Asia (BISA), New Delhi, India

Dr. R.K. Naresh
Sardar Vallabhbhai Patel University of Agriculture & Technology, Meerut, Uttar Pradesh, India

Chapter - 5

Climate Change and Sustainable Food Security

Dr. R.K. Gupta and Dr. R.K. Naresh

Climate change is a major challenge for agriculture, water availability, food security and rural livelihoods for millions of people including the poor in India. Adverse impact will be more on small holding farmers. Climate change is expected to have adverse impact on the living conditions of farmers who are already vulnerable and food insecure. Rural communities, particularly those living in already fragile environments, face an immediate and ever-growing risk of increased crop failure, loss of livestock and forest products. They would have adverse effects on water availability, food security and livelihoods of small farmers in particular. In order to have climate change sensitive and pro-poor policies, there is a need to focus on small farmers. Agriculture adaptation and mitigation could provide benefits for small farmers. The coping strategies would be useful to have long term adaptation strategies.

The importance of collective action in climate change adaptation and mitigation is recognized and very important for technology transfer in agriculture and natural resource management among small holders and resource dependent communities. The rise in global food prices (Koning *et al.*, 2008; Swinnen and Squicciarini, 2012) has raised concerns about food security and how to feed the hungry world Godfray *et al.*, (2010). The year 2012 was marked by extreme weather and globally, confirming the impacts of climate change on food security Parry *et al.*, (2005). The world population of 7.2 billion and the atmospheric CO_2 concentration of 400 ppmv in 2013 are increasing at the annual rate of 75 million people and 2.2 ppmv, respectively (GHG Bulletin, 2011). Indeed, there exists a strong correlation between the human population and CO_2 emission: growth in world population by one billion increases CO_2-C emission from fossil fuel consumption by 1.4 Pg (1 Pg = 1015, g = 1 Gt) (Lal *et al.*, 2013). That being the case, the projected increase in the world population to 9.6 billion by 2050 and 11 billion by 2100 will increase CO_2-C emission from 10 Pg C/year in 2012 to 13.4 Pg C/year by 2050 and 15.3 Pg C/year by 2100. In comparison, the Energy Information Authority (EIA, 2013) estimated the world's energy-related CO_2-C emissions at 8.5 Pg in 2010, 9.9 Pg in 2020, 11.3 Pg in 2030 and 12.4 Pg in 2040, with an average annual rate of increase of 1.3% per year. The ramifications of the growing competition for soil, water and energy Pimentel, (2011) are exacerbated by anthropogenic forces. Important among these are climate change and its uncertainties, soil resource degradation, water quality and renewability, activity, use efficiency of energy-based inputs, the stability and sustainability of agricultural productivity and global food security. The imbalance in food-soil-people stands out in several regions, including South Asia (SA) Swaminathan, (2012). The strong nexus between fossil fuel combustion, climate change and food security is widely recognized (Lal *et al.*, 2013). Yet, there are numerous uncertainties regarding the impacts of climate Lovejoy *et al.*, (2012) and variability on agriculture and food security Zhang, (2011) because of the confounding interaction with the CO_2 fertilization effect as moderated by water and nutrient availability and growing risks of extreme weather Godfray *et al.*, (2010).

India accounts for about 2.3% of world's geographical area and 4.2% of its water resources but has to support almost 18% of world's human population and 15% of the livestock. Agriculture remains the most important sector of Indian economy with 18% share in gross domestic product (GDP) at 2011–12 prices, 11% of exports and 53.3% share in total employment or workforce in 2013–14. As per the Agriculture Census 2010–11, the total number of operational holdings in the country has almost doubled from 71.01 million in 1970–71 to 138.35 million in 2010–11 (Fig 5.1). During the same period, the average size of operational holding has declined to 1.15 ha from 2.28 ha in 1970–71 (GOI, 2014).The small and marginal holdings taken together (below 2.0 ha) constituted 85.01% with operated area of 44.58% in 2010–11 against 83.29% with 41.14% of operated area in 2005–06. In a total of 138.35 million operational holdings the highest one belonged to Uttar Pradesh (23.33 million) and total of 159.59 million hectares operated area in 2010-11,the contribution was made by Uttar Pradesh (17.62million ha.).The average size of operational holding in the country was1.15 ha in 2010-11and in Uttar Pradesh was 0.76 ha (Chand *et al.*, 2011)

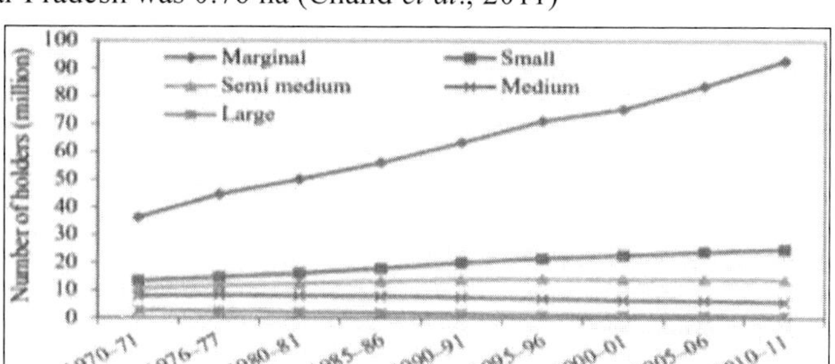

Fig 5.1: Number of operational holdings in India. Adapted from GOI (2014)

Food Production

The country's production of cereals is next only to China and USA. Among the important cereals, India is the second largest producer globally of rice and wheat, and the top producer of pulses. India is also the largest producer of milk and second largest producer of groundnut, vegetables, fruits, sugarcane, and cotton. As per the fourth advance estimates for 2013–14, total food grain production in India has been estimated at 264.77 Mt (DAC, 2015). Total production of pulses and oilseeds estimated at record levels of 19.27 and 32.88 Mt, respectively. Total production of rice in the country is estimated at 106.54 Mt which is a new record, higher by 1.30 Mt than the production of rice during 2012–13. Production of wheat, estimated at record level of 95.91 Mt is also higher than the production of 93.51 Mt during 2012–13. The production of coarse cereals is estimated at 43.05 Mt. Similarly, the production of milk (137.7 Mt), eggs (73.4 billion nos.), and wool (47.9 million kg) was also higher during 2013–14 compared to previous years. Uttar Pradesh has accounted for 19% share in the country's total food grain output during 2014-15. Total sugarcane production in India 28.3 million tonne in the fiscal ending September 2015, and Uttar Pradesh production was 7.35 million tonne and contribute 7.8% to India's GDP. Small and marginal farmers also make larger contribution to the production of high value crops. They contribute around 70% to the total production of vegetables, 55% to fruits against their share of 44% in land area (Birthal, 2011). Their share in cereal production is 52% and 69% in milk production. Thus, small farmers contribute to both diversification and food security.

Food Demand

The consumption pattern of food grains, which contribute to major nutritional intake, is decreasing; it was 64% during the base year 2000, and may decrease to 57 and 48% during the projected years of 2025 and 2050, respectively (Amarasinghe *et al.*, 2007). Contributions from non-grain crops and other animal products are increasing. The contribution from non-grain crops was 28% during 2000, and is projected to be 33 and 36% during 2025 and 2050, respectively. Contribution from other animal products was 8% during 2000, and is projected to increase to 12 and 16% during 2025 and 2050, respectively. Taking these trends into consideration the total food grain demand is estimated to be 291 Mt by 2025 and 377 Mt by 2050, whereas the total production is estimated to be 292 Mt by 2025 and 385Mt by 2050, which is 2.0% more than the demand. However, production deficits are projected for other cereals, oilseeds, and pulses. The projected deficit is 33 and 3% in 2025 and 43 and 7% in 2050 for other cereals and pulses, respectively. Another study by Singh (2009) found that the food consumption levels in India will increase from the current level of 2400 kcal/capita per day to about 3000 kcal/capita per day in 2050 and the demand for cereals will rise to 243 Mt in 2050. The rain-fed crop yields are expected to increase to 1.8 t/ha in 2030 and 2.0 t/ha in 2050. The irrigated cereal yields are projected to increase from 3.5 to 4.6 t/ha during the same period. The cereal production in India is thus projected to increase by 0.9% per year between 1999–2001 and 2050, and is expected to exceed the demand by 2050 even if the projected growth is about 0.9% per year.

Mittal (2008) concluded that an increase demand for food is seen with low yield growths especially in cereals any surplus in cereal production in the country is likely to diminish in the years to come. This is more alarming in the case of edible oil and pulses and the policy focus is needed for productivity enhancement (Table 5.1). Negative signs in respect of total cereals, pulses, oilseeds and sugars indicate the demand for the commodity outstrips supply and thus imply a deficit of the commodity in future. The deficit however is projected to increase over the years particularly for the below food items.

Table 5.1: Supply-Demand Gap for selected food item (Million tonnes)

Gap (Supply Minus Demand)		
Food Items	2021	2026
Rice	8.98	9.13
Wheat	27.33	32.04
Total Cereals	-2.94	-16.97
Pulses	-24.92	-39.31
Edible oils	-17.68	-26.99
Sugar	-39.67	-74.13

Source: Surabhi Mittal (2008)

Food Security and Climate Change

The concept of food security has been undergoing an evolutionary change during the last 50 years. In the 1950s, food security was considered essentially in terms of production. More recently, it is becoming evident that even if availability and access are satisfactory, the biological absorption of food in the body is related to the consumption of clean drinking water as well as to environmental hygiene, primary health care, and primary education. There are large disparities among sub-regions and countries in the region. India still has the second-highest estimated number of undernourished people in the world. The number of undernourished people in India is 194.6 (15.2%) million during 2014–16 compared to 210.1 million (23.7%) during 1990–92; "Global Hunger Index-2015" released by International Food Policy Research Institute ranks India at 25 with a score of 29 among 117 countries. About 54% of pregnant women suffer from anemia in India. The prevalence of vitamin A deficiency in the population is also very high in India (62% in 2003). The Indian Government has been responding with a number of measures to overcome hunger and malnutrition. International Rice Research Institute (IRRI), we might lose 15% of the world's rice crop with every degree of global warming. Agricultural production in India is closely linked to the performance of summer monsoon (June–September) which contributes about 75% of the annual precipitation (Fig 5.2). Apart from the inter-annual variability in summer monsoon rainfall, occurrence of many of the hydro-meteorological events is found to influence Indian agriculture at different spatial scales production during kharif (rainy season) during different years.

Apart from the summer monsoon rainfall, India receives about 15% of annual precipitation during the winter months of December–March. This precipitation is very important for Rabi (winter) crops. Though rainfall and its distribution have profound influence on the Indian food grain production, otherclimatic elements like radiation and temperature are also exerting considerable effect. Projected climate change may exacerbate the extreme climatic events and aggravate the risks of drought, flooding, pest infestation, and water scarcity to agro-ecosystems already under great stress (Beddington *et al.*, 2012). Climate change may affect food systems in several ways (Gregory *et al.*, 2005), although not all effects of climate change may be adverse to agronomic/food production (Lal, 2013).

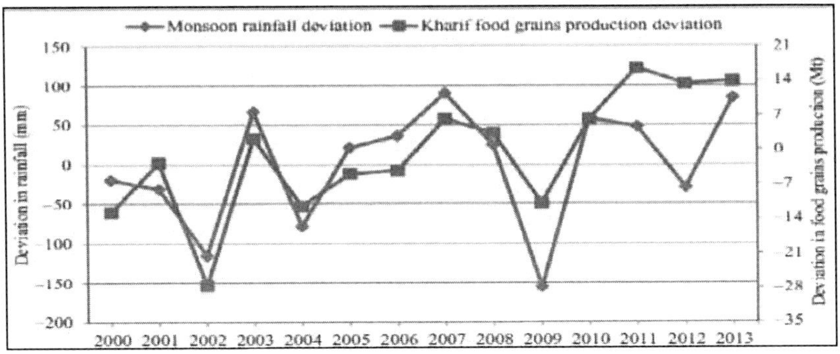

Source: Adapted from Srinivasa Rao *et al.*, (2013)

Fig 5.2: Monsoon rainfall versus deviation in food grain

Birthal *et al.*, (2014) projected the effects of climate change on crop yields for three timescales (2035, 2065, and 2100) at minimum and maximum changes in temperature and rainfall. In general, the production of pulses will be affected more by climate change than other crops. By the year 2100, with a significant change in climate, the yield of chick pea and pigeon pea will be lower by around 25% vis-a-vis without climate change (Table 5.2). The climate impacts on cereals will vary widely in rainy season as well as winter seasons. In the winter season, wheat yield will be less by about 22%, almost 3 times that of barley. Similarly, among rainy season cereals, rice will be affected more than maize and sorghum by the climate change. Rice yield will decline by over 15% with significant changes in climate as compared to loss of 7% in sorghum and of 4% in maize. Groundnut also stands to lose, but rapeseed-mustard is likely to gain by a small margin. If the climate does not change significantly, yield losses will be much smaller. However, the climate impacts will not be so severe in the short-run (2035). Maximum changes in temperature and rainfall are 1.3°C and 7% by 2035, 2.5°C and 26% by 2065, and 3.5°C and 27% by 2100, respectively.

Table 5.2: Projected changes in crop yields (%) at maximum changes in temperature and rainfall by 2035, 2065, and 2100

Crop	2035	2065	2100
Rainy season			
Rice	-7.1	-11.5	-15.4
Maize	-1.2	-3.7	-4.2
Sorghum	-3.3	-5.3	-7.1
Groundnut	-5.6	-8.6	-11.8
Winter Season			
Wheat	-8.3	-15.4	-22.0
Barley	-2.5	-4.7	-6.8
Chickpea	-10.0	-18.6	-26.2
Rapeseed-mustard	0.3	0.7	0.5

Source: Adapted from Birthal *et al*., (2014)

The direct impacts of climate change would be small on rainy season crops but the crops will become vulnerable due to increased incidence of weather extremes such as changes in rainy days, rainfall intensity, duration and frequency of drought and floods, diurnal asymmetry of temperature, change in humidity, and pest incidence and virulence. Winter crop production may become comparatively more vulnerable due to larger increase in temperature, asymmetry of day and night temperature, and higher uncertainties in rainfall (Rajeevan, 2013). The effects of climate change on food production are not limited to crops. It will affect food production and food security via its direct or indirect impact on other components of the agricultural production systems, especially livestock production which is closely linked with crop production. Livestock in India are raised under mixed crop–livestock systems deriving a substantial share of their encrgy requirements from crop by-products and residues. Any decline in crop area or production will reduce fodder supplies. Heat stress on animals will reduce rate of feed intake. The higher temperatures and changing rainfall patterns may cause increased spread of the existing vector-borne diseases and macro-parasites, alter disease pattern, give rise to new diseases, and affect reproduction behavior. All these factors will affect performance of the livestock (Birthal *et al.*, 2014).

By 2065, India's population is likely to cross 1.7 billion mark demanding more and diversified foods. With climate change, ensuring food security with more food production under limited resources will be a big challenge. It is, however, possible for farmers and other stakeholders to adapt to climate change and reduce the losses. Simple adaptations, such as change in crop variety, planting dates, rainwater conservation, adoption of resilient intercropping systems, particularly in rain-fed areas could help in reducing impacts of climate change. For example, losses in wheat production can be reduced from 4–5 to 1–2 Mt, if a large percentage of farmers could change to timely planting (Aggarwal, 2008). Ye and Ranst, (2009) concluded that declining food production under changing and variable climate from 18% surplus in 2005 to 22%–32% deficits by 2030–2050, with a doubling of the rate of soil degradation.

Kumar and Parikh (2001) shown for rice and wheat crop that projected large-scale changes in the climate would lead to significant reductions in their crop yields, which in turn would adversely affect agricultural production by 2060 and may affect the food security of more than one billion people in India. Kumar *et al.*, (2011) mentioned that decline in irrigated area for maize, wheat and mustard in northeastern and coastal regions; and for rice, sorghum and maize in Western Ghats of India may cause loss of production due to climate change. Hundal and Prabhjyot-Kaur (2007) concluded that an increase in minimum temperature up to 1 to 3°C above normal has led to decline in productivity of rice and wheat by 3% and 10% respectively in Punjab. Geethalakshmi *et al.*, (2011) concluded that productivity of rice crop has declined up to 41% with 40C increase in temperature in Tamil Nadu. Saseendran *et al.*, (2000) analyzed the projected results for duration 1980-2049; found that increment in temperature up to 50C can lead to continuous decline in the yield of rice and every one degree increment of temperature will lead up to 6% decline in yield in Kerala (India).

Srivastava *et al*., (2010) found that climate change will reduce monsoon sorghum productivity up to 14% in central zone (CZ) and up to 2% in south central zone (SCZ) by 2020; and this model also suggested that yields are likely to be affect even more in 2050 and 2080 scenarios; climate change impacts on winter crop are projected to reduce yields up to 7%, 11% and 32% by 2020, 2050 and 2080 respectively in India. Asha *et al*., (2012) found that the yields of sorghum, maize, pigeon pea (arhar), groundnut, wheat, onion and cotton has decreased up to 43.03, 14.09, 28.23, 34.09, 48.68, 29.56 and 59.96 kilogram per hectare respectively in rain-fed area; they also mentioned that almost 100% and 92.22% small and sample farmers respectively reported that the reduction in the rainfall was the major reason for reduction in the yield levels over the period followed by the pest and disease to the extent of 72.22%; and changes in temperature and seasonal patterns were quoted as the reason for the reduction in the yield by 42.22% of the sample respondents in Dharwad district in Karnataka (India). Vinesh *et al*., (2011) revealed that climate change has shifted and shortened crop duration in major crops rice and sugarcane, and it has significantly affected cane productivity in Uttar Pradesh and Uttarakhand. Kapur *et al*., (2009) projected surface warming and shift in rainfall may decrease crop yields by 30% by the mid 21st century, due to this reason there may be reduction in arable land resulting into pressures on agriculture production in India.

Anthropogenic activities have drastically transformed the biosphere. Humans appropriate about 40% of terrestrial photosynthesis and are the cause of the 50% of the global N flux from synthetic fertilizers and fossil fuel combustion and >50% of the renewable fresh water withdrawal Koning *et al*., (2008). Therefore, the continuing increase in population and consumption growth until the year 2100 Krishna *et al*., (2011) imply even more exploitation of the planet. Thus, there is an urgency to reduce the impact of agro-ecosystems on the environment Kumar, (1998). The record-breaking weather events of 2011 and 2012, along with the severity of wildfires and tornados, in conjunction with alternating droughts and floods, are indicative of anthropogenic climate changes. Thus, any strategy to achieve food production sustainability must also address climate change. It is feared that more than 100 million people may perish as a result of climate change by 2030 Kumar and Parik (2001). Climate change may influence food production directly (Fig 5.3) by affecting crop growth (e.g., drought, floods, length of growing season, suboptimal or supra-optimal temperatures) or indirectly by altering the ground water and the incidence of pests and pathogens (Lal *et al*., 2001; Lal, 2007).

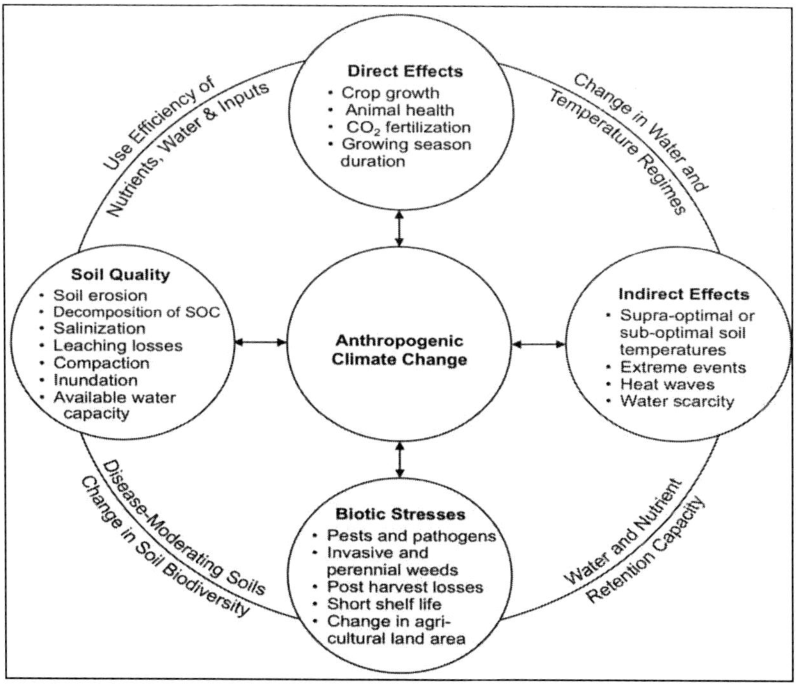

Fig 5.3: Direct and indirect effects of human-induced climate change on food security, which can be negative or positive, depending on the biome and regional characteristics

Climate Variability and Agriculture

Surface air temperatures over the country showed a warming trend, which is similar to the warming of global surface air temperatures. The annual average surface temperatures have increased over the years with a trend of 0.56°C/100 years which is close to the global warming trend. There is a substantial increase in surface temperatures since mid 1970s. Further, surface temperature warming is noted over most of the country except over a few pockets over northwest India. It is further noted that night time temperatures have increased sharply during the recent years, highlighting the role of greenhouse gases. These warming trends of surface temperature also may have an impact on agriculture.

Fig 5.4 shows that most districts with very high and high vulnerability to climate change are in Rajasthan, Gujarat, Maharashtra, Madhya Pradesh, Karnataka and Uttar Pradesh. Wheat and rice, two crops central to nutrition in India, have been found to be particularly sensitive to climate change. Lobell *et al.*, (2012) found that wheat growth in northern India is highly sensitive to temperatures greater than 34°C. The Intergovernmental Panel on Climate Change (IPCC) report of 2007 echoed similar concerns on wheat yield: a 0.5°C rise in winter temperature is likely to reduce wheat yield by 0.45 tonnes per hectare in India. Acute water shortage conditions, together with thermal stress, will affect rice productivity even more severely.While there has been considerable progress in understanding the sensitivities of crop production to yield; there are relatively few models which assess the impact of climate change on access to food.

Climate change amplifies the economic drivers of food insecurity. Variation in the length of the crop growing season and higher frequency of extreme events due to climate change and the consequent growth of output adversely affect the farmer's net income. India is particularly vulnerable because its rural areas are home to small and marginal farmers who rely on rain-fed mono-cropping, which provides barely a few months of food security in a normal year.

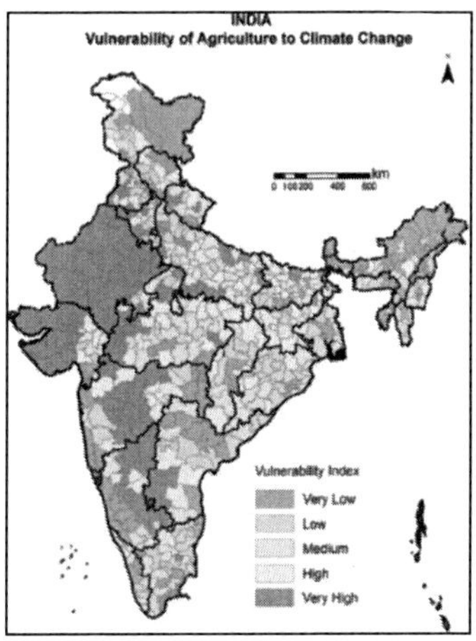

Fig 5.4: Vulnerability of Indian agriculture to Climate Change (2021-2050) Source: Rama Rao *et al.*, (2013)

Indian agricultural production and consequently, the country's Gross Domestic Product (GDP) show a strong link with ISMR (Gadgil and Gadgil 2006). In addition, temperature variations also play an equally critical role in crop production. The contribution of agriculture to the gross domestic product (GDP) has declined from 57% in 1950–51 to around 17% due primarily to growth in other sectors of the economy.

There has been substantial growth in agricultural production due to technological developments. However, substantial year to variability about the trend is also noted. The year to year variability of weather and climate is primarily responsible for the year to year fluctuations in the agricultural production. The most striking feature of the variation of the crop yield with the anomaly of ISMR (Fig 5.5) is the asymmetry in response to deficits versus excess of ISMR. It is seen that the magnitude of the impacts on crop yield increases rapidly with increasing deficits of ISMR. However, for positive anomalies of ISMR, the rate of increase of the impact on crop yield with ISMR is very small. This asymmetry in response to monsoon variability has important implications for the food security of the country. It implies that even when over a long period, the mean monsoon rainfall does not decrease, the negative impact of the deficit monsoon seasons cannot be made up by increases in crop yield in good monsoon seasons (Gadgil and Gadgil, 2006; Gadgil, 2012).

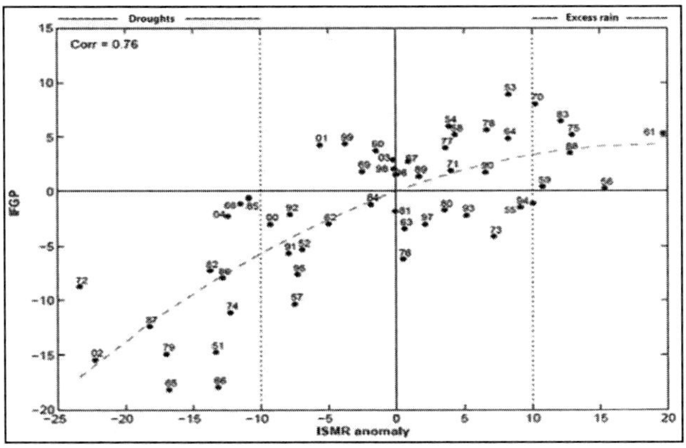

Fig 5.5: Relationship between ISMR anomaly and Food grain production in the country (Gadgil and Gadgil, 2006)

When we consider crop production at an aggregate spatial scale, climate variability is more relevant than weather related impacts. Since variability in agricultural output is largely driven by the year to-year fluctuations in strength of the summer monsoon rains, it is important to examine how global warming and future climate change may affect the Indian monsoon climate. On agricultural perspective, in addition of quantum of rainfall, the length of the monsoon season, rainy days, length of weak/break monsoon spells, frequency of heavy rains etc are also important. Surface temperature variability also will play an important role in affecting agricultural production in the country (Kumar *et al.*, 2011).

Impacts of Climate Change on Agriculture

Global warming is projected to have a significant impact on conditions affecting agriculture, including temperature, CO_2, glacier runoff, precipitation, and the interaction of these elements. These conditions in turn determine the carrying capacities of different ecosystems to produce enough biomass including food for human population and domesticated animals. The overall effect of climate change on agriculture will depend on the balance of these effects.

Climate change is predicted to have severe consequences on agriculture and the rural poor in South Asia. It has been identified as one of the most disaster-prone regions in the world. This is critical in view of the fact that increased frequency of extreme weather events like droughts and floods can have severe impacts in these countries (IPCC, 2007a). Given that approximately three-fifths of the cultivated area in South Asia is rainfed, the onset, duration, spatial extent, and total precipitation during the monsoon is the most critical factor in determining the livelihoods of a large majority of the population in rural areas. Moreover, resilience is typically low in rural areas as the existing asset base is limited and services are often insufficient. Global warming is likely to affect all these factors (Sterrett, 2011).

India is the second largest producer of rice and wheat in the world. In addition, 4% of the soybean (9.1 MT) and 23% of maize (19.7 MT) in the world is grown in India (FAOSTAT, 2010). Crop growth and yield are largely determined by weather during the growing season. With slight variations from normal weather, the effectiveness of externally supplied inputs decreases and consequently yields are affected. Increasing CO_2 concentration in the atmosphere and predicted higher atmospheric temperature are also expected to have an effect on crop production through changes in plant growth and transpiration (Mall *et al.*, 2006). Increased temperature reduces the grain yield due to faster development, increased respiration losses, and decreased time of grain growth in food crops. Yield reduction in wheat, rice, and corn in the region due to temperature rise during winter is expected (Challinor *et al.*, 2008).

The possible scenarios and impacts are unique to each country and are largely influenced by its geographical profile. Bhutan and Nepal have fragile mountainous ecosystem; Bangladesh and Sri Lanka have low-lying coastal areas, while India and Pakistan depend on cultivation in river plains, deltas, arid and semiarid lands. Key impacts of climate change would be an increase in temperature particularly during post monsoon and winter, increased frequency of floods during the monsoon, decreases in winter precipitation, and lower number of rainy days all affecting crop growth and production.

Increases in atmospheric temperature and annual rainfall result in frequent floods and cyclones in Bangladesh, whereas increased temperature would shift the cultivating zone further into higher elevations in the mountain ranges of Nepal, Bhutan, and India. Coastal region inundation or intrusion of seawater is a likely cause of a loss of the cultivated area in Bangladesh. With extreme weather events and encroaching salt water, the impacts on rice yield may vary between -6 to + 14% depending on different climate change scenarios (Mahmood *et al.*, 2004). On an average during the period 1962–88, Bangladesh suffered a loss of about 0.5 MT of rice production annually as a result of floods, accounting for nearly 30% of the country's average annual grain imports.

Increased frequency of monsoonal storms and flooding in the Himalayas could aggravate the occurrence of landslides in Nepal. In addition to the danger to life and property, some of the generated sediments may be deposited in the agricultural lands or in irrigation canals and streams, which will contribute to deterioration in crop production and in the quality of agricultural lands. There will be an increase of 0.6 1°C in mean temperature in coastal areas in Pakistan; increase in summer and winter rainfall in northern Pakistan and 10–15% decrease in the coastal belt and arid plains. The warming trend in India over the past 100 years (1901–2007) was observed to be 0.51°C, with accelerated warming of 0.21°C every 10 years since 1970 (Kumar, 2009).

Spatial and temporal variation projected changes in the temperature and rainfall, are likely to lead to differential impacts in the different regions (Byjesh *et al.*, 2010). The IPCC reports, and a few other global studies, indicate a probability of a 10–40% loss in crop production in India with increase in temperature by 2080–2100 (Majumdar, 2008). Studies conducted in India (Aggarwal, 2008) have confirmed similar declining trends in agricultural productivity due to climate change. For every 1°C increase in temperature, the yields of wheat, soybean, mustard, groundnut, and potato are expected to decline by 3–7% (Aggarwal, 2009b) and in rice by 6% (IWMI, 2007). Projections indicate the potential loss of 4–5 MT of wheat with every rise of 1°C temperature throughout the growing period with current land use in India alone (Aggarwal, 2008). Losses were also significant in other crops, such as mustard, peas, tomatoes, onion, garlic, and other vegetables and fruit crops (Aggarwal, 2009a). Various districts in the western Rajasthan, southern Gujarat, Madhya Pradesh, Maharashtra, northern Karnataka, northern

Andhra Pradesh and southern Bihar are also highly vulnerable to climate change. Sorghum yields are predicted to vary from + 18 to -22% depending on a rise of 2–4°C in temperature and increase by 20–40% of precipitation (Mall *et al.*, 2006). Rain-fed areas are likely to be more vulnerable in terms of extreme events (Mall *et al.*, 2006). Aberrations in the South–West monsoon could include a delay in the onset of the monsoon, long dry spells, and early withdrawal, etc., adversely affecting the productivity (Lal, 2001). This increase in variability could make it more difficult for resource-poor farmers to take decisions on investing on inputs and new technologies (Pandey *et al.*, 2007).

Climate change is expected to accelerate the hydrological cycle. In India and Pakistan, winter precipitation is projected to decline and this is likely to result in a greater demand for water during the Rabi season. During the monsoon season, the intensity of rains is projected to increase, which will imply more frequent and severe floods and a lesser recharge of groundwater (Jain, 2012). The flow of Indus river basin is also likely to affect cotton production in Pakistan, the main cash crop of the country. Further, water for agriculture is becoming increasingly scarce, and climate change-induced higher temperatures will increase crops' water requirements, so shortages will become more serious in the coming decades. By 2025, 15–20 million of the world's 79 million ha of irrigated rice lowlands, which provide three quarters of the world's rice supply, are expected to suffer some degree of water scarcity (IWMI, 2007). It is also estimated that to eliminate hunger and undernourishment of the world's population by 2025, the additional water requirements may be equivalent to all freshwater withdrawn and used today for agricultural, industrial, and domestic purposes (SIWI, 2005).

Agricultural sector is one of the sensitive areas which would be influenced by the projected global warming and associated climate change. In spite of the uncertainties about the precise magnitude of climate change on regional scales, an assessment of the possible impacts of changes in key climatic elements on our agricultural resources is important for formulating response strategies. Climate change may affect agricultural crops in four ways (Hulme, 1996). First, changes in temperature and precipitation will alter the distribution of agro-ecological zones. (i) An increase of potential evapo-transpiration is likely to intensify drought stress, especially in the semi-arid tropics and subtropics. (ii) Carbon dioxide effects are expected to have a positive impact due to, greater water use efficiency and higher rate of photosynthesis. (iii) Water availability (or runoff) is another critical factor in determining the impact of climate change. A number of studies suggested that precipitation and the length of the growing season are critical in determining whether climate change positively or negatively affects agriculture. (iv) Agricultural losses can result from climatic variability and the increased frequency of extreme events such as droughts and floods or changes in precipitation and temperature variance. Higher frequency of droughts is likely to increase pressure on water supplies for numerous reasons ranging from plant transpiration to allocation.

Attri and Rathore (2003) used CERES-wheat dynamic simulation model and climate change scenarios projected by the middle of the current century. They found increase in wheat yield between 29–37% and 16–28% under rainfed and irrigated conditions especially in different genotypes under a modified climate. An increase in temperature by 3^0C or more shall cancel out the positive effects of CO_2. Mall *et al.*, (2004) used the CROPGRO-soybean model to simulate the impact of climate change on soybean production in India. Climate change scenarios for the selected regions of the Indian subcontinent were developed using the results from three climate models. For the crop growth model, the probable changes in surface air temperature during the growing season were estimated at the selected sites in the region following standard rationalization techniques. Probable changes in precipitation, cloudiness and solar radiation under the climate changes scenarios were not taken into consideration in this analysis in view of the significant uncertainties associated with non-linear, abrupt and threshold rainfall events projected by the climate models over the Indian subcontinent. In this study, all the GCM projected climate change scenarios (at the time of doubling of CO_2 concentrations) predicted decreased yields for almost all locations. Mean decline in yields across different scenarios ranged from 14% in Pune (West India) to 23% in Gwaliar (Central India). Decline in soybean yield is found to be less in west and south India as compared to other parts of the country. The mean yield was found to be significantly affected under UKMO climate model generated climate scenarios for both current and doubled CO_2 atmosphere. These studies have indicated that the direct impacts of climate changes would be small on kharif crops but kharif agriculture will become vulnerable due to increased incidence of weather extremes such as changes in rainy days, rainfall intensity, duration and frequency of drought and floods, diurnal asymmetry of temperature, change in humidity, and pest incidence and virulence. Rabi crop production may become comparatively more vulnerable due to larger increase in temperature, asymmetry of day and night temperature and higher uncertainties in rainfall. The impacts of the climate change on Indian agriculture would be small in near future, but in long run the Indian agriculture may be seriously affected depending upon season, level of management, and magnitude of climate change.

Night time temperatures above some threshold can have stronger negative impacts on rice yields compared to radiation and day time temperatures over India and other parts of south Asia (Peng *et al.*, 2004). The recent studies highlight the importance of changing temperatures on the yields of major crops. The study by Kumar *et al.*, (2011) suggested higher temperatures reduce the yields regardless of rainfall consistent with the findings by others. Notwithstanding the positive benefits that may occur in crop yields under increasing CO_2 (Aggarwal and Mall, 2002), the substantial increase in temperatures could result in a net loss of productivity in the coming decades.

Projected Climate Change over India

The only way to understand the impact of global warming on the Indian monsoon and to assess future monsoon climate is to use climate models based on carefully constructed scenarios of emission of greenhouse gases. Kumar *et al.*, (2011) presented a comprehensive assessment of the present and expected future monsoon climate using observations and IPCC AR4 global climate model projections.

He study suggested that monsoon seasonal rainfall over India in the latter half of the 21st Century may not be different in abundance to that experienced today. However, their intensity and duration of wet and dry spells may change appreciably. However, this assessment on precipitation has large uncertainty. On the other hand, Indian temperatures during the late 21st century will very likely exceed the highest values experienced in the 130-year instrumental record of Indian data. The assessment on temperatures has more confidence than for rainfall due to large spatial scale driving the thermal response of climate to greenhouse gases. The study revealed that monsoon climate changes, especially temperature could heighten human and crop mortality, thus posing a socio-economic threat to the Indian sub-continent.

Kumar *et al.*, (2011) used the IPCC AR4 models. However, latest climate projections for the fifth assessment report of the IPCC are now available. These projections are made using the newly developed representative concentration pathways (RCPs) under the Coupled Model Inter comparison Project 5 (CMIP5). For the IPCC AR5 projections, the scientific community has developed a set of new-emission scenarios termed as representative concentration pathways (RCPs), which represent pathways of radiative forcing and not detailed socio-economic narratives or scenarios. There are four RCP scenarios: RCP2.6, RCP4.5, RCP6.0 and RCP8.5. The naming convention reflects socio-economic pathways that reach a specific radiative forcing by 2100. For example, RCP8.5 leads to a radiative forcing of 8.5 Wm^{-2} by 2100. Chaturvedi *et al.*, (2012) examined the CMIP5 model projections on the future climate of the Indian monsoon. The study provides multi-model and multi-scenario temperature and precipitation projections for India for the period 1860-2099. They have used 18 CMIP5 coupled models to develop future projections. The study examined ensemble mean climate instead of individual coupled models as they found out that ensemble mean climate is closer to observed climate than any individual model. The salient results from this study are as follows:

1) Under the business-as-usual scenario, mean warming over India is likely to be in the range 1-7-2.0 ^{0}C by 2030s and 3.3-4.8 ^{0}C by 2080s relative to pre-industrial times.
2) All-India precipitation under the business-as-usual scenario is projected to increase from 4% o 5% by 2030s and from 6% to 14% towards the end of the century (2080s) compared to the 1961-1990 baseline.
3) The precipitation projections are generally less reliable than temperature projections.
4) There is consistent positive trend in frequency of extreme precipitation days (e.g., greater than 40 mm/ day) for decades 2060s and beyond.

Fig 5.6 shows the CMIP5 model ensemble-based time series of temperature and precipitation anomalies (historical and projections) from 1861-2099 relative to the 1961-1990 baseline for the RCP scenario (Chaturvedi *et al.*, 2012).

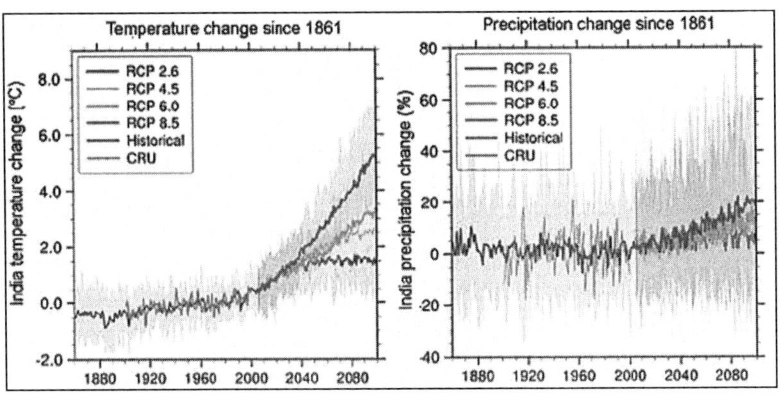

Fig 5.6: CMIP5 model based time series of temperature and precipitation anomalies (historical and projections) from 1861-2099 relative to the 1961-1990 baseline for the RCP scenario (Chaturvedi *et al.*, 2012).

Fig 5.7 suggests that under RCP2.6 the ensemble mean temperature increases by approximately 2°C over the period 1880's to 2070-2099 and 4.8°C in RCP8.5. For RCP4.5 and RCP6.0, which represent the moderate scenarios, the projected increase in temperature ranges from 2.9°C to 3.3°C.? All- India precipitation projections have larger uncertainties as evident from the large spread of the precipitation change projections in the fig, which ranges from -20% to 60% towards the end of the century. The model projections also suggest that all-India ensemble mean annual precipitation rise of 7%, 9.4%, 9.4% and 18.7% for RCP2.6, RCP4.5, RCP6.0 and RCP8.5 respectively by 2099 compared to the 1961-1990 baseline.

Fig 5.7 shows the percentage increase in the number of days when precipitation exceeds various thresholds. The period of 1861–1870 was selected as the reference period. The analysis projects consistently increasing trends in frequency of extreme precipitation days (e.g. 40 mm/day) for decades 2060's and beyond. The changes in extreme precipitation events have not emerged as a clear signal in 2010s in this model. The decline in heavy rainfall events in the 2050's is likely related to circulation changes on decadal variability in the model.

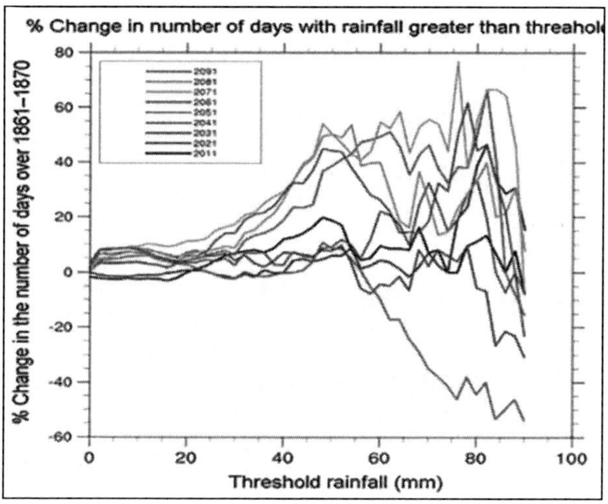

Fig 5.7: Projected change in the frequency of extreme rainfall days for future decades relative to the 1861 1870 baseline based on the MIROCESM-CHEM model for RCP4.5 scenario (Chaturvedi *et al.*, 2012).

References

1. Aggarwal PK. Global climate change and Indian agriculture: impacts, adaptation and mitigation. Indian J. Agric. Sci. 2008; 78 (11):911–919.

2. Aggarwal PK, Mall RK. Climate Change and Rice Yields in Diverse Agro-Environments of India: II. Effect of Uncertainties in Scenarios and Crop Models on Impact Assessment. Climatic Change. 2002; 52(3):331–43.
3. Aggarwal PK. Global Climate Changes and Indian Agriculture-Case Studies from ICAR Network Project. IARI, New Delhi, 2009a, 148
4. Aggarwal PK. Vulnerability of Indian agriculture to climate change-current stage of knowledge. In: National Workshop Review of Important of Work Program Towards Indian Network of Climate Change Assessment, Ministry of Environment & Forest, New Delhi. 2009b
5. Amarasinghe UA, Shah T, Anand BK. India's Water Supply and Demand From 2007; 2025-2050: Businessas-Usual Scenario and Issues. www.iwmi.cgiar.org/NRLP %20 Proceeding 2% 20 Paper % 202.pdf 9. 2007
6. Asha LKV, Gopinath M, Bhat ARS. Impact of climate change on rain-fed agriculture in India', International J Environ Sci Dev. 2012; 3(4):368-371.
7. Attri SD, Rathore LS. Simulation of impact of projected climate change on wheat in India. Int. J. Climatol. 2003; 23:693–705.
8. Beddington JR, Asaduzzaman M, Clark ME. What next for agriculture after Durban. Sci. 2012; 335:289-290.
9. Birthal PS, Joshi PK, Narayanan AV. Agricultural Diversification in India: Trends, contribution to growth and small farmer participation, ICRISAT, mimeo. 2011
10. Birthal PS, Khan MJ, Negi DS, Agarwal S. Impact of climate change on yields of major food crops in India. Implications for food security. Agric. Econ. Res. Rev. 2014; 27(2):145-155.
11. Byjesh KS, Naresh Kumar S, Aggarwal PK. Simulating impacts, potential adaptation and vulnerability of maize to climate change in India. Mitig. Adapt. Strat. Global Change. 2010; (15):413–431.
12. Challinor AJ, Wheeler TR. Use of a crop model ensemble to quantify CO_2 stimulation of water-stressed and well-watered crops. Agric. Forest Meteorol. 2008; 148:1062–1077.
13. Challinor AJ, Wheeler TR, Craufurd PQ, Ferro CAT, *et al.*, Adaptation of crops to climate change through genotypic responses to mean and extreme temperatures. Agric. Ecosyst. Environ. 2008; 119:190–204.

14. Chand R, Lakshmi Prasanna PA, Aruna Singh. Farm size and productivity: Understanding the strengths of smallholders and improving their livelihoods. Economic and Political Weekly. 2011; 46:26-27

15. Chaturvedi RK, Joshi J, JayaramanM, Bala G, Ravindranath NH. Multi-model climate change projections for India under representative concentration pathways. Curr Sci. 2012; 103:791–802.

16. DAC. Agricultural Statistics at a Glance 2014. Department of Agriculture and Cooperation. Ministry of Agriculture, Government of India 2015, 2015, 452.

17. FAOSTAT. FAO Statistical Year Book 2010. Food and Agricultural Organization of United Nations, Rome, Italy. 2010

18. Geethalakshmi V, Lakshmanan A, Rajalakshmi D, Jagannathan R, Sridhar G, Ramara AP. Climate change impact assessment and adaptation strategies to sustain rice production in cauvery basin of Tamil Nadu. Current Sci. 2011; 101:103

19. Godfray HCJ, Beddington JR, Crute IR, Haddad L, Lawrence D. Food security: The challenge of feeding 9 billion people. Sci. 2010; 327:812–818.

20. GOI. Agriculture Census 2010–11. All India Report on Number and Area of Operational Holdings. Ministry of Agriculture, Government of India, 2014, 87.

21. Hulme, Mike ed. Climate Change and Southern Africa. Norwich, United Kingdom: Climatic Research Unit, University of East Anglia. 1996

22. Hundal SS, Prabhjyot-kaur. Climatic variability and its impact on cereal productivity in Indian Punjab. Current Sci. 2007; 92(4):506-512

23. Jain SK. Sustainable water management in India considering likely climate and other changes. Curr. Sci. 2012; 102(2):177-188.

24. IPCC. Intergovernmental Panel on Climate Change (IPCC). Fourth Assessment Report: Climate Change, Geneva. 2007a

25. IPCC. Summary for Policymakers. Climate Change: The Physical Science Basis. Contribution of Working Group I to the Fourth Assessment Report of the Intergovernmental Panel on Climate Change, Cambridge University Press, Cambridge UK. 2007b

26. IPCC. Climate Change: Climate Change Impacts, Adaptation and

Vulnerability, Summary for policymakers. Intergovernmental Panel on Climate Change. 2007c

27. IWMI. Rice: feeding the Billions. In: Water for Food, Water for Life: A Comprehensive Assessment of Water Management in Agriculture, Earthscan; International Water Management Institute, London; Colombo. 2007

28. Kapur D, Khosla R, Mehta PB. Climate change: India's options', Economic and Political Weekly, 2009, 36:31

29. Koning NBJ, van Ittersum, MK, Becx GA, *et al.*, Long-term global availability of food: Continued abundance or new scarcity? J. Life Sci. 2008; 55:229–292.

30. Krishna Kumar K, Kamala K, Rajagopalan B, Hoerling MP, *et al.*, The once and future pulse of Indian monsoonal climate. Climate Dynam. 2011; 36(11):2159-2170.

31. Kumar K. Impact of climate change on India's Monsoon climate and development of high resolution climate change scenario for India. Paper presented at MoEF, New Delhi, October 14, 2009.

32. Kumar KK, Patwardhan SK, Kulkarni A, Kamala K, *et al.*, Simulated projections for summer monsoon climate over India by a high-resolution regional climate model (PRECIS). Curr. Sci. 2011; 101(3):312-325.

33. Kumar KSK, Parikh J. Socio-economic impacts of climate change on Indian agriculture. Int. Review Environ Strategies. 2001; 2 (2):277-293.

34. Kumar P. Food Demand and Supply Projections for India. Agricultural Economics Policy Series Indian Agricultural Research Institute, New Delhi, 1998, 98-01.

35. Lal M. Future climate change: implications for India Summer monsoon and its variability. Curr. Sci. 2001; 81:1205-1207.

36. Lal M, Nozawa T, Emori S, Harasawa H, *et al.*, Future climate change: implications for Indian summer monsoon and its variability. Curr. Sci. India. 2001; 81:1196-1207.

37. Lal M. Implications of climate change on agricultural productivity and food security in South Asia. Key vulnerable regions and climate change – Identifying thresholds for impacts and adaptation in relation to Article 2 of the UNFCCC, Springer, Dordrecht, in press. 2007

38. Lal R, Stewart BA. Principles of Sustainable Soil Management in

Agroecosystems. In Advances in Soil Science; Lal, R., Stewart, B.A., Eds; Taylor and Francis: Boca Raton, FL, USA, 2013, 568

39. Lovejoy S, Schertzer D, Varon D. Do GCM's predict the climate or macro-weather? Earth Syst. Dyn. Discuss. 2012; 3:1259-1286.

40. Mall RK, Lal M, Bhatia VS, Rathore LS, Singh R. Mitigating climate change impact on soybean productivity in India: a simulation study. Agri. Forest Meteoro. 2004; 121:113–125.

41. Mall RK, Singh R, Gupta A, Srinivasan G, Rathore L.S. Impact of climate change on Indian agriculture: A review. Climatic Change. 2006; 78: 445-478.

42. Majumdar PP. Implications of climate change for sustainable water resources management in India, Physics and Chemistry of the Earth, PartsA/B/C. 2008; 33:354-358.

43. Mahmood R, Hubbard KG, Carlson C. Modification of growing season surface temperature records in the Northern Great Plains due to land use transformation: verification of modelling results and implications for global climate change. Int. J. Climatol. 2004; 24:311–327.

44. Mittal S. Demand-Supply Trends and Projections of Food in India. Working Paper No. Indian Council for Research on International Economic Relations, 2008, 209-20

45. Pandey S, Bhandari H, Hardy B, editors. Economic costs of drought and rice farmers' coping mechanisms: a cross-country comparative analysis. Los Baños (Philippines): International Rice Research Institute, 2007, 203

46. Parry M, Rosenzweig C, Livermore M. Climate change, global food supply and risk of hunger. Phil. Trans. Roy. Soc. London B. 2005; 360:2125–2138.

47. Peng S, Huang J, Sheehy JE, Laza RC, *et al.*, Rice yields decline with higher night temperature from global warming. Proc Natl Acad Sci USA. 2004; 101:9971–9975.

48. Pimentel D. Food for thought: A review of the role of energy in current and evolving agriculture. Crc Crit. Rev. Plant Sci. 2011; 30:35-44

49. Rajeevan M. Climate change and its impact on Indian agriculture. In: Shetty, P.K., Ayyappan, S., Swaminathan, S. (Eds.), Climate Change and Sustainable Food Security. National Institute of Advanced Studies, India, 2013, 1-12.

50. Rama Rao, Raju CA, Subba Rao BMK, Rao AVM, *et al.*, Atlas on Vulnerability of Indian Agriculture to Climate Change. Central Research Institute for Dryland Agriculture, Hyderabad, 2013, 116

51. Samra JS, Singh G. Heat Wave of March Impact on Agriculture, 32 pp. Indian Council of Agricultural Research. 2004.

52. Saseendran SA, Singh KK, Rathore LS, Singh SV, Sinha SK. 'Effects of climate change on rice production in the tropical humid climate of Kerala, India', Climatic Change, Kluwer Academic Publishers, Netherland. 2000; 44(4):495-514.

53. Singh RB. Towards a Food Secure India and South Asia. Making Hunger a History. http://www.apaari.org/wpcontent/uploads/2009/08/towards-a-food-secureindiamakinghunger-history.pdf. 2009.

54. SIWI, IFPRI, IUCN, IWMI. Let it Reign: The NewWater Paradigm for Global Food Security. Final Report to CSD-13 Stockholm InternationalWater Institute, Stockholm. 2005

55. Song MW, Li AZ, Cai LQ. Zhang RS. Effects of different tillage methods on soil organic carbon pool. J Agro-Enviro Sci. 2011; 27(2):6222-626

56. Srivastava A, Kumar SN, Aggarwal PK. Assessment on vulnerability of sorghum to climate changes in India. Agri. Ecosyst Environ. 2010; 138:160-169.

57. Sterrett C, Review of Climate Change Adaptation Practices in South Asia. Oxfam Research Reports. Climate Concern, Melbourne, Australia. 2011

58. Swaminathan MS. Combating hunger. Sci. 2012; 338:1009

59. Swinnen J, Squicciarini P. Mixed messages on prices and food security. Sci. 2012; 335:405–406.

60. Vinesh K, Sharma YS, Chauhan YS. Impact of climate change on the growth and production of saccharum officianarum and magnifera Indiaca. Int. J Sci Tech Mgt. 2011; 2(1):92-107.

61. Ye L, van Ranst E. Production scenarios and the effect of soil degradation on long-term food security in China. Global Environ. Change. 2009; 19:464-481

62. Zhang J. China's success in increasing per capita food production. J Exp. Bot. 2011; 62: 3707-3711.

Chapter - 6
Climate Change and Water Availability

Authors

Dr. R.K. Naresh

Sardar Vallabhbhai Patel University of Agriculture & Technology, Meerut, Uttar Pradesh, India

Dr. S.S.Tomar

Rajamata Vijayaraje Scindia Krishi Vishwa Vidyalala, Gwalior ZARS, A B, Road Morena, Madhya Pradesh, India

Chapter - 6

Climate Change and Water Availability

Dr. R.K. Naresh and Dr. S.S. Tomar

Climate is a very decisive factor in natural resource water availability of a region. Warming of the climate system in recent decades is evident from increase in average air and ocean temperatures, widespread melting of snow and ice, and rising sea level. The hydrological cycle is intimately linked with changes in atmospheric temperature and radiation balance. A warmer climate may lead to intensification of the hydrological cycle, resulting in higher rates of evaporation and increase of liquid precipitation. These processes, in association with a shifting pattern of precipitation, may affect the spatial and temporal distribution of runoff, soil moisture, groundwater reserves etc. and may increase the frequency of droughts and floods. Agricultural demand, particularly for irrigation water is considered more sensitive to climate change. A change in field-level climate may alter the need timing of irrigation and water productivity. Crop production can increase if resources are expanded or irrigation is intensified, but these may increase the rate of natural resources degradation. Since climate change impacts on soil water balance will lead to changes of soil evaporation and plant transpiration, consequently, the crop growth period may shorten in the future impacting on water productivity.

Approximately 75% of the Himalayan glaciers are on retreat and will disappear by 2035. Moreover in Uttar Pradesh, India by 2050 the rainfall could drop by 10%, which would reduce drainage by 17%. Majority of the fresh water resources has already been depleted and there is reduction in agricultural production globally with escalation in population and food demand. Global food supply will depend on how well agriculture adopts to climate change. The apparent impacts of the global climate change in India include erratic monsoon, high intensity floods, increased frequency of draughts, decreasing crop yields among others. Rapid industrial development, urbanization and the increasing demand for irrigation water to feed the burgeoning population of India are already placing immense pressure on water resources. There is a direct link between the rise in global temperature and damage to eco-systems. About 130 million hectares (mha) land in India is undergoing different levels of degradation, namely water erosion (32.8 mha), wind erosion (10.8 mha), desertification (8.5 mha), water logging (8.5 mha). The greenhouse gases (GHGs) in atmosphere are increasing at the rate of 0.5%, 0.6% and 0.25 ppbv yr^{-1} for CO_2, CH_4 and N_2O, respectively. Contribution of these three GHGs to global warming is to an extent of 20% due to agricultural activities and 14% due to in land use changes and attendant deforestation. Major agricultural activities that contribute to emission of GHGs include ploughing, application of manures and fertilizers, biomass burning and crop residue removal. Historic global C loss due to agriculture is estimated at 55 to 100 Pg from soil C pool and 100 to 150 Pg from the biotic C pool. Trends indicate that agricultural productivity will decline up to 25 percent which could be as much as 50 percent in rain-fed agriculture. Small and marginal farmers with small land holdings will be more vulnerable to climate change. With the temperature increasing and precipitation fluctuations, water availability and crop production are likely to decrease in the future.

India accounts for about 2.3% of world's geographical area and 4.2% of its water resources but has to support almost 18% of world's human population and 15% of the livestock (Table 6.1). Agriculture remains the most important sector of Indian economy with 18% share in gross domestic product (GDP) at 2011–12 prices, 11% of exports and 53.3% share in total employment or workforce in 2013–14. As per the Agriculture Census 2010–11, the total number of operational holdings in the country has almost doubled from 71.01 million in 1970–71 to 138.35 million in 2010–11. During the same period, the average size of operational holding has declined to 1.15 ha from 2.28 ha in 1970–71 (GOI, 2014). The small and marginal holdings taken together (below 2.0 ha) constituted 85.01% with operated area of 44.58% in 2010–11 against 83.29% with 41.14% of operated area in 2005–06. In a total of 138.35 million operational holdings the highest one belonged to Uttar Pradesh (23.33 million) and total of 159.59 million hectares operated area in 2010-11, the contribution was made by Uttar Pradesh (17.62 million ha.). The average size of operational holding in the country was1.15 ha in 2010-11and in Uttar Pradesh was 0.76 ha (Chand *et al.*, 2011). Thus, the small holding character of Indian agriculture is much more prominent today than even before.

Table 6.1: Annual water requirement for different uses (in BCM)

Uses	Year					
	2010		2025		2050	
	Low	High	Low	High	Low	High
Irrigation	543	557	561	611	628	807
Domestic	42	43	55	62	90	111
Industry	37	37	67	67	81	81
Power	18	19	31	33	63	70
Inland navigation	7	7	10	10	15	15
Environment – ecology	5	5	10	10	20	20
Evaporation loss	42	42	50	50	76	76
Total	694	710	784	843	973	1180
Population (million)	1286	1333	1286	1333	1346	1581

Source: Jain (2011)

Climate change projections to 2100 AD for India indicate an overall increase in temperature by 2–4°C (Table 6.2) with no substantial change in precipitation (Kavikumar, 2010). At an all India level, there is no definite trend that can be established in the monsoon rainfall during last 100 years, but there are some regional patterns. Areas of increasing trend in monsoon rainfall are found along the west coast, north Andhra Pradesh and adjoining areas, north-east India and parts of Gujarat and Kerala (-6 to -8% of normal over 100 years). The warming trend in India over the past 100 years (1901–2007) was observed to be 0.51°C with accelerated warming of 0.21°C every 10 years since 1970 (Kumar, 2009).

The spatial distribution of temperature changes indicates a significant warming trend along the west coast, central India, the interior Peninsula, and over northeast India. However, a cooling trend has been observed in the northwest and in some parts in southern India. Records over the past 130 years do not show any significant long-term trend in the frequencies of large-scale droughts or floods in the summer monsoon season. The frequency of cyclonic storms that form over Bay of Bengal has remained almost constant over the period 1887–1997. The magnitude of the impact of climate change is likely to vary in different parts of India. Parts of western Rajasthan, Southern Gujarat, Madhya Pradesh, Maharashtra, Northern Karnataka, Northern Andhra Pradesh, and Southern Bihar are likely to be more vulnerable in terms of extreme events (Mall *et al*., 2006).

Table 6.2: Projected changes in climate in India (207 e0-99)

Region	Jan.–Mar.	Apr.–Jun.	Jul.–Sep.	Oct.–Dec.
Change in Temperature (°C)				
Northeast	4.95	4.11	2.88	4.05
Northwest	4.53	4.25	2.96	4.16
Southeast	4.16	3.21	2.53	3.29
Southwest	3.74	3.07	2.52	3.04
Change in Precipitation (%)				
Northeast	-9.3	20.3	21.0	7.5
Northwest	7.2	7.1	27.2	57.0
Southeast	-32.9	29.7	10.9	0.7
Southwest	22.3	32.3	8.8	8.5

Source: Kavikumar (2010).

Water Demand

Though India is industrializing rapidly, producing sufficient food remains the country's primary water challenge (although reducing the impact of destructive floods is also a critical concern in some regions). Population is assumed to be growing at 1.0 percent per annum and GDP at 6.8 percent per annum between 2005-2030. During the same period, the share of agriculture in GDP decreases from 19 percent to 10 percent. But because the vast majority (more than 95 percent) of agricultural production is and will continue to be for domestic consumption, the growing population coupled with an average increase in wealth mean that increasing caloric intake of the national population will be one of the key trends underlying the water resource challenge.

Overall, the unconstrained demand implied by this production growth, driven by the rapid increase in demand for food and feed crops, particularly rice and wheat, would mean that in 2030 agriculture will account for almost 1,200 billion m^3 or 80 percent of total water demand—almost double agriculture's water demand in 2005. Projected municipal and domestic water demand will also double by 2030, to 108 billion m^3 (7 percent of total demand), while projected demand from industry will quadruple to 196 billion m^3 (13 percent), pushing overall demand growth close to 3 percent per annum. This demand, weighed against today's reliable, accessible supply, would create severe projected deficits for most of India's river basins (Fig 6.1).

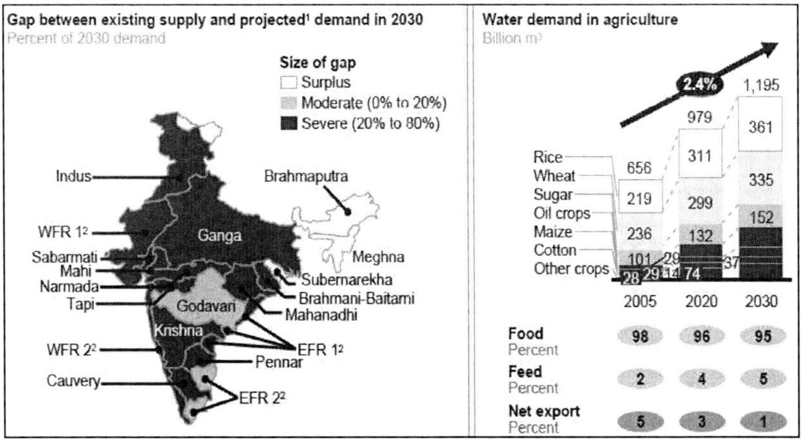

Source: 2030 Water Resources Group

Fig 6.1: India-water suppEly and demand gap

1 The unconstrained projection of water requirements under a static policy regime and at existing levels of productivity and efficiency 2 WFR = western-flowing coastal rivers; EFR = eastern-flowing coastal rivers

In the base case projection for India, of the projected 685 million metric tons of food production in 2030, 175 million tons are expected to be rain-fed, leaving 510 million tons of irrigated production.

Underlying these numbers is the assumption that existing rain-fed lands would reduce slightly in extent given some conversion to irrigated lands, and that additional crop demand will primarily come from additional irrigated lands. For irrigated production in 2030, projected demand from rice will be 361 billion m^3 or 30 percent of irrigation demand, followed by wheat (335 billion m^3) sugarcane (152 billion m^3) and oil crops (137 billion m^3). These requirements of course do not necessarily reflect relative weight of yields due to different water intensity. For example, the existing requirement of rice for irrigation withdrawals is 3.7 billion m^3 per million tons of production, while sugarcane requires 3.3 billion m^3 per million tons of production. Withdrawals also differ between basins. For example, in the Ganga, sugarcane requires 3.7 billion m^3 per million tons of production, in the Krishna basin only 3.0 billion m^3. These differences in the water requirements also reflect the current mix of irrigation technologies applied to each crop and in each basin, the source of irrigation (surface water or groundwater), and differences in the local climate.

Water Supply

At 2,518 billion m^3, the total water resource base for India, including surface and groundwater, is substantial but highly variable: during the monsoon season 50 percent of annual precipitation falls in less than one month and 90 percent of river flows occur in only 4 months of the year (Briscoe and Malik, 2006). The ability of the current infrastructure to buffer that variability is low, making it difficult for accessible, reliable supply to meet projected demand. With only 200 m^3 of water storage capacity per person, compared to 2,200 m^3 per person in China and some 6,000 m^3 per person in the United States (ICOLD, database), India's accessible, reliable supply of water amounts to 744 billion m^3, or 29 percent of its total water resource. Data on groundwater availability is uncertain, but we estimate that out of a total 400 billion m^3 of renewable groundwater, approximately 230 billion m^3 are accessible reliably. Yet, the importance of groundwater for India cannot be underestimated. For example, groundwater has played a large role in the success of rapidly increasing grain yields in India's "Green Revolution," when higher yielding seeds accompanied by fertilizer inputs multiplied the yields of India's agriculture. A key enabler of this yield increase was the use of both shallow and deep tube wells, allowing farmers more control of irrigation water (Briscoe and Malik, 2006).

Water supply is not distributed uniformly across the subcontinent. The Ganga basin accounts for a large fraction of the accessible supply, with almost 311 million m^3, split between groundwater (35 percent) and surface water (65 percent). This is also reflected in the individual basin gaps, with the Ganga having by far the largest demand, and thus representing the largest gap in 2030 at 350 million m^3 or 53 percent of local demand, followed by the Indian tributaries of the Indus basin (106 million m^3, or 52 percent of local demand) and the Krishna basin (90 million m^3, or 64 percent of local demand). These three basins combined account for over 70 percent of the total gap in India.

Within the total supply, groundwater supply differs substantially by region as well. In some basins, such as the western rivers, the true renewable groundwater supply is much less than what is actually pumped, leading to massive overdraft, declining water tables and elevated pumping costs; while in basins such as the Eastern Ganges additional groundwater supply could be increased sustainably. In addition, the core challenge of availability of groundwater is compounded by rapidly deteriorating water quality in many areas of the country, which could reduce the quantities that can be considered accessible supply, even for agricultural purposes.

Water Resources of India

Total utilizable water resource in the country has been estimated to be about 1123 billion cubic metres (BCM) (690 BCM from surface and 433 BCM from ground water), which is just 28% of the water derived from precipitation (Table 6.3). About 80% of the water (688 BCM) is being diverted for irrigation, which may increase to 1072 BCM by 2050. On the basis of the available water resources, the total irrigation potential from surface and ground water resources is estimated to be 139.9 Mha. The major source for irrigation is groundwater. Annual groundwater recharge is about 433 BCM of which 212.5 BCM is used for irrigation and 18.1 BCM for domestic and industrial use (CGWB 2011). By 2025, demand for domestic and industrial water usage may increase to 29.2 BCM.If the international yardstick of 1700 m^3 per capita availability of water is taken as a criterion, then the whole country became "Water stressed" in the year 2007 with per capita water availability declining to 1656 m^3 from 5200 m^3 in 1951. The reported per capita availability of water was 1545 m^3 in 2011. It is estimated that the per capita availability of water is likely to reach a level of around 1100 m^3 in 2050 but in reality we may reach this threshold much earlier considering the wasteful ways of water usage. It may be noted that per capita availability of <1000 m^3 indicates "Water scarcity".If the per capita availability is scrutinized basin wise than a significant number of basins in the country were either in the "water scarcity zone" or worse in 2010.

Table 6.3: Water resources of India

Estimated Annual Precipitation (including snow fail)	**4000 BCM**
Estimated Utilisable Water	
Surface	690 BCM*
Gronnd	433 BCM
Total	1123 BCM
irrigation Potential	
***Major and and Medium (Surface Water)**	
Ultimate	58.5 Mha
Created (upto 2009-10)	44.4 Mha
Utilized (upto 2009-10)	34.4 Mlla
***Minor irrigationgation**	
Ultimate	81.4 Mha
Created (upto 2009-10)	62.S Mha
Utilized (upto 2009-10)	52.5 Mha
Total (Major and Medium + Minor)	
Ultimate	139.9 Mha
Created (upto 2009-10)	107.2 Mha
Utilized (upro 2009-10)	96.9 Mha
Area Irrigated (2009-10)	
Gross	86.42 Mha
Net	63.25 Mha

*BCM= Biliion Cubic Meter = 1 km^3

Source: CWC (2013)

Ground Water Development

Development of ground water resources which was a low key affair in the early stages is now contributing to more than 60% of the irrigation demand and playing a critical role in ensuring food security to the millions in the country. However, it's over exploitation is now a very serious cause of concern and the potential of judicious exploitation of ground water resources is now confined only to some states in the eastern part of the country. Out of the 5842 assessed blocks, 4277 are safe while the rest are in the semi critical (523), critical (169) and a large number of overexploited (802) blocks (Fig 6.2).

India has a dubious distinction of having the highest ground water abstraction (Fig 6.3). Considering the availability of the ground water resources, Government of India (GOI) has rightly decided to focus on Eastern India and has implemented a programme aptly titled "Bringing Green Revolution to Eastern India (BGREI)". In view of the seriousness of the problem, the Government is, therefore, putting a lot of emphasis on artificial recharge of ground water. A project for setting up 225 thousand recharge structures at an estimated cost of approximately Rs. 20,000 crores has been launched with this objective and a clear defined goal that it has to be implemented in a participatory mode with involvement of local communities, locally elected bodies like Panchayats, and Non Government Organizations (NGOs).

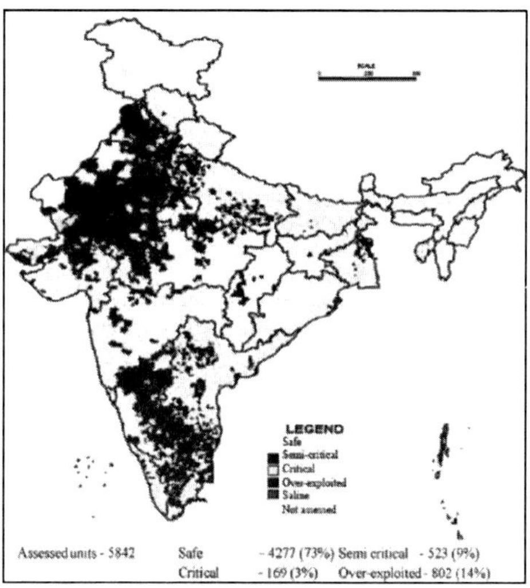

Fig 6.2: Block–wise status of ground water exploitation

Water Resources of Uttar Pradesh

Total geographical area of the state is 24,170 thousand hectare (which is 7.33% of total area of India) out of which 16,573 thousand hectare is under cultivation. Gross cropped area is 25,414 thousand ha with the cropping intensity of 153%.

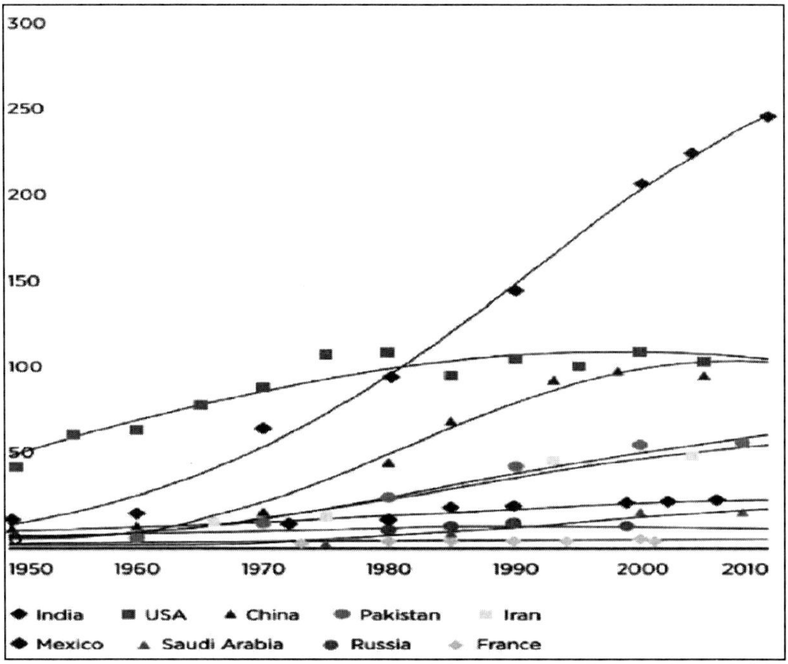

Fig 6.3: Groundwater abstraction trends in selected countries (in km^3/year)

In Uttar Pradesh size of holding is around 0.83 ha and per capita land area is 0.14ha, which is less than a half of the national average of 0.32 ha. Uttar Pradesh is largest producer of wheat, potato, sugarcane and milk whereas third largest producer of rice. Agriculture still constitutes the backbone of the state economy, more so, because it provides livelihood to about two third population of the state.

The state is endowed with ample alluvial soil along with diverse agro climatic profile which can support the cultivation of variety of crops. Due to large cultivated area, its share in national agricultural production is quite impressive but low crop productivity has hindered the realisation of ultimate potential.

The state is endowed with bountiful water resources which were considered abundant but because of increasing demand for various purposes namely irrigation, drinking and domestic, power (thermal and hydro), industrial and other uses, its scarcity is becoming apparent which shall get more pronounced with increasing population. Water is a prime natural resource, a basic human need and a precious asset. Planning and development of water resources of the state need to be governed by the development perceptions of the state. Being a border state of India, Uttar Pradesh's northern frontiers adjoin Nepal which earlier extended upto Tibet before the creation of Uttaranchal whose Shivalik ranges near Tibet Border. UP's boundaries touch Haryana, Delhi and Rajasthan in the west, Madhaya Pradesh in the south and Bihar in the east. In geophysical terms, the Shivalik range of the Himalayas in the north, the river Yamuna and the Vindhyas in the west, southwest and the south and the Gandak River in the east demarcate the terrain called Uttar Pradesh as it is. Topographically, U.P. consisted of Terai Bhabhar, plains and the southern plateau, Terai Bhabhar.

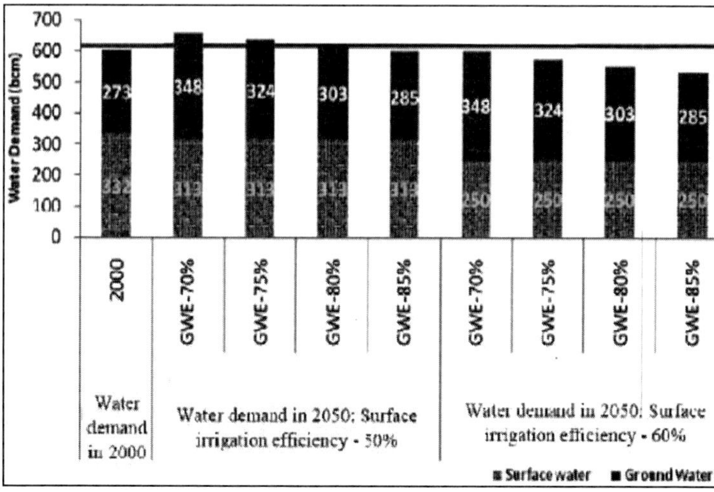

Fig 6.4: Irrigation efficiency and water demand scenarios. (Amarasinghe *et al.*, 2007)

The food production in UP is commensurate with the self sufficiency of the country. One of the major contributors for this sufficiency is irrigation. To meet this high irrigational requirement, water resources are being increasingly developed. Ground water contributes to about 71% of the irrigation needs of the State. The indiscriminate development of ground water has resulted in depletion of groundwater storage and lowering of water level in certain areas on one hand (Fig 6.4). On other side the surface water development in areas having shallow water level has resulted in water logging and soil salinisation.

Agricultural Response to Water Scarcity

Farmers are typically very adaptable to changes in market opportunities and access to productive inputs including water. The successful response of agriculture to fast population increase in the second half of the last century and to progressive shortage of land and water are a case in point: over the past 30 years, the world's total agricultural production doubled, while the expansion of cultivated land was only about 15 percent and all of this growth occurred in land equipped for irrigation (Fig 6.5). In regions of land scarcity, such as South Asia, the growth in commodity production was almost completely based on increases in yields and cropping intensities. In contrast, in South America, the pressure on land is less, and 40 percent of the growth in production was due to an aerial increase in farmed land. Adaptation is evident in water-scarce regions such as the Near East and North Africa, where irrigation efficiencies are often 20 percent higher than in water-rich areas of Southeast Asia, Latin America or sub-Saharan Africa.

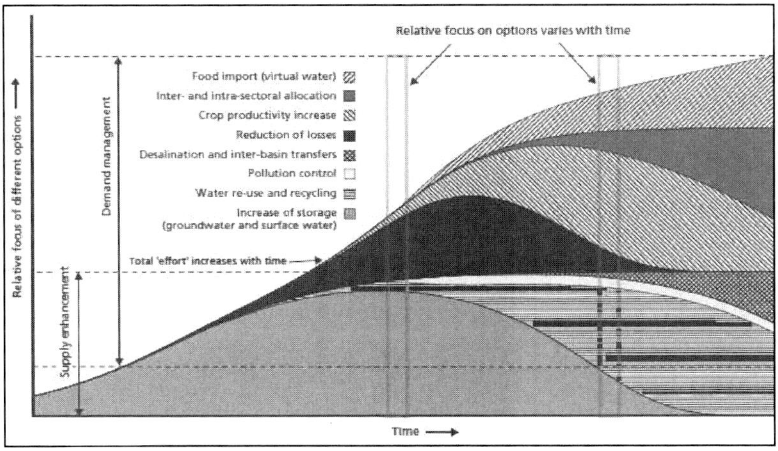

Fig 6.5: A schematic representation of the relative focus on different options for the agricultural sector to cope with increasing levels of water scarcity over time

Climate change Impacts on soil Water Balance

The total gross irrigated area has nearly trebled from 22.6 m ha in 1950–51 to 99.1 m ha in 1998–99. Out of this, 34.3 m ha is from major and medium projects, 12.7 m ha from minor schemes using surface water and 52.2 m ha from groundwater. As against the national average of 38% of the total cropped area being irrigated, Punjab has the distinction of achieving the highest level of irrigation (92%), followed by Haryana (79%) and Uttar Pradesh (66%) (CWC, 2002). Umapathi and Ramashesha (2001) found that area under water-intensive crops like paddy and sugar cane has increased by 8.0 and 0.7 lakh ha, respectively, over a period of 11 years in Tamil Nadu. They recommended that area under water-intensive crops should not exceed the area under non-water intensive crops.

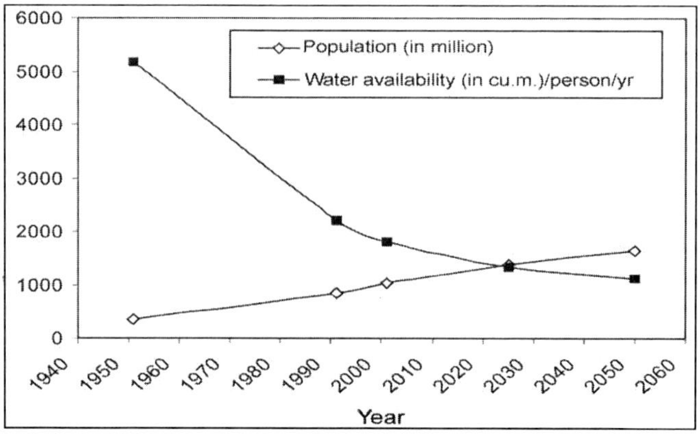

Source: Ministry of Water Resources Govt. of India, 2003.

Fig 6.6: Observed and projected decline in per capita average annual freshwater availability and growth of population from 1951 to 2050.

Fig 6.6 clearly indicates the 'two-sided' effect on water resources – the rise in population will increase the demand for water leading to faster withdrawal of water and this in turn would reduce the recharging time of the water-tables. As a result, availability of water is bound to reach critical levels sooner or later. Agricultural demand, particularly for irrigation water, which is a major share of total water demand of the country, is considered more sensitive to climate change. A change in field-level climate may alter the need and timing of irrigation. Increased dryness may lead to increased demand, but demand could be reduced if soil moisture content rises at critical times of the year 2. Naresh *et al.*, (2017a) projected that most irrigated areas in India would require more water around 2025 and global net irrigation requirements would increase relative to the situation without climate change by 3.5–5% by 2025, and 6–8% by 2075. In India, roughly 52% of irrigation consumption across the country is extracted from groundwater; therefore, it can be an alarming situation with decline in groundwater and increase in irrigation requirements due to climate change.

Managing Water for Alleviating Drought Stress

Enhancing the "green water" supply in the root zone is essential to the sustainability of agriculture in arid, semi-arid and sub-humid regions. The strategy is to improve yield per unit of rainfall in dry-land agriculture (more crop per drop). It is argued that supplemental irrigation, based on water harvesting and application through micro-irrigation technologies (e.g., drip sub-irrigation), is a practical adaptation to climate change and variability and the ground water-based irrigation. Nonetheless, the effectiveness of irrigation and sustainable use of limited water resources involves improving intrinsic water-use efficiency (WUE) and crop yield. The latter is defined as the ratio of the instantaneous rates of CO_2 assimilation (A) and transpiration (T) at the stomata. While understanding plant physiological attributes are critical, including the possible gains by the CO_2 fertilization effects, soil quality and its management to enhance plant-available water capacity is also essential. The problem of low WUE is strongly confounded by soil degradation. Thus, restoration of soil quality to improve green water is a prudent strategy. In addition to having lower opportunity cost, the use of green water to increase agronomic production has lesser environmental externalities compared with the use of blue water involving irrigation through canal or ground water.

The externalities of blue water use can be minimized through the uptake of precision irrigation techniques. Precision irrigation is defined as "site-specific irrigation management that relies on variable application of water". It is emerging as a potential solution to increase the productivity and reduce the environmental impact of irrigated agriculture. The strategy is to match irrigation to soil type, the inherent permeability and water/nutrient retention capacity by using irrigation techniques in conjunction with a variable rate of water applications. Micro-irrigation (drip or trickle) is water saving, but an expensive, option. In comparison, precision irrigation through the sub-division of control zones and electronic controllers may be efficient and economical. Precision irrigation can be used in conjunction with the concept of regulated deficit irrigation (RDI) or partial root zone drying (PRD). Subjected to drought stress, roots generate abscisic acid (ABA), which moderates stomata closure. Withholding irrigation also stimulates deeper rooting, reduces irrigation needs, and enhances resilience.

Three challenges of using micro-irrigation are:

i) Monitoring variable/heterogeneous soil moisture content by using electromagnetic induction and near-infrared systems mounted on unmanned aerial vehicles;

ii) Measuring plant-water status and regulating irrigation through it by using high resolution and high frequency remote sensing (e.g., MODIS) technology; and

iii) Applying variable rate of water irrigation. There are numerous proven options of enhancing the WUE, and their adoption must be promoted through soil governance. Improving soil fertility by establishing fertilizer trees can also improve WUE in drought-prone agro-ecosystems.

Rainwater harvesting and management is a useful strategy to store the excess water and reuse it during the dry season. This strategy is especially relevant to rain-fed agriculture and is relevant to smallholder farmers to intensify production systems. Rice, a semi-aquatic plant and a major food staple of the world, has a high water requirement and a large yield gap, which must be closed. Flooded paddies in an arid environment, such as in western parts of the Indo-Gangetic Plains, have been responsible for the rapid depletion of the ground water. Water scarcity, exacerbated by extreme events and climate change, necessitates the identification of water-saving strategies. Therefore, aerobic rice is a viable option for increasing WUE even with a slight decline in agronomic yield. Crop yields of aerobic rice can be enhanced by N management. Among several challenges that remain to be addressed in promoting aerobic rice is soil sickness caused by aerobic mono-cropping. Micronutrient deficiency and effective weed control are other factors to be considered under continuous aerobic rice cultivation. Strategies for soil water management are outlined in (Fig 6.7), and those of improving productivity on the plant scale are conceptualized.

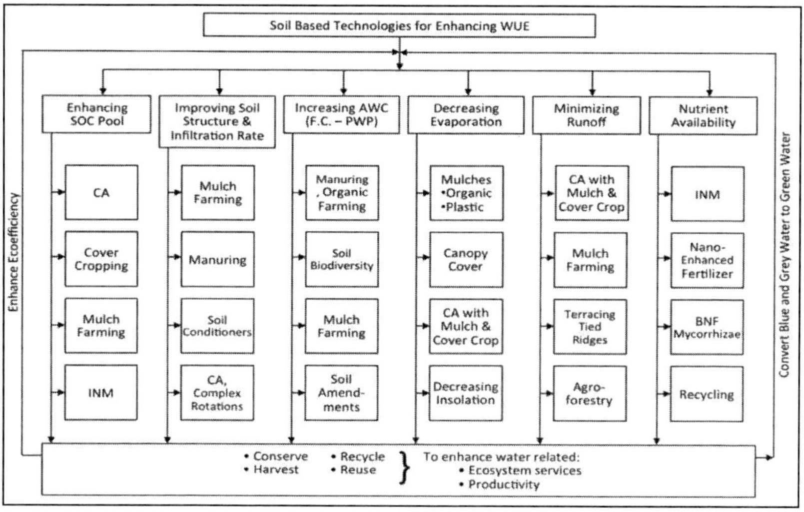

Fig 6.7: Principles and practices of enhancing the water use efficiency.

Notes: CA, conservation agriculture; INM, integrated nutrient management; BNF, biological nitrogen fixation; F.C., field moisture capacity; PWP, permanent wilting point; WUE, water-use efficiency; AWC, available water capacity.

Utilization of Wastewater for Groundwater Recharge

The artificial recharge to groundwater with wastewater depends on a number of factors such as porosity, permeability, hydraulic conductivity, transmissivity and the most important stratigraphical arrangement. In majority of the cases the subsurface sequencing of the layers and the aquifer depth creates problems. Thus the selection of the possible locations for the construction of artificial recharge facility should be established taking care of factors such as suitable site, infiltration capacity, hydro-geological conditions, systematic planning and environmental considerations. Based on these considerations Soil-Aquifer Treatment (SAT) facility can be established.

Infiltration Capacity of the Vadose Zone

Infiltration capacity of the Vadose Zone is a very important factor that governs the rate of saturation of the zone of aeration. Therefore the lithological zones with an infiltration rate of 0.30 m/day or more are most suitable.

Hydrogeological Consideration

Integration of data obtained from field, laboratory analysis and simulation methods are generally used to develop an understanding of the hydro-geological system as a basis for predicting potential consequences. Some of the important considerations that are important for such projects are as follows

i) The aquifer characteristics at the recharging site must have good hydro-geological conditions such as storage coefficient, availability of storage space and permeability.

ii) Young alluvium buried channels, alluvial fans and sand dunes etc. are some of the best places for recharge.

iii) In hilly terrain and hard rock areas, fractured, weathered to semi weathered and cavernous rocks are the best sites for recharge.

Planning and Management Consideration

Planning and management of artificial recharge projects need careful consideration of several water management objectives, water routing capabilities, economics, off-site effects and different other factors. Water requirement for irrigation purposes can be fulfilled by geo-purified wastewater, if it is carried out with systematic planning. A large amount of wastewater is generated in urban areas. After partial treatment this water can be used for artificial recharge of groundwater aquifer, which can be exploited for irrigational purposes. For creating partial treatment facility near the recharge and discharge area planning should be based on hydro-geological consideration and based on easily and economically feasible techniques.

Appropriate Spacing

It is also important that there should be appropriate spacing between the two recharge structures. Because when the recharge water begins to merge with the existing groundwater, a mound will develop that drastically reduces the recharge rate. Further if many recharge structures are placed side by side the hydraulic gradients of these recharges will become zero due to superposition, due to this the mound will continue to grow and the recharge rate will drop more quickly (Chatdarang, 2001). Therefore two recharge structures should be at least 1000 m away from each other for getting good results.

Environmental Consideration

It is very important to carry out a detailed environmental impact assessment of each selected location and project before the implementation, which includes mitigation and monitoring planning. The most important factor for environmental consideration is that in any circumstances the selected site should not be within or upstream of a groundwater drinking community and no recharge should be planned where groundwater is flowing into the river.

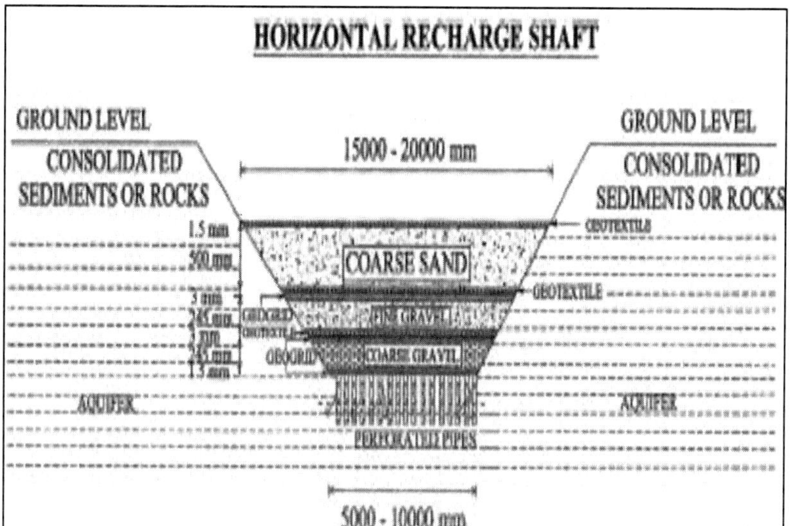

Fig 6.8: Design of horizontal shaft structure shaft structure equipped with artificial set of lithology suitable for hard Rock areas and suitable for hard rock areas, older alluvial places with low surface runoff.

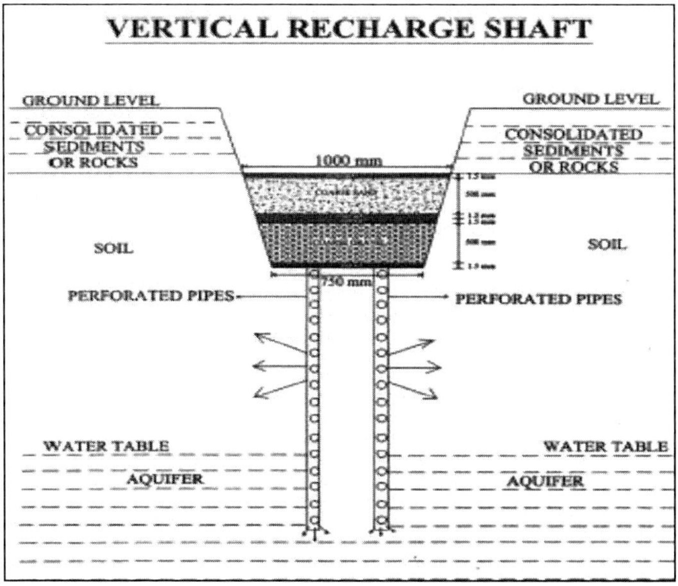

Fig 6.9: Design of vertical shaft artificial set of lithology older alluvial plains and places plains and places with high surface runoff.

Artificial Recharge Methods

Artificial recharge methods were specially design for uses in the arid and semi arid regions of the world. These recharge methods include three important components.

Area of the watershed to produce runoff

A storage facility (soil profile, surface reservoirs or groundwater aquifers); and.

A target area to beneficially use the water (agriculture, domestic or industry).

According to O'Hare *et al*., (1986) some of the important factors, which are considered for artificial recharge, are

a) Availability of wastewater
b) Quantity of source water available
c) Resultant water quality (after reactions with native water and aquifer materials)
d) Clogging potentials

e) Underground storage space available

f) Depth of underground storage space

g) Transmission characteristics of the aquifer

h) Applicable methods (injection or infiltration)

i) Legal/institutional constraints

j) Costs (h) Cultural/social considerations.

Adaptation of Advance Agriculture Practices

Adaptation of advance technologies can help in making the framing more easy and productive task. The good agricultural practice involves providing adequate water to plants so that excess standing water can be prevented. Excessive amounts of water can cause poor aeration of the root system leading to inhibition of plant development; therefore avoiding water stress is particularly important in arid regions. Further use of GPS tractors, along with sprayers can minimize the farming cost. These tractors can accurately drive themselves through the field after receiving the correct work plan in their computer system (Fig 6.8). Moreover the pressurized irrigation with sprinklers is extremely useful in increasing the water use efficiency in modernizing the agriculture system (Fig 6.9). Furthermore adaptation of mobile technique can play a big role in monitoring and controlling crop irrigation systems. With these instruments a farmer can control his irrigation systems from a phone or computer instead of driving to each field. The major challenge is to accelerate the adaptation of modern techniques in irrigation with which the crop productivity cannot be increased. Fig 6.10 shows the framework for agricultural water management in the context of the climate change.

Crop Breeding

Adaptation of plant breeding can improve the quality, performance and diversity of crops with the objective of developing plants better adaptable to human needs. Studies show that for overcoming abiotic stresses in crops, the crop breeding technique has been proved to be an effective means of increasing food production and mitigating climate change effects. Crop breeding techniques not only enhance the crop quality and yield and but also increase the tolerance of environmental pressures like soil salinity, high temperature and drought conditions. Therefore large scale adaptation of crop breeding is essential to ensure food security.

Climate Forecasting for Crops

Erratic climatic conditions and their variability with time play an important role in the crop production and overall yield. Most of the crop failures worldwide are associated with either a lack or excess of rainfall. Precise climate forecasting can reduce the risks of crop failure and also help in the pre and post decision making processes for better agricultural yield. Several studies have evaluated the potential benefits of using seasonal climate forecasts on the decision making process in agriculture. Further the nature of the forecasting also influences the ability of farmers to respond like farmers are more concerned about within-season characteristics of rainfall rather than the total seasonal rainfall amounts. The value of forecasts diminishes if information is received after the number of pre-planting decisions are made, therefore the forecasting should be in time and specific.

Managing Food Security

Managing food security and its sustainable development is one of the biggest challenges worldwide. Majority of the world's poor population lives in South Asia and Sub- Saharan Africa countries, nearly 1.7 billion people (Chen and Ravallion, 2007) and out of this, approximately for 860 million people there food security. Countries are not able to provide sufficient quantities of nutritious food to these people so that they can live healthily. It is strongly believed that there is enough food in the world to feed everyone adequately but the problem is distribution and management. Therefore development and implementation of a food security plan is necessary that must include procedures for handling threats, product tampering, and product storage and distribution plan along with a monitoring procedure. Moreover there should be a corrective action that prevents products from entering commerce.

Reduction in Farming Cost

Artificial recharge can play a major role in minimizing the farming cost. There are a number of examples that proof its utility in irrigation sector. In 2000, the International Water Management Institute (IWMI) carried out a study in India on the project "Madhya Ganga Canal Project (MGCP)", which occupies lower Ganga canal command area in central Ganga plain. The study was carried out on Lakhaoti Branch canal of the MGCP, which spreads over 205.6 thousand hectares and covers the area of western Uttar Pradesh (UP) in India. Under this experiment excess river water was used to recharge the groundwater via earthen canals.

The experiment was successful in raising the groundwater table to 6.6 meters and brought down the cost of pumping for irrigation from Rs. 4500 per ha meter to an economical level of Rs. 2700 per ha meter. The conclusion of this study is that the prevailing practices of supplying water only during summer/dry seasons need to be changed. In fact during monsoon such supplies should be carried out to help farmers to grow water intensive crops, this way both irrigation and recharge of groundwater will take place.

The construction of structures like "ooranies" is another example. These are big or small types of dugout tanks and ponds constructed thousands of years ago in very large numbers (more than 500,000) in the Peninsular India. These "ooranies" were constructed to fulfill the demand for water for agriculture and domestic purposes. There are several shallow dug wells near these tanks which are recharged with tank water. Further many wells which are used for drinking purposes, located near the tank and tank beds are artificially recharged from the tank into these wells to provide a clean water supply throughout the year with natural filtering (DHAN foundation, 2002). The structures like "ooranies" if formed in large scales especially in the arid and semiarid areas, the problems of water scarcity for domestic and irrigational purposes can be solved up to some extent despite drastic changes in climate.

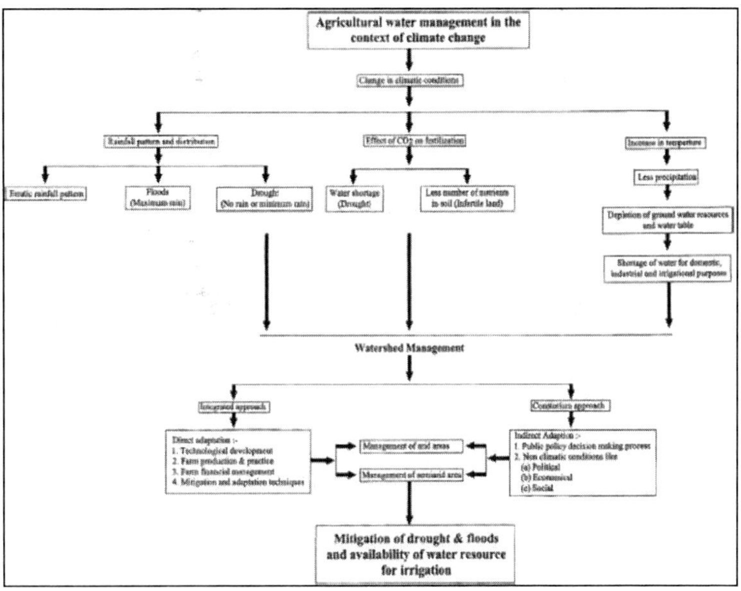

Fig 6.10: Proposed framework for agricultural water management in the context of climate change.

Projections of Precipitation

The projections of precipitation indicate a 3% to 7% overall increase in all-India summer monsoon rainfall in the 2030's with respect to the 1970's.

However, on a seasonal scale, except for the Himalayan region, all other regions are likely to have lower rainfalls in the winter period as well as pre-summer period. Spatial patterns of monsoon rainfall indicate a significant decrease in the monsoon rainfall in future except in some parts of the southern peninsula.

Himalayan Region

The projected precipitation show a net increase in 2030's with respect to the simulated rainfall of 1970's in the Himalayan region by 60 to 206 mm. The increase in annual rainfall in 2030's with respect to 1970's ranges from 5 to 13%. All seasons in the Himalayan region indicate an increase in rainfall, with the monsoon months of June, July, August and September showing the maximum increase in rainfall by 12 mm. The winter rain in the months of January and February are also projected to increase by 5mm in 2030's with respect to 1970's, with minimum increase in October, November and December.

Costal Region-West Coast

The west coast projections indicate that in 2030's the annual rainfall will vary from 935± 185.33 mm to 1794±247.1 mm. The trend of rainfall in 2030's is showing an increase with respect to the 1970's in this region as well. The increase in rainfall is by 6 to 8 %, an increase that is ranging from 69 to 109 mm. Though June, July and August show an average increase of 8mm rainfall in 2030's with respect to 1970's, however, the winter rainfall is projected to decrease on an average by 19 mm during the period January and February in 2030's with reference to 1970's. The period March, April and May also show a decrease in rainfall with respect to 1970's.

Costal Region- East Coast

In the eastern coast of India, the projected annual rainfall varies from a minimum of 858±10mm to a maximum of 1280±16mm. The increase in 2030's with respect to the 1970's is estimated to be 2 to 54 mm, an increase of 0.2% to 4.4 % respectively. The maximum increase in rainfall is projected to happen in March, April and May in 2030's, with rainfall set to increase by 14 mm on an average with respect to the same period in 1970's. The winter rainfall is projected to decrease by 6 mm with respect to 1970's.

North- Eastern Region

The projected mean annual rainfall is varying from a minimum of 940±149mm to 1330 ±174.5 mm. The increase with respect to 1970's is by 0.3% to 3%. The north-east also show a substantial decrease in rainfall in the winter months of January and February in 2030's with respect to 1970's with no additional rain projected to be available during the period March to May and October to December. In fact, recent data indicates the same pattern. However, the monsoon rainfall during June, July and August is likely to increase by 5 mm in 2030's with reference to 1970's a rise of 0.6%.

Projections of Mean Annual Surface Temperature

The annual mean surface air temperature rise by 2030's ranges from 1.7°C to 2°C as in the three simulations. The seasons may be warmer by around 2°C towards 2030's. The variability of seasonal mean temperature may be more in winter months

Himalayan Region

The annual temperature is projected to increase from 0.9±0.6 °C to 2.6±0.7°C in 2030's. The net increase in temperature is ranging from 1.7°C to 2.2°C with respect to the 1970's. Seasonal air temperatures also show rise in all seasons. However, winter temperatures during October, November and December in the show a decrease by 2.6°C in 2030's with respect to 1970's.

West Coast

The annual temperatures are set to increase from a minimum of 26.8°C to a maximum of 27.5°C in the 2030's. The rise in temperature with respect to the 1970's correspondingly ranges between 1.7 to 1.8°C. Temperatures are also projected to rise for all seasons for all the three simulations from 1.5 to 2.2°C, with the rainfall period of June, July, August and September showing the minimum rise amongst all seasons.

East Coast

In the east coast, the surface annual air temperature is set to rise from 28.7°C to 29.3°C. The standard deviation in temperatures varies from 0.6 to 0.7 respectively. The rise in temperature with respect to 1970's is of the order of 1.6 to 2.1 °C. The maximum increase in temperature is for March, April and May in all simulations and is ranging from 1°C to 3.3 °C.

North-East

Surface air temperature is projected to rise by 25.8 to 26.8 °C in 2030's with a standard deviation ranging from 0.8 to 0.9. The rise in temperature with respect to 1970's is ranging from 1.8 to 2.1 °C.

Water Resources in India-The regional distribution

The annual precipitation (including snowfall), which is the main source of water in the country, is estimated at 4,000 km^3. The resource potential of the country, which occurs as natural runoff in rivers, is about 1,869 km^3, as per the basin-wise latest estimate of Central Water Commission, considering both surface and groundwater as one system. The majority of river runoff occurs during 3–4 months in a year during the monsoon. Inland water resources of the country are classified as rivers and canals, reservoirs, tanks and ponds, beels, lakes, derelict water and brackish water (Mall *et al*., 2007)

Rivers

The Ganga–Brahmaputra–Meghna system is the major contributor to the total water resource potential of the country. Its share is about 60% in total water resource potential of the various rivers. As per the 2001 census, the per capita freshwater availability is about 1,820 m^3. Due to various constraints of topography, uneven distribution of resources over space and time, it has been estimated that only about 1,122 km^3 of the total potential of 1,869 km^3 – of which 690 km^3 comes from surface water resources – can be put into beneficial use. In the majority of river basins, the present utilization is significantly high and is in the range of 50%–90% of utilizable surface resources. However, in rivers such as the Narmada and Mahanadi, the percentage utilization is quiet low and is 23% and 34%, respectively. Per capita gross water availability in the Brahmaputra and Barak basins is of the order of 14,057 m^3 per annum and in Sabarmati basin, the water availability is as low as 307 m^3 (MoWR, 2003).

Average water yield per unit area of the Himalayan Rivers is almost double that of South Peninsular river systems, indicating the importance of snow and glacier melt contribution from the Himalayan region.

The average intensity of mountain glaciation varies from 3.4% for the Indus to 3.2% for the Ganga and

1.3% for the Bramhaputra. The tributaries of this river system show maximum intensity of glaciation (2.5%–10.8%) for Indus, followed by the Ganga (0.4%–10%), and the Brahmaputra (0.4%–4%). The average annual and seasonal flow of these systems give a different picture, thereby demonstrating that rainfall contributions are greater in the eastern region, while the snow and glacier melt contributions are more important in the western and central Himalayan region. In recent decades the hydrological characteristics of the watersheds in the Himalayan region seem to have undergone substantial change as a result of land use (e.g. deforestation, agricultural practices and urbanization) leading to more frequent hydrological disasters, enhanced variability in rainfall and runoff, extensive reservoir sedimentation and pollution of lakes, etc. (Ramakrishnan, 1998).

Most of the rivers in the south like the Cauvery, the Narmada and the Mahanadi are fed through groundwater discharges and are supplemented by monsoon rains. So, these rivers have very limited flow during non-monsoon periods. The flow rate in these rivers is independent of the water source and depends upon the precipitation rate in the region. Therefore, in spite of being smaller, the rivers flowing into the west coast have a higher flow rate due to higher precipitation over that region.

Glaciers

Major glacier-fed Himalayan Rivers, along with glaciated catchments, have regional importance-the water from the glacier melt sustains stream flow in these rivers through the dry season. The "frozen water" in the Himalayas is crucial for the people inhabiting the mountain areas as well as those inhabiting the downstream regions. The Indus and the Ganga – the two major rivers in western Himalayan region – directly impact the lives of a large population living in the northern part of India, and even beyond the national boundaries.

The Indus basin has 7,997 glaciers with a total glacier cover of 33,679km^2 and total ice volume of 363.10km^3. The Ganga basin has 968 glaciers with a total glacier cover of 2,857km^2 and total ice volume of 209.37km^3. The contribution of snow to the runoff of major rivers in the eastern Himalayas is about 10% (Sharma, 1993) but more than 60% in the western Himalayas (Vohra, 1981).

It is estimated that Himalayan Mountains cover a surface area of permanent snow and ice of about 97,020km^2, with a volume of 12,930km^3. In these mountains, it is estimated that 10%–20% of the total surface area is covered with glaciers while an additional area ranging from 30%–40% has seasonal snow cover. These glaciers, providing snow and glacial melt waters, keep our rivers perennial. Bahadur (1999) reported that a very conservative estimate puts the snow and ice melt water contribution to Himalayan streams at no less than 500km^3 per year, while another study reports about 515km^3 per year from the upper Himalayas. The most useful facet of glacier runoff is that glaciers release more water in a drought year and less water in a flood year.

Ground Water

The other source of water is the groundwater resource, which has two components-static and dynamic. The static fresh groundwater reserve (i.e, aquifer zones below the zone of groundwater table fluctuation) of the country has been estimated at 10,812 billion m^3. The dynamic component, which is replenished annually, has been assessed as 432 billion m^3. As per the national water policy, development of groundwater resources is to be limited to utilization of dynamic component of groundwater. The present development policy, therefore, forbids utilization of the static reserve to prevent groundwater mining.

The total annual replenishable groundwater resource is about 43 million hectare metres (Mham). After making a provision of 7Mham for domestic, industrial and other uses, the available groundwater resource for irrigation is 36Mham, of which the utilizable quantity is 32.6Mham (Mall *et al.*, 2007).

The Central Ground Water board has established about 15,000 network monitoring stations in the country to monitor the water level and its quality. The water level in major parts of the country generally does not show any significant rise or decline. However, certain blocks in 289 districts in the states of Andhra Pradesh, Assam, Bihar, Chhattisgarh, NCT Delhi, Jharkahand, Gujarat, Haryana, Karnataka, Kerala, Madhya Pradesh, Maharashtra. Orissa, Punjab, Rajasthan, Tamil Nadu, Tripura, Uttar Pradesh and West Bengal show a significant decline in groundwater.

Observed Changes in Glaciers in India

The glaciers in the region show fluctuations in retreat rates during the last century, possibly due to the mixed influence of variable topography, temperature and snowfall regime. The Gangotri glacier, which was earlier receding at the rate of around 26 m/year between 1935 and 1971, has shown a gradual decline in the rate of recession. It had come down to around 17 m/year between 1971 and 2004, and in recent years has shown a recession rate of about 12 m/year during 2004–2005 (Kumar *et al.*, 2008). The rate of recession of the Pindari glacier has come down to 6.5 m/year in comparison to the earlier reported rate of 26 m/year between 1996 and 2007. Similarly, the rate of recession of the Milam glacier has been observed as 16.5 m/year in the last 150 years (Bali *et al.*, 2009).

The snout of the Donagiri glacier has shown signs of moderate recession, and the Satopanth glacier, which had been receding at the rate of 22.86 m/year earlier, has lately shown a recession rate of 6.5 m/year during 2005–2006 (Nainwal *et al.*, 2004). Between 1962 and 1995, the volume of ice in the Dokriani glacier has reduced by approximately 20%, and the frontal area has vacated by 10%, whereas the glacier has receded by 550m at an average rate of 16.6 m/yr. However, the yearly monitoring of the snout position of the glacier during 1991–1995 revealed an average rate of recession of 17.4 m/yr, and it has vacated an area of 3,957 m^2. The estimated retreat of the Dokriani glacier in 1998 was 20m compared to an annual average of 16.5m over 1993–1998 (Mirza *et al.*, 2002).

Studies conducted during last three decades by the National Institute of Hydrology, Roorkee, reveal that in Ladakh, Zanskar and the Great Himalayan ranges of Jammu and Kashmir are generally receding, and the glacier volume changes range between 3.6% and 97%, with the majority of glaciers showing a degradation of 17%–25% (Annual Report, NIH, 2008-09). The 23km long Durung Drung glacier in the Zanskar valley is highly affected by western disturbances (ablation rate variations between 0.75cm/day and 2.67 m/day during July and August);

References

1. Adger WNNW, Arnell Tompkins, EL. Successful adaptation to climate change across scales. Global Environ. Change Part A. 2005; 15:77–86.
2. Bali R, Agarwal KK, Ali SN, Rastogi SK, Krishna K. Monitoring recessional pattern of Central Himalayan Glaciers: some optimistic observations. Proceedings of the Indian Science Congress. 2009; 96:79-80.
3. Bahadur J. The Himalayan Environment [edts. Dash SK, Bahadur J]. New Age International (pvt ltd.), New Delhi, 1999, 258-267.
4. Chatdarang V. Artificial recharge for conjunctive use in irrigation: The San Joaquin valley, California: (M. Tech Thesis); Massachusetts Institute of Technology, June, USA, 2001.
5. Chaturvedi RK, Gopalakrishnan Ranjith, Jayaraman Mathangi, Bala Govindasamy *et al.*, Impact of climate change on Indian forests: a dynamic vegetation modeling approach, Mitigation and Adaptation Strategies for Global Change, 2010.
6. DHAN foundation. Revisiting tanks in India, National Seminar on conservation and development of Tanks. New Delhi India, 2002.
7. Gupta SK, Deshpande RD. Water for India in 2050: first-order assessment of available options. Rev. Pap. Curr. Sci. 2004; 86(9):1216-1224.
8. IPCC. Fresh water resources and their management climate change; Impact, adaptation and vulnerability. Contribution of working group-II to the fourth assessment report of the intergovernmental panel on climate change. Cambridge University Press: Cambridge, UK, 2007.
9. IPCC. Climate change and water, Intergovernmental panel on climate change technical report IV. June, 2008.

10. Kavikumar KS. Climate sensitivity of Indian agriculture: role of technological development and information diffusion. In: Lead Papers of National Symposium on Climate Change and Rainfed Agriculture, February 18–20, Indian Soc. Dryland Agric., Central Research Institute for Dryland Agriculture, Hyderabad, India, 2010, 192.

11. Kumar K, Dumka RK, Miral MS, Satyal GS, Pant M. Estimation of retreat rate of Gangotri glacier using rapid static and kinematic GPS survey. Current Science. 2008; 94(2):258-262.

12. Kumar BM, Takeuchi K. Agro-forestry in the Western Ghats of Peninsular India and the Satoyama landscapes of Japan: a comparison of two sustainable land use systems. Sustain. Sci. 2009; 4:215-232.

13. Kurylyk Barret L, MacQuarrie, Kerry TB. The uncertainty associated with estimating future groundwater recharge: a summary of recent research and an example from a small unconfined aquifer in a northern humid-continental climate. J. Hydrol. 2013; 492:244–253.

14. Lovejoy TE, Hannah L. Climate Change and Biodiversity, Yale University Press, New Haven, Connecticut, 2005, 418.

15. Mall RK, Singh R, Gupta A, Srinivasan G, Rathore L. Impact of climate change on Indian agriculture: a review. Climate Change. 2006; 78:445-478.

16. Mall RK, Bhatla R, Pandey SK. Water Resources in India and Impact of Climate Change, Jalvigyan Sameeksha. 2007; 22:157.

17. Mirza MMQ, Warrick RA, Ericksen NJ, Kenny J. The Implications of Climate Change on Flood Discharges of the Ganges, Brahmaputra and Meghna Rivers in Bangladesh, Climatic Change. 2002; 57(3):287-318.

18. Misra AK. Global climatic variations and its impact on Indian hydrological conditions. Int. J Eng. Sci. Res. Technol. 2013; 2(2):148-158.

19. Misra AK. Climate change and challenges of water and food security. Int J Sustainable Built Envir. 2014; 3:153-165.

20. Nainwal HC, Naithani AK, Kumar D, Negi BDS, Prasad C. Glacial morphology of the Gangotri group of glaciers in Garhwal Himalaya, Uttaranchal, India. In Srivastava D, Gupta KR, Mukerji S. Editors, Geological Survey of India, Spl. Pub. 2004; 80:87-103.

21. O'Hare MP, Fairchild DM, Hajali PA, Canter LW. Artificial recharge of Groundwater. Proceedings of the second International Symposium on Artificial Recharge of Groundwater, 1986.

22. Ramakrishnan S. Groundwater, Scitech publications Pvt. Ltd., Chennai, 1998.

23. Vohra CP. The Climate of the Himalayas, The Himalaya: aspects of change, [Lall JS (edt.)], Oxford University Press, New Delhi, India, 1981, 138-151.

Chapter - 7
Climate Change and Soil Health

Authors

Dr. S.S. Dhaliwal
Department of Soil Science, Punjab Agricultural University, Ludhiana, Punjab, India

Dr. Vineet Kumar
Indian Institute of Farming System Research Modipuram, Meerut, Uttar Pradesh, India

Chapter - 7

Climate Change and Soil Health

Dr. S.S. Dhaliwal and Dr. Vineet Kumar

Soil health indicators are a composite set of measurable physical, chemical and biological attributes which relate to functional soil processes and can be used to evaluate soil health status, as affected by management and climate change drivers.

Defining soil health in relation to climate change should consider the impacts of a range of predicted global change drivers such as rising atmospheric carbon dioxide (CO_2) levels, elevated temperature, altered precipitation (rainfall) and atmospheric nitrogen (N) deposition, on soil chemical, physical and biological functions (French *et al.*, 2009). Many studies have progressed our understanding of relationships between particular soil properties and climate change drivers, e.g. responses to temperature, CO_2 or rainfall; however, Wixon and Balser (2009) note that "a comprehensive explanation of the factors at the heart of the issue is currently lacking".

Determination of how predicted changes in climate relate to soil health will thus depend on our capacity to clearly define soil health properties and their relationship with specific soil functions, including complexity associated with interactive effects of climate change.

Note that both the terms "soil health" and "soil quality" are used synonymously although it is realised that the former term gives greater emphasis on soil biodiversity and ecological functions that make soil a dynamic living resource with capacity for self-organization. Indicators, calculated values or estimated statistics relative to a threshold level are being increasingly used across biological, environmental, economical, social, institutional and political disciplines to assess current condition or trend of soil health (Dalal *et al.*, 2003a). Indicators may be used as an indirect measure of soil function, serving to assess soil quality or health and its direction of change with time, by linking functional relationships among measurable attributes and monitoring for sustainable land management, including environmental impacts (Dalal *et al.*, 2003a). Kinyangi (2007) proposes that soil health assessment involves an evaluation process consisting of a series of actions: (1) selection of soil health indicators, (2) determination of a minimum data set, (3) development of an interpretation scheme of indices, and (4) on-farm assessment and validation. While the use of integrative soil health tests is increasing (e.g. Idowu *et al.*, 2009), limited information exists to evaluate the applicability of soil health indicators for monitoring soil functions within the context of climate change. A range of frameworks identifying soil health indicators to measure change and implement strategies to adapt to climate change exist (Kibblewhite *et al.*, 2008; Nuttall 200). Potential linkages between soil health indicators, land management and climate change from these reviews are summarised in Fig 7.1. Elevated CO_2 concentration, increasing temperature, atmospheric N deposition and changes in total and seasonal distribution of rainfall and extreme events such as droughts and floods will impact on soil biological processes, C and N cycling, and consequently on soil structure and erosion events, nutrient availability and plant diseases, and hence on ecosystem functionality and agricultural productivity.

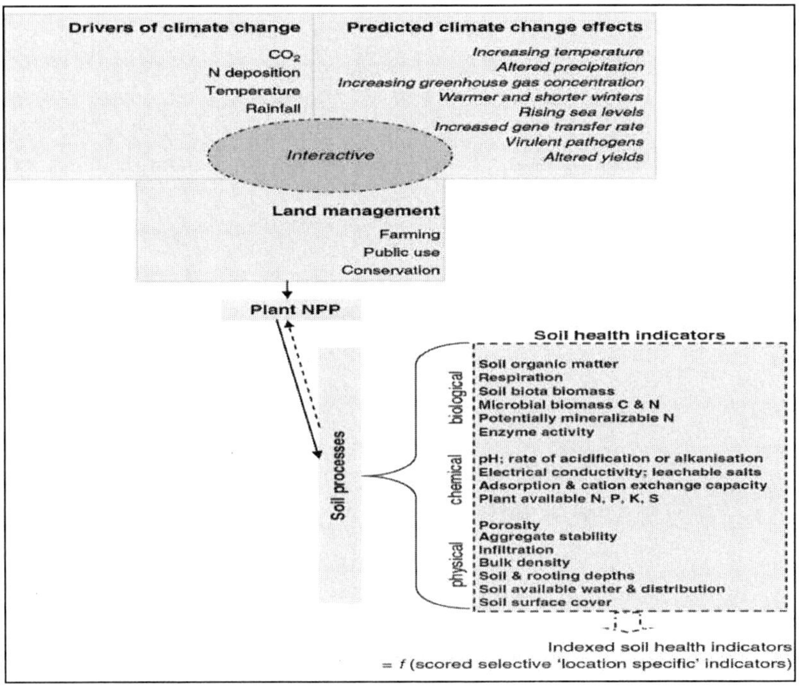

Fig 7.1: Schematic representation of the potential links between climate change, land use and management change and soil health indicators (Modified from Dalal and Moloney, 2000; Karlen *et al.*, 2003)

7.1 Soil Physical Properties, Soil Health and Climate Change

Soil physical properties provide information related to water and air movement through soil, as well as conditions affecting germination, root growth and erosion processes. Many soil physical properties thus form the foundation of other chemical and biological processes, which may be further governed by climate, landscape position and land use. A range of soil physical properties are highlighted as potential soil health indicators, and key soil physical indicators in relation to climate change include soil structure, water infiltration, bulk density, rooting depth, and soil surface cover, which are discussed below.

Soil Structure (Aggregate Stability, Porosity)

Aggregate stability, the resistance of soil aggregates to external energy such as high intensity rainfall and cultivation, is determined by soil structure, as well as a range of chemical and biological properties and management practices (Moebius *et al*., 2007). It is considered a useful soil health indicator since it is involved in maintaining important ecosystem functions in soil including organic carbon (C) accumulation, infiltration capacity, movement and storage of water, and root and microbial community activity; it can also be used to measure soil resistance to erosion and management changes (Moebius *et al*., 2007; Rimal and Lal 2009). Because of its association with the storage of soil organic carbon (SOC) and water, its measurement can be useful to guide climate adaptation strategies, especially in areas that are likely to experience high and intense rainfall and consequently increased erosion events. Since aggregate stability is measured in many different ways, standardised procedures are required within a soil health monitoring framework under climate change scenarios (Salvador Sanchis *et al*., 2008). Porosity, a measure of the void spaces in a material as a fraction (volume of voids to that of the total volume), and pore size distribution provide a direct, quantitative estimate of the ability of a soil to store root-zone water and air necessary for plant growth (Reynolds *et al*., 2002). Pore characteristics are strongly linked to soil physical quality; bulk density and macroporosity are functions of pore volume, while soil porosity and water release characteristics directly influence a range of soil physical indices including soil aeration capacity, plant available water capacity and relative field capacity (Reynolds *et al*., 2009).

Soil water balance and ecosystem conditions under present-day and projected climatic scenarios use porosity as a model parameter (Porporato *et al*., 2005). Since root development and soil enzyme activities are closely related to soil porosity and pore size distribution (Piglai and De Nobili 1993) and because future climate change scenarios (e.g. elevated CO_2 and temperature, and variable and extreme rainfall events) may alter root development and soil biological activities, soil porosity and pore size distribution and consequently soil functions are likely to be affected in unexpected directions; this aspect needs attention in future studies on the relationship of soil health and climate change. Moreover, pore size distribution and aeration status, besides other factors, are the key factors in governing methane (CH_4) fluxes, both CH_4 emission and uptake (Dalal *et al.*, 2008), and nitrous oxide (N_2O) emissions from soil (Dalal *et al.*, 2003a,). However, data on relationships of greenhouse gas emissions and soil porosity and pore size distribution in response to climate change are limited and hence urgently required to guide development of climate adaptive strategies.

Infiltration, Soil Available Water and Distribution

Soil water infiltration, the rate at which water enters the soil surface and moves through soil depth, is gaining increasing interest within soil water modelling, as field-based measurements continue to improve (Joel and Messing 2001). Since infiltration rate may change significantly with soil use, management and time, it has been included as an indicator of soil health for assessments of land use change impacts (O'Farrell *et al.*, 2010). The availability of water for plant growth and important soil processes is governed by a range of soil properties including porosity, field capacity, lower limit of plant available water (thus excluding osmotic potential) and hence plant available water capacity, macropore flow and texture (Jarvis 2007). Plant available water capacity has been used as part of integrative soil health tests to assess management impacts, although variation in methodology may complicate the interpretation of results (Idowu *et al.*, 2009). Furthermore, the soil available water and distribution may respond rapidly to climate change, especially to variable and high intensity rainfall or drought events, and thus management strategies, such as the planting of cover crops, conservation tillage and incorporation of organic matter, that maintain or even enhance water infiltration and available water in soil may help in mitigating the impacts of severe rainfall and drought events or severe erosion events (Salvador Sanchis *et al.*, 2008).

Bulk Density

Bulk density is routinely assessed in agricultural systems to characterise the state of soil compactness in response to land use and management

(Ha°kansson and Lipiec 2000). It is considered as a useful indicator for the assessment of soil health with respect to soil functions such as aeration and infiltration (Reynolds *et al.*, 2009). Since bulk density is in general negatively correlated with soil organic matter (SOM) or SOC content (Weil and Magdoff 2004), loss of organic C from increased decomposition due to elevated temperatures (Davidson and Janssens 2006) may lead to increase in bulk density and hence making soil more prone to compaction via land management activities and climate change stresses, for example, from variable and high intensity rainfall and drought events (Birka´s *et al.*, 2009).

Table 7.1: Soil health and relations to prpcesses and functions under projected climate change scenarios

Soil health indicators		Soil processes affected	Landscape scale (direct determination or estimated from pedotransfer functions)	Relevance to assess climate change impacts	Inclusion in a minimum data set
Physical	Soil structure	Aggregate stability, organic matter turnover	Aggregation, surface seal, indication of water and chemical retention and transportation	Medium	Frequent
	Porosity	Air capacity, plant available water capacity, relative field capacity	Soil crusting, reduced seed germination, aeration, water entry	High	Occasional/ frequent
	Infiltration	Soil water availability and movement	Potential for leaching, productivity, erosion	High	Occasional
	Bulk density	Soil structural condition, compaction	Volumetric basis for soil reporting	Low	Frequent
	Soil depth and rooting	Plant available water capacity, subsoil salinity	Productivity potential, uncertain whether trends can be discerned over long time periods	Medium	Occasional
	Soil/plant available water and distribution	Field capacity, permanent wilting point, macropore flow, texture	Water and chemical retention and transportation; yield	High	Frequent
	Soil protective cover	Soil water and nutrient movement, soil stabilisation, C and N fixation	Soil physical movement, organic matter input and movement	Medium	Frequent
Chemical	pH	Biological and chemical activity thresholds	Soil acidification, salinisation, electrical conductivity, soil structural stability	Medium	Frequent
	EC	Plant and microbial activity thresholds	Soil structural decline; leachable salts	Medium	Frequent
	Plant available N, P, K	Plant available nutrients and potential for loss	Capacity for crop growth and yield; environmental hazard (e.g. algal blooms)	Medium	Frequent
Biological	Soil organic matter				
	Light fraction or Macro-organic matter	Plant residue decomposition, organic matter storage and quality,	Loss of organic matter, soil aggregate formation	High	Frequent
	Mineralisable C and N	macroaggregate formation	Total organic C, soil respiration rate, nutrient supply	High	Occasional
		Metabolic activity of soil organisms, net inorganic N flux from mineralisation and immobilisation	Microbial activity, nutrient supply		
	Soil total C and N	C and N mass and balance	Soil structure, nutrient supply	High	Occasional/ frequent
	Soil respiration	Microbial activity	Microbial activity	High	Occasional
	Microbial biomass C and N	Microbial activity	Soil structure, nutrient supply, pesticide degradation	High	Frequent/ Occasional
	Microbial quotients	Substrate use efficiency	Substrate quality	High	Occasional
	Microbial diversity	Nutrient cycling and availability		High	Occasional
	Other microbiological indicators, enzyme activity	Soil structure, labile carbon, K_m, V_{max}, K_i, Q_{10}	Biochemical activity, nutrient supply	High	Occasional

[a]Adapted from Dalal and Moloney 2000; Gregorich et al. 1994; Haynes 2008; Idowu et al. 2009; Kinyangi 2007; Reynolds et al. 2009; Stenberg 1999

Rooting Depth

Rooting depth is considered an important indicator of soil health, since changes in this property is likely to affect plant available water capacity, subsoil salinity, SOC content or other properties to indicate physicochemical constraints in the soil profile (Birka's *et al.*, 2009). Under prolonged drought, the impact of subsoil constraints such as salinity and high chloride concentrations (Rengasamy 2010) is likely to be greater on plant available water and hence plant productivity. Also, Birka's *et al.*, (2008) included rooting depth as a soil health parameter for monitoring of soil condition and plant growth under extreme drought and variable rainfall events to indicate the potential for adaptability and mitigation of climate stresses through alteration of rooting depth.

Soil Surface Cover

Soil surface cover (e.g. a layer of crop residues or biological soil crust) provides a range of important ecological functions including protection of soil surface by dissipating raindrop impact energy, soil stabilisation, and reduction in erodible surface area, water and nutrient retention, C fixation and, in some instances, N fixation and support of native seed germination (Box and Bruce 1996). Dalal and Moloney (2000) note that soil surface cover provides an integrated indicator of soil physical management, organic matter input and effects associated with erosion and runoff, although correct timing of monitoring of soil surface cover change is required to evaluate impacts of management and climate change on soil health.

Soil structural conditions such as soil crust and soil seal formation, primarily related to sodicity, are also indicators that may be used to characterise soil health under climate change. The formation of soil crusts and seals can affect a range of soil processes, including water infiltration, oxygen diffusion, runoff, surface water evaporation and wind erosion. A range of methods exist to measure their thickness and strength, although research effort is needed to relate these properties with soil processes (Assouline 2004) affecting ecosystem functions and plant productivity, as well as to evaluate their role in mitigating adverse climate change impacts, thereby assisting in climate change adaptation.

7.2 Soil Chemical Properties, Soil Health and Climate Change

pH

Soil pH, a function of parent material, time of weathering, vegetation and climate, is considered as one of the dominant chemical indicators of soil health, identifying trends in change for a range of soil biological and chemical functions including acidification, salinisation, crop performance, and nutrient availability and cycling and biological activity (Dalal and Moloney 2000). Soil pH has thus been included in integrative soil health tests to assess impacts of land use change and agricultural practices (Gil *et al.*, 2009; Idowu *et al.*, 2009). While Brinkman and Sombroek (1999) suggested that most soils would not be subjected to rapid pH changes resulting from drivers of climate change such as elevated temperatures, CO_2 fertilisation, variable precipitation and atmospheric N deposition (McCarthy *et al.*, 2001), it is likely, however, that these drivers of climate change will affect organic matter status, C and nutrient cycling, plant available water and hence plant productivity, which in turn will affect soil pH (Reth *et al.*, 2005).

Electrical Conductivity

Soil electrical conductivity (EC), a measure of salt concentration, is considered an easily measured, reliable indicator of soil quality/health (Arnold *et al.*, 2005). It can inform trends in salinity, crop performance,nutrient cycling (particularly nitrate) and biological activity and, along with pH, can act as a surrogate measure of soil structural decline especially in sodic soils (Arnold *et al.*, 2005). Electrical conductivity has been used as a chemical indicator to inform soil biological quality in response to crop management practices (Gil *et al.*, 2009). Using elevation gradient as a surrogate for increasing temperatures and decreasing precipitation under climate change scenarios, Smith *et al.*, (2002) found that EC decreased and pH increased in a semi-arid environment. Pariente (2001) examined the dynamics of soluble salts concentration in soils from four climatic regions (Mediterranean, semi-arid, mildly arid and arid) and found a non-linear relationship between the soluble salts content and rainfall, with sites that received <200 mm rainfall contained significantly high soluble contents and vice versa. Clearly, there is a need for comprehensive assessment of the influence of drivers of climate change on soil EC as an important soil health indicator in different ecosystems.

Sorption Capacity and Cation Exchange Capacity

Sorption capacity and cation exchange capacity (CEC) are considered important determinants of soil chemical quality, particularly the retention of major nutrient cations Ca, Mg and K, and immobilisation of potentially toxic cations Al and Mn; these properties can thus be useful indicators of soil health, informing of a soil's capacity to absorb nutrients, as well as pesticides and chemicals (Ross *et al.*, 2008). This may qualify CEC as a useful soil health indicator that is also required to be monitored in response to climate change. Since CEC of coarse-textured soils and low-activity clay soils is attributed to that of SOM (Weil and Magdoff 2004), the increasing decomposition and loss of SOM due to elevated temperatures (Davidson and

Janssens 2006) may lead to the loss of CEC of these soils. Low CEC of soil may result in increased leaching of base cations in response to high and intense rainfall events, thus transporting alkalinity from soil to waterways.

7.3 Soil Health Indicators under Climate Change

Plant Available Nutrients

In their identification of basic soil properties to meet requirements of indicators for screening soil quality/health, Doran *et al.*, (1999) list extractable nutrients N, phosphorus, potassium, since "they provide information on plant available nutrients and potential loss from soil. providing indication of productivity and environment quality". Measurement of extractable nutrients may provide indication of a soil's capacity to support plant growth; conversely, it may identify critical or threshold values for environmental hazard assessment (Dalal and Moloney 2000). Nutrient cycling, especially N, is intimately linked with soil organic C cycling (Weil and Magdoff 2004), and hence drivers of climate change such as elevated temperatures, variable precipitation and atmospheric N deposition are likely to impact on N cycling and possibly the cycling of other plant available nutrients such as phosphorus and sulphur, although direction and exact magnitude of change in plant available nutrients need to be investigated in detail.

Soil Biological Properties, Soil Health and Climate Change

"While the chemistry (and physics) of the soil system provides the context. It is the soil biota which is adaptive to changes in environmental circumstances" (Kibblewhite *et al.*, 2008). Under conditions of climate change, biological indicators form an integral component in soil health assessment, since, by virtue, they involve complex adaptive systems (i.e. the biota) by integrating key soil processes in ways that other indicators do not (Ritz *et al.*, 2009).

Ritz *et al.*, (2009) note that published information on potential biological indicators has shown an almost exponential increase since the 1970s; however, many of these publications cover a range of processes, for example, microbial, invertebrate or ecological processes. Consequently, a comprehensive explanation of the factors including formal criteria identifying a collective set of biological indicators is limited. Key biological indicators selected for the scope of this study include SOM and its constituents, soil C, respiration and soil microbial biomass. Other microbiological indicators will also be briefly discussed.

Soil Organic Matter

SOM comprises an extensive range of living and non-living components; it has been widely acknowledged that SOM is one of the most complex and heterogeneous components of soils, which vary in their properties, functions and turnover rates (Weil and Magdoff 2004). It is considered an important attribute of soil quality (or soil health) due to the many functions it provide and/or support, including the contribution to the charge characteristics of soils, a sink for and source of C and N, and to a variable extent regulates phosphorus and sulphur cycling, possesses an ability to complex with multivalent ions and organic compounds, provides microbial and faunal habitat and substrates, as well as affecting aggregate stability, trafficability, water retention and hydraulic properties (Haynes 2008). Main indicators for evaluating SOM status include: SOC, since it comprises about 50% of SOM; organic N, since it is closely associated with organic C and is the most important nutrient for plant productivity; and readily mineralisable C and N (Haynes 2008).

As SOM drives the majority of soil functions, decreases in SOM can lead to a decrease in fertility and biodiversity, as well as a loss of soil structure, resulting in reduced water holding capacity, increased risk of erosion and increased bulk density and hence soil compaction (Weil and Magdoff 2004). Land use and management practices that lead to build up of SOM will help in absorbing CO_2 from the atmosphere, thus mitigating global warming. By increasing water storage, SOM can play an important role in the mitigation of flooding impacts following extreme rainfall events, while storing water in the event of droughts thus increasing soil resilience.

SOM indicators have been used in long-term soil experiments for the assessment of climate change (Richter *et al.*, 2007; Rinnan *et al.*, 2007), although the response of SOM to elevated temperature remains scientifically controversial and without consensus (Agren and Wetterstedt 2007). In general, increases in temperature have been reported to enhance decomposition of SOM, but rising temperature and precipitation, CO_2 fertilisation and atmospheric N deposition may support high plant productivity and organic matter input to soil and consequently increase SOM. According to Davidson and Janssens (2006) and Kuzyakov and Gavrichkova (2010), it is the accessibility and availability of SOM to microorganisms that govern SOM losses rather than the rate-modifying climate factor (i.e. temperature).

Soil Carbon

Soil C can be used as an indicator of change for a number of reasons: it is familiar, direct, linked to ecosystem performance and it has "memory", that is, changes across time; however, it may not be reliable as a solitary indicator of soil quality/ health, as it may not encompass all ecosystem traits (Janzen 2005). While soil contains C in diverse forms and residence times, considerable research attention has focused on the SOC form, since it (1) has been largely modified through human activities, and (2) is predicted to decline with increase in mean global temperatures, which would have adverse effects on important soil functions and processes and soil quality/health (Lal *et al.*, 2007). However, in his review on the role of soil C in the changing global C cycle, Janzen (2005) proposed that "if soil C is to be used as an indicator of ecosystem performance, we may need to rethink how it is measured and interpreted (Fig 7.2). Since, although we have been measuring soil C for a long time, our methods was not developed to measure changing C stocks". Since then substantial progress has been made, especially in geospatial and geostatistical methodologies to measure soil C across the landscape and over time (Allen *et al.*, 2010; Worsham *et al.*, 2010).

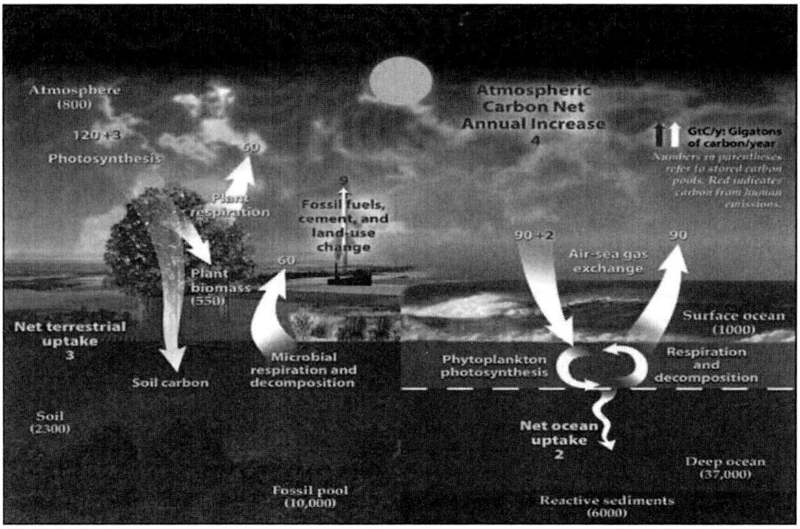

Fig 7.2: Carbon Cycle Diagram from the DOE

Light Fraction and Macro-Organic Matter (Labile Organic Matter)

Light (or low-density) fraction and macro-organic components of SOM consist mainly of mineral-free particulate plant and animal residues, which serve as readily decomposable substrate for soil micro-organisms, as well as a labile nutrient reservoir (Wagai *et al.*, 2009). Since light fraction and macro-organic matter are responsive to management practices, they may act as early indicators to measure the effectiveness of changing management practice in adaptive response to climate change (Gregorich *et al.*, 1994). For example, the labile soil organic C is rapidly depleted as the temperature rises (Davidson and Janssens 2006). In addition, elevated CO_2 in the future may reduce sequestration of root-derived soil C, a major source of labile, light fraction C (Heath *et al.*, 2005).

Potentially Mineralisable C and N

The amount of mineralisable organic matter in soil is an indicator of organic matter quality, acting as the interface between autotrophic and heterotrophic organisms during the nutrient cycling process (Gregorich *et al.*, 1994). While potentially mineralisable C and N may be measured in both the field and laboratory, it is generally treated as a relative rather than an absolute value due to inconsistencies in methods (Haynes 2008). However, mineralisable organic matter may be a useful indicator to assess soil health under climate change, since it affects nutrient dynamics within single growing seasons, and may be used to compare management regimes and C sequestration over extended periods of time (Gregorich *et al.*, 1994).

7.4 Soil Respiration

Soil respiration is often used as a biological indicator for soil health, since it is positively correlated with SOM content (and often with microbial biomass and activity) and can be determined as either CO_2 production or O_2 consumption, e.g. "soil" or "basal" respiration, using a range of in situ or laboratory methods (Haynes 2008). Soil respiration, particularly its temperature response (often expressed as the Q10 function), is widely acknowledged to be a critical link between climate change and the global C cycle (Wixon and Balser 2009), although the nature of this relationship is under current scientific debate (A° gren and Wetterstedt 2007; Balser *et al.*, 2006; Kuzyakov and Gavrichkova 2010). Recent studies have also shown that soil respiration is relatively responsive to changes in the seasonal timing of rainfall, which is predicted to change according to global and regional climate models (Chou *et al.*, 2008).

Soil Microbial Biomass

Microbial biomass, the living component of SOM, is considered the most labile C pool in soils and a sensitive indicator of changes in soil processes, with links to soil nutrient and energy dynamics, including mediating the transfer between SOC fractions (Haynes 2008;Saha and Mandal 2009). While it has been compared to related soil properties to determine change in SOM quality (Gregorich *et al.*, 1994), its use as an indicator of soil quality or soil health is currently limited due to unavailability of benchmark values, difficulty in interpretation and cost-effectiveness of measurement procedures (Dalal 1998). However, soil microbial biomass, similar to labile C, has been shown to be responsive to short-term environmental changes (Haynes 2008), with recent studies revealing significant decline in the soil microbial biomass during long-term simulated climatic warming experiments (Rinnan *et al.*, 2007). When combined with 13^{C} isotope labelling technique, the shift in microbial biomass 13^{C} may provide a more sensitive measure of changes in soil C processes in response to climate and land use changes than the total microbial biomass C (Pateron *et al.*, 2009).

The use of Microbial and Metabolic "Quotients"

In addition to individual biological indicators, ecophysiological indices or "quotients" have also been used to assess environmental change. Indices of physiological performance (e.g. respiration, growth/death, C uptake) against the total organic C or total microbial biomass C per unit time are expressed as microbial quotient (microbial C/organic C), and respiratory or metabolic quotient (CO_2-C respired/hr/microbial C), respectively, and have been used to assess management impacts on organic C dynamics (Moscatelli *et al.*, 2005). While the effectiveness of these quotients to assess ecosystem disturbance and development has been questioned (Moscatelli *et al.*, 2005), there are cases where these quotients have provided useful information on substrate quality and ecosystem response to stresses across different soil types and environments (Weil and Magdoff 2004). Their potential for use as soil health indicators to assess climate change impacts is also promising, with research under controlled Free Air CO_2 Enrichment and N fertilisation trials, reporting that both microbial and metabolic quotients were sensitive to changes due to elevated CO_2 and N fertilisation (Moscatelli *et al.*, 2005).

7.5 Enzyme Activity

Soil enzyme activities may serve to indicate change within the plant-soil system, since these (1) are closely linked to the cycling of nutrients and soil biology, (2) are easily measured, (3) integrate information on both the microbial status and the physicochemical soil conditions, and (4) show rapid response to changes in soil management (Garcı´a-Ruiz *et al.*, 2009). Studies of individual enzyme activities report strong temporal and spatial variability, thus often leading to conflicting results (Garcıa-Ruiz *et al.*, 2009). Furthermore, Dorodnikov *et al.*, (2009) showed that by altering the quantity and quality of belowground C input by plants, elevated CO_2 may stimulate microbial enzyme activities, abundance of microbial enzymes and C turnover possibly affecting microbial community functioning in soil, and furthermore, the extent of stimulation of microbial enzyme activities may depend on soil aggregate size. In addition, atmospheric N deposition may affect extracellular enzymes, which are involved in SOC decomposition and nutrient cycling processes (Frey *et al*., 2004). It is still to be known how soil microbial enzyme activities involved in organic C turnover, nutrient cycling and greenhouse gas emissions will respond to the interactive effects of multiple global change drivers (such as climate change, land use change), thereby supporting the generalised view of their use as indicators of changes in soil health (Fig 7.3).

7.6 Other Microbiological Indicators as Integrative Indicators of Soil Health

Many other soil microbial properties may be considered with respect to soil health indicators; their inclusion within research literature is gaining popularity as more holistic definitions of soil health include soil and plant attributes and functionality, and as advances in microbiological techniques offer promising developments in the detection of soil health status. Other microbial indicators noted in the literature, although not included in the scope of this chapter, include microbial autotrophic nitrification, arbuscular mycorrhiza, community studies of soil macro- and microfauna, and soil disease suppressiveness and plant disease incidence (French *et al*., 2009; Kibblewhite *et al*., 2008; Ritz *et al*., 2009).

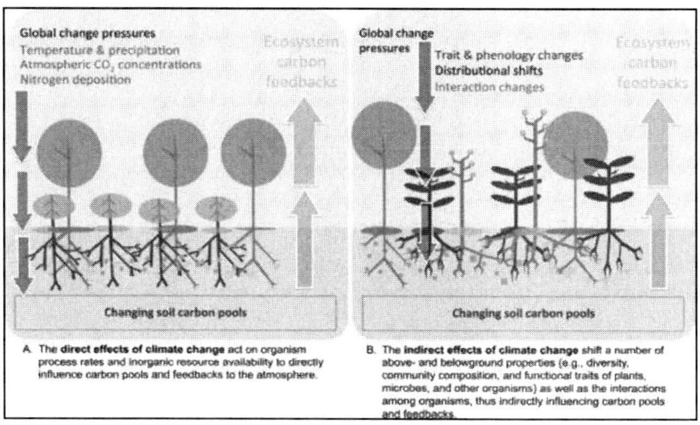

Fig 7.3: The direct effects of global change on carbon feedbacks to the atmosphere have received considerable experimental attention (A); however, there has been less of a focus on understanding the magnitude of indirect effects of global change on the composition and function of ecosystems (B). The ecosystem-scale responses to the indirect effects of global change on community interactions (e.g., via changes in species distributions and/or traits) may be as large, or even larger, than the direct effects. Combined, the direct and indirect effects of global change on ecosystems may magnify, counterbalance, or reverse ecosystem carbon feedbacks to the atmosphere.

7.7 Genetic and Functional Biodiversity of Soils, Soil Health and Climate Change

Soil community structure and function is a current theme for debate among soil scientists, particularly whether ecosystem functioning is influenced by biodiversity, and how the loss of microbial functional groups influences ecosystem functioning (Hunt and Wall 2002; Kibblewhite *et al.*, 2008). Hunt and Wall (2002) note that much of the focus has been on aboveground systems to assess potential effects of climate change on plant biodiversity and biochemistry, although recent studies suggest that climatic stresses, such as warming, will have a profound effect on the rhizosphere, soil heterotrophic community structure and soil processes, including soil respiration, N mineralisation and ecosystem C functioning (Bardgett *et al.*, 2008; Briones *et al.*, 2009). Changing climatic factors is also of concern due to possible evolutionary changes, which allow the spread of virulence factors and genes that aid in environmental survival (French *et al.*, 2009). As molecular technique advancements continue to expand our understanding of microbial diversity and the conditions driving changes in microbial diversity and community structure (Arias *et al.*, 2005; Prosser 2002), it is likely that the adoption of genetic and functional biodiversity indicators of soil health will increase, particularly within food web and nutrient cycling models, to improve predictions of climate change impacts on soil health (see Chap. 8).

7.8 **Selection of Soil Health Key Indicators:** Requirements and Conceptual Framework for In Situ Soil Health Assessment under Climate Change.

Many soil health indicators discussed above need prioritisation with regard to assessment of soil health under sustainable management practices and predicted climate change scenarios. Recent reviews have focused upon measurement of individual indicators or a suite of indicators to assess changes in soil health within a soil physical, chemical and/or biological context (Moebius *et al.*, 2007; Zagal *et al.*, 2009), although Stockdale and Watson (2009) note that work is needed to test existing indicator frameworks to guide management decisions, rather than simply to compare with the status quo.

The indicators of soil health are interlinked (Fig 7.1) and so are the drivers of global change (land use change, elevated temperatures, elevated atmospheric CO_2 concentration, increasing atmospheric N deposition, variability in the amount, intensity and distribution of rainfall, extreme climatic events, and their interactions). Conceptualisation of interactions between climate change, land management and soil health indicators, as presented in Fig 7.1, further highlights the complexity and interdependence of many of these drivers to influence soil health, primarily through their influence on plant primary productivity, microbial and faunal biomass diversity and activity, including their products (intracellular and extracellular enzymes), as well as the overall C and N cycles (Bardgett *et al.*, 2008; Kibblewhite *et al.*, 2008; Wixon and Balser 2009).

Table 7.1 presents a summary of major soil health indicators, their processes and functions under projected climate change scenarios, their relevance to climate change impacts and their frequency of inclusion within a minimum data set for soil health assessment. Most studies agree that a minimum data set for the assessment of soil health should include key indicators that: (1) are sensitive to changes due to management and climate variations, (2) integrate soil physical, chemical and biological properties, (3) are relatable to important soil functions, (4) are applicable to field conditions and (5) are accessible to many users (Idowu *et al.*, 2009; Karlen *et al.*, 2003; Lagomarsino *et al.*, 2009; Pattison *et al.*, 2008; Rao and Siddaramappa 2008; Schindelbeck *et al.*, 2008). The selection of key indicators of soil health under climate change scenarios, as discussed in this chapter, should be able to reflect the mitigation and adaptive capability of soil and its resilience to climate change over short to long term.

References

1. Agren GI, Wetterstedt M. What determines the temperature response of soil organic matter decomposition? Soil Biol Biochem. 2007; 39:1794-1798.

2. Allen DE, Pringle MJ, Page KL, Dalal RC. A review of sampling designs for the measurement of soil organic carbon in Australian grazing lands. Rangel J, 2010; 32:227–246.

3. Aon MA, Cabello MN, Sarena DE, Colaneri AC *et al.*, I. Spatio-temporal patterns of soil microbial and enzymatic activities in agricultural soils. Appl Soil Ecol. 2001; 18:239-254.

4. Arias ME, Gonza´lez-Pe´rez FJ, Gonza´lez-Vila ASB. Soil health-a new challenge for microbiologists and chemists. Int Microbiol. 2005; 8:13–21.

5. Arnold SL, Doran JW, Schepers J, Wienhold B. Portable probes to measure electrical conductivity and soil quality in the field. Commun Soil Sci Plant Anal. 2005; 36:2271–2287.

6. Assouline S. Rainfall-induced soil surface sealing: a critical review of observations, conceptual models, and solutions. Vadose Zone J, 2004; 3:570–591.
7. Balser TC, McMahon KD, Bart D, Bronson D, *et al.*, Bridging the gap between micro-and macro-scale perspectives on the role of microbial communities in global change ecology. Plant Soil. 2006; 289:59–70.
8. Bardgett RD, Freeman C, Ostle NJ. Microbial contributions to climate change through carbon cycle feedbacks. ISME J, 2008; 2:805–814.
9. Birka´s M, Stingli A, Szemo KA, Kalma RT, Bottlik L. Soil condition and plant interrelations in dry years. Cereal Res Commun 36, 2008, 15-18.
10. Birka´s M, Dexter A, Szemo KA. Tillage-induced soil compaction, as a climate threat increasing stressor. Cereal Res Commun. 2009; 37:379–382.
11. Blanco-Canqui H, Lal R. Mechanisms of carbon sequestration in soil aggregates. Crit Rev Plant Sci. 2004; 23:481–504.
12. Box JE, Bruce RR. The effect of surface cover on infiltration and soil erosion. In: Agassi M (ed) Soil erosion, conservation and rehabilitation. Marcel Dekker Inc, New York, 1996.
13. Brinkman R, Sombroek W. The effects of global change on soil conditions in relation to plant growth and food production. In: Bazzaz F, Sombroek W (Eds) Global climate change and agricultural production. Food and Agriculture Organization of the United Nations, 1999.
14. and John Wiley and Sons Ltd., Rome, Italy.
15. Briones MJI, Ostle NJ, McNamara NP, Poskitt J. Functional shifts of grassland soil communities in response to soil warming. Soil Biol Biochem. 2009; 41:315–322.
16. Chou WW, Silver WL, Jackson RD, Thompson AW, Allen-Diaz B. The sensitivity of annual grassland carbon cycling to the quantity and timing of rainfall. Glob Chang Biol. 2008; 14:1382–1394.
17. Dalal RC. Soil microbial biomass-what do the numbers really mean? Aust J Exp Agric, 1998, 649-665.
18. Dalal RC, Moloney D. Sustainability indicators of soil health and biodiversity. In: Hale P, Petrie A, Moloney D, Sattler P (Eds) Management for sustainable ecosystems. Centre for Conservation Biology, Brisbane, 2000, 101-108.

19. Dalal RC, Lawrence P, Walker J, Shaw RJ *et al*., A framework to monitor sustainability in the grains industry. Aust J Exp Agric. 1999; 39:605-620.

20. Dalal RC, Eberhard R, Grantham T, Mayer DG. Application of sustainability indicators, soil organic matter and electrical conductivity, to resource management in the northern grains region. Aust J Exp Agric. 2003a; 43:253–259.

21. Dalal RC, Wang WJ, Robertson GP, Parton WJ. Nitrous oxide emission from Australian agricultural lands and mitigation options: a review. Aust J Soil Res. 2003b; 41:165–195.

22. Dalal RC, Allen DE, Livesley SJ, Richards G. Magnitude and biophysical regulators of methane emission and consumption in the Australian agricultural, forest, and submerged landscapes: a review. Plant Soil. 2008; 309:43–76.

23. Dang YP, Dalal RC, Mayer D, McDonald M *et al*., High subsoil chloride concentrations reduce soil water extraction and crop yield on Vertisol in northeastern Australia. Aust J Agric Res. 2008; 59:321–330.

24. Davidson EA, Janssens IA. Temperature sensitivity of soil carbon decomposition and feedbacks to climate change. Nature. 2006; 440:165-173.

25. DeVries W, Breeuwsma A. The relationship between soil acidification and element cycling. Water Air Soil Pollut. 1987; 35:293–310.

26. Doran JW. Soil health and global sustainability: translating science into practice. Agric Ecosyst Environ. 2002; 88:119–127.

27. Doran JW, Zeiss MR. Soil health and sustainability: managing the biotic component of soil quality. Appl Soil Ecol. 2000; 15:3–11.

28. Doran JW, Jones AJ, Arshad MA, Gilley JE. Determinants of soil quality and health. In: Pankhurst CE, Doube BM, Gupta VVSR (eds) Biological indicators of soil health. CAB International, Oxon, 1999.

29. Dorodnikov M, Blagodatskaya E, Blagodatsky S, Marhan S, Fangmeier A, Kuzyakov Y. Stimulation of microbial extracellular enzyme activities by elevated CO_2 depends on soil aggregate size. Glob Change Biol. 2009; 15:1603-1614.

30. French S, Levy-Booth D, Samarajeewa A, Shannon KE, Smith J, Trevors JT. Elevated temperatures and carbon dioxide concentrations: effects on selected microbial activities in temperate agricultural soils. World J Microbiol Biotechnol. 2009; 25:1887–1900.

31. Frey SD, Knorr M, Parrent JL, Simpson RT. Chronic nitrogen enrichment affects the structure and function of the microbial community in temperate hardwood and pine forests. For Ecol Manage. 2004; 196:159–171.

32. Garcı´a-Ruiz R, Ochoa V, Vinegla B, Hinojosa MB *et al*., Soil enzymes, nematode community and selected physico-chemical properties as soil quality indicators in organic and conventional olive-oil farming: influence of seasonality and site features. Appl Soil Ecol. 2009; 41:305–314.

33. Gil SV, Meriles J, Conforto C, Figoni G *et al.* Field assessment of soil biological and chemical quality in response to crop management practices. World J Microbiol Biotechnol. 2009; 25:439-448.

34. Gregorich EG, Carter MR, Angers DA, Monreal CM, Ellert BH. Towards a minimum data set to assess soil organic matter quality in agricultural soils. Can J Soil Sci. 1994; 74:367–385.

35. Hadas A. Soil tilth-the desired soil structural state obtained through proper soil fragmentation and reorientation processes. Soil Tillage Res. 1997; 43:7-40.

36. Ha°kansson I, Lipiec J. A review of the usefulness of relative bulk density values in studies of soil structure and compaction. Soil Tillage Res. 2000; 53:71–85.

37. Haynes RJ. Soil organic matter quality and the size and activity of the microbial biomass: their significance to the quality of agricultural soils. In: Huang Q, Huang PM, Violante A (eds) Soil mineral-microbe-organic interactions: theories and applications.Springer, Berlin, 2008, 201–230.

38. Heath J, Ayres E, Possell M, Bardgett RD *et al*., Rising atmospheric CO_2 reduces sequestration of root-derived soil carbon. Sci. 2005; 309:1711–1713.

39. Hunt HW, Wall DH. Modelling the effects of loss of soil biodiversity on ecosystem function. Glob Change Biol. 2002; 18:33–50.

40. Idowu OJ, Van Es HM, Abawi GS, Wolfe DW *et al*., Use of an integrative soil health test for evaluation of soil management impacts. Renew Agric Food Syst. 2009; 24:214–224.

41. Janzen HH. Soil carbon: a measure of ecosystem response in a changing world? Can J Soil Sci. 2005; 85:467–480.

42. Jarvis NJ. A review of non-equilibrium water flow and solute transport

in soil macropores: principles, controlling factors and consequences for water quality. Eur J Soil Sci. 2007; 58:523–546.

43. Joel A, Messing I. Infiltration rate and hydraulic conductivity measured with rain simulator and disc permeameter on sloping arid land. Arid Land Res Manage. 2001; 15:371–384.

44. Karlen DL, Doran JW, Andrews SS, Wienhold BJ. Soil Quality – Humankinds' foundation for survival. J Soil Water Conserv. 2003; 58:171–179.

45. Kibblewhite MG, Ritz K, Swift MJ. Soil health in agricultural systems. Philos Transact R Soc. 2008; 363:685–701.

46. Kinyangi J. Soil health and soil quality: a review. Draft publication. Available Accessed 15 Feb 2010, 2007.

47. Knorr W, Prentice IC, House JI, Holland EA. Long-term sensitivity of soil organic carbon turnover to warming. Nature. 2005; 433:298–301.

48. Kuzyakov Y, Gavrichkova O. Time lag between photosynthesis and carbon dioxide efflux from soil: a review of mechanisms and controls. Global Change Biology. 2010; 16:3386–3406.

49. Lagomarsino A, Moscatelli MC, Tizio AD, Mancinelli R, Grego S, Marinari S. Soil biological indicators as a tool to assess the short-term impact of agricultural management on changes in organic C in a Mediterranean environment. Ecol Indic. 2009; 9:518–527.

50. Lal R. Tillage systems in the tropics: management options and sustainability implications. Food and Agriculture Organization of the United Nations Soils Bulletin 71, FAO, Rome, 1995.

51. Lal R (ed). Soil quality and soil erosion. Soil and Water Conservation Society, Ankeny, 1999.

52. Lal R, Follett RF, Stewart BA, Kimble JM. Soil carbon sequestration to mitigate climate change and advance food security. Soil Sci. 2007; 172:943–956.

53. McCarthy JJ, Canziani OF, Leary NA, Dokken DJ, White KS. Climate Change: impacts, adaptation, and vulnerability. Contribution of Working Group II to the Third Assessment Report of the Intergovernmental Panel on Climate Change. Cambridge University Press, Cambridge, 2001.

54. Moebius BN, Van Es HM, Schindelbeck RR, Idowu OJ, Clune DJ,

Thies JE. Evaluation of laboratory-measured soil properties as indicators of soil physical quality. Soil Sci. 2007; 172:895–912.

55. Moscatelli MC, Lagomarsino A, Marinari S, De Angelis P, Grego S. Soil microbial indices as bioindicators of environmental changes in a poplar plantation. Ecol Indic. 2005; 5:171–179.

56. Nelson KL, Lynch DH, Boiteau G. Assessment of changes in soil health throughout organic potato rotation sequences. Agric Ecosyst Environ. 2009; 131:220–228.

57. Nuttall JG. Climate changes-identifying the impacts on soil and soil health. Department of Primary Industries, Future Farming Systems Research Division, Victoria, 2007.

58. O'Farrell PJO, Donaldson JS, Hoffman MT. Vegetation transformaion, functional compensation, and soil health in a semi-arid environment. Arid Land Res Manage. 2010; 24:12–30.

59. Pankhurst CE, Doube BM, Gupta VVSR. (Eds). Biological indicators of soil health. CAB International, Oxon, 1997.

60. Pariente S. Soluble salts dynamics in the soil under different climatic conditions. Catena. 2001; 43:307–321.

61. Paterson E, Midwood AJ, Millard P. Through the eye of the needle: a review of isotope approaches to quantify microbial processes mediating soil carbon balance. New Phytol. 2009; 184:19–33.

62. Pattison AB, Moody PW, Badcock KA, Smith LJ *et al.*, Development of key soil health indicators for the Australian banana industry. Appl Soil Ecol. 2008; 40:155–164.

63. Piglai M, De Nobili M. Relationship between soil porosity and root development and soil enzyme activity in cultivated soils. Geoderma. 1993; 56:243–256.

64. Porporato A, Daly E, Rodriguez-Iturbe I. Soil water balance and ecosystem response to climate change. Am Nat. 2005; 164:625–632?

65. Post WM, Kwon KC. Soil carbon sequestration and land-use change: processes and potential. Glob Change Biol. 2000; 6:317–328.

66. Prosser JI. Molecular and functional diversity in soil micro-organisms. Plant Soil. 2002; 244:9–17.

67. Rao BKR, Siddaramappa R. Evaluation of soil quality parameters in a tropical paddy soil amended with rice residues and tree litters. Eur J Soil Biol. 2008; 44:334–340.

68. Rengasamy P. Soil processes affecting crop production in salt-affected soils. Funct Plant Biol. 2010; 37:613–620.

69. Reth S, Reichstein M, Falge E. The effect of soil water content, soil temperature, soil pH value and root mass on soil CO2 efflux. Plant Soil. 2005; 268:21–33.

70. Reynolds WD, Bowman BT, Drury CF, Tan CS, Lu X. Indicators of good soil physical quality: density and storage parameters. Geoderma. 2002; 110:131–146.

71. Reynolds WD, Drury CF, Tan CS, Fox CA, Yang XM. Use of indicators and pore volume function characteristics to quantify soil physical quality. Geoderma. 2009; 152:252–263.

72. Richter DD, Hofmockel M, Callaham MA, Powlson DS, Smith P. Long-term soil experiments: keys to managing Earth are rapidly changing ecosystems. Soil Sci Soc Am J, 2007; 71:266-279.

73. Riley J. Multidisciplinary indicators of impact and change: key issues for identification and summary. Agric Ecosyst Environ. 2001c; 87:245-259.

74. Rimal BK, Lal R. Soil and carbon losses from five different land management areas under simulated rainfall. Soil Tillage Res. 2009; 106:62–70.

75. Rinnan R, Michelsen A, Baath E, Jonasson S. Fifteen years of climate change manipulations alter soil microbial communities in a subarctic heath ecosystem. Glob Change Biol. 2007; 13:28–39.

76. Ritz K, Black HIJ, Campbell CD, Harris JA, Wood C. Selecting biological indicators for monitoring soils: a framework for balancing scientific and technical opinion to assist policy development. Ecol Indic. 2009; 9:1212–1221.

77. Ross DS, Matschonat G, Skyllberg U. Cation exchange in forest soils: the need for a new perspective. Eur J Soil Sci. 2008; 59:1141–1159.

78. Saha N, Mandal B. Soil health-a precondition for crop production. In: Khan MS, Zaidi A, Musarrat J (Eds) Microbial strategies for crop improvement. Springer, Heidelberg, 2009, 161–168.

79. Salvador Sanchis MP, Torri D, Borselli L, Poesen J. Climate effects on soil erodibility. Earth Surf Process Landf. 2008; 33:1082–1097.

80. Schindelbeck RR, Van Es HM, Abawi GS, Wolfe DW, *et al.*,

Comprehensive assessment of soil quality for landscape and urban management. Landsc Urban Plan. 2008; 88:73–80.

81. Schjønning P, Elmholt S, Christensen BT. Soil quality management-concepts and terms, 2004.

82. In: Schjønning P, Elmholt S, Christensen BT (eds) Managing soil quality: challenges I modern agriculture. CAB International, Oxon.

83. Smith JL, Halvoson JJ, Bolton H Jr. Soil properties and microbial activity across a 500 m elevation gradient in a semi-arid environment. Soil Biol Biochem. 2002; 34:1749–1757.

84. Sparling GP. Soil microbial biomass, activity and nutrient cycling as indicators of soil health. In: Pankhurst CE, Doube BM, Gupta VVSR (Eds) Biological indicators of soil health. CAB International, Oxon, 1997, 79–96.

85. Stenberg B. Monitoring soil quality of arable land: microbiological indicators. Acta Agriculturae Scandinavica., Sect. B. Soil Plant Sci. 1999; 49:1–24.

86. Stockdale EA, Watson CA. Biological indicators of soil quality in organic farming systems. Renew Agric Food Syst. 2009; 24:308–318.

87. Wagai R, Mayer LM, Kitayama K. Nature of the "occluded" low-density fraction in soil organic matter studies: a critical review. Soil Sci Plant Nutr. 2009; 55:13–25.

88. Weil RR, Magdoff F. Significance of soil organic matter to soil quality and health. In: Weil RR, Magdoff F (Eds) Soil organic matter in sustainable agriculture. CRC press, Florida, 2004, 1-43.

89. Wixon DL, Balser TC. Complexity, climate change and soil carbon: a systems approach to microbial temperature response. Syst Res Behav Sci. 2009; 26:601–620.

90. Worsham L, Markewitz D, Nibblelink N. Incorporating spatial dependence into estimates of soil carbon contents under different land covers. Soil Sci Soc Am J. 2010; 74:635–646.

91. Zagal E, Munoz C, Quiroz M, Co´rdova C. Sensitivity of early indicators for evaluating quality changes in soil organic matter. Geoderma. 2009; 151:191–198.

Chapter - 8
Climate Change and Biodiversity

Authors

Dr. Mukesh Kumar

Department of Genetics & Plant Breeding, Sardar Vallabhbhai Patel University of Agriculture & Technology, Meerut, Uttar Pradesh, India

Dr. Purushattom

Department of Microbiology & Pathology, Sardar Vallabhbhai Patel University of Agriculture & Technology, Meerut, Uttar Pradesh, India

Chapter - 8

Climate Change and Biodiversity

Dr. Mukesh Kumar and Dr. Purushottam

Biodiversity is crucial to ecosystem health, and healthy ecosystems are central to human well-being. Healthy ecosystems interlinked with working landscapes and other open spaces form the ecological infrastructure of the country and are the foundation for clean air and water, fertile soil and food. It does depend on healthy ecosystems for economic and livelihood activities, including agriculture and a number of income generating and subsistence level activities. These natural ecosystems are under pressure from land use change and related processes causing degradation, as well as invasive alien species. Accelerated climate change (resulting in increasing temperature, rising atmospheric CO_2 and changing rainfall patterns) is exacerbating these existing pressures. Well-functioning ecosystems provide natural solutions that build resilience and help society adapt to the adverse impacts of climate change. This includes, for example, buffering communities from extreme weather events such as floods and droughts, reducing erosion and trapping sediment, increasing natural resources for diversifying local livelihoods, providing food and fiber, and providing habitats for animals and plants which provide safety nets for communities during times of hardship. Sustainably managed and/or restored ecosystems help in adapting to climate change at local or landscape level. Adaptation responses involving natural resources require a coherent framework that can be applied at national level, and then refined for sub-national purposes.

Biodiversity describes the variety of life on earth in terms of genes, species and ecosystems, and the ecological and evolutionary processes that maintain this diversity. It is a measure of ecosystem health.

Ecosystems provide ecosystem services: provisioning: including water, food, and timber; regulating: water filtration, climate regulation (carbon storage), pollination, and protection from natural disasters (e.g. mangroves and dunes against storm surges); cultural: recreation and relaxation, spiritual values, tourism, and education; and supporting: soil formation, photosynthesis (oxygen), and nutrient cycling.

Climate Change Effects on Biodiversity

The multiple components of climate change are anticipated to affect all the levels of biodiversity, from organism to biome levels (Fig 8.1). They primarily concern various strengths and forms of fitness decrease, which are expressed at different levels, and have effects on individuals, populations, species, ecological networks and ecosystems. At the most basic level of biodiversity, climate change is able to decrease genetic diversity of populations due to directional selection and rapid migration, which could in turn affect ecosystem functioning and resilience. However, most studies are centered on impacts at higher organizational levels, and genetic effects of climate change have been explored only for a very small number of species. Beyond this, the various effects on populations are likely to modify the 'web of interactions' at the community level (Gilman *et al.*, 2010). In essence, the response of some species to climate change may constitute an indirect impact on the species that depend on them. A study of 9650 inter-specific systems, including pollinators and parasites, suggested that around 6300 species could disappear following the extinction of their associated species.

Climate change has led to phenological shifts in flowering plants and insect pollinators, causing mismatches between plant and pollinator populations that lead to the extinctions of both the plant and the pollinator with expected consequences on the structure of plant–pollinator networks. Other modifications of inter-specific relationships (with competitors, prey / predators, host / parasites or mutualists) also modify community structure and ecosystem functions. At a higher level of biodiversity, climate can induce changes in vegetation communities that are predicted to be large enough to affect biome integrity. The Millenium Ecosystem Assessment forecasts shifts for 5–20% of Earth's terrestrial ecosystems, in particular cool conifer forests, tundra, scrubland, savannahs and boreal forest. Of particular concern are 'tipping points' where ecosystem thresholds can lead to irreversible shifts in biomes (Leadley *et al.*, 2010).

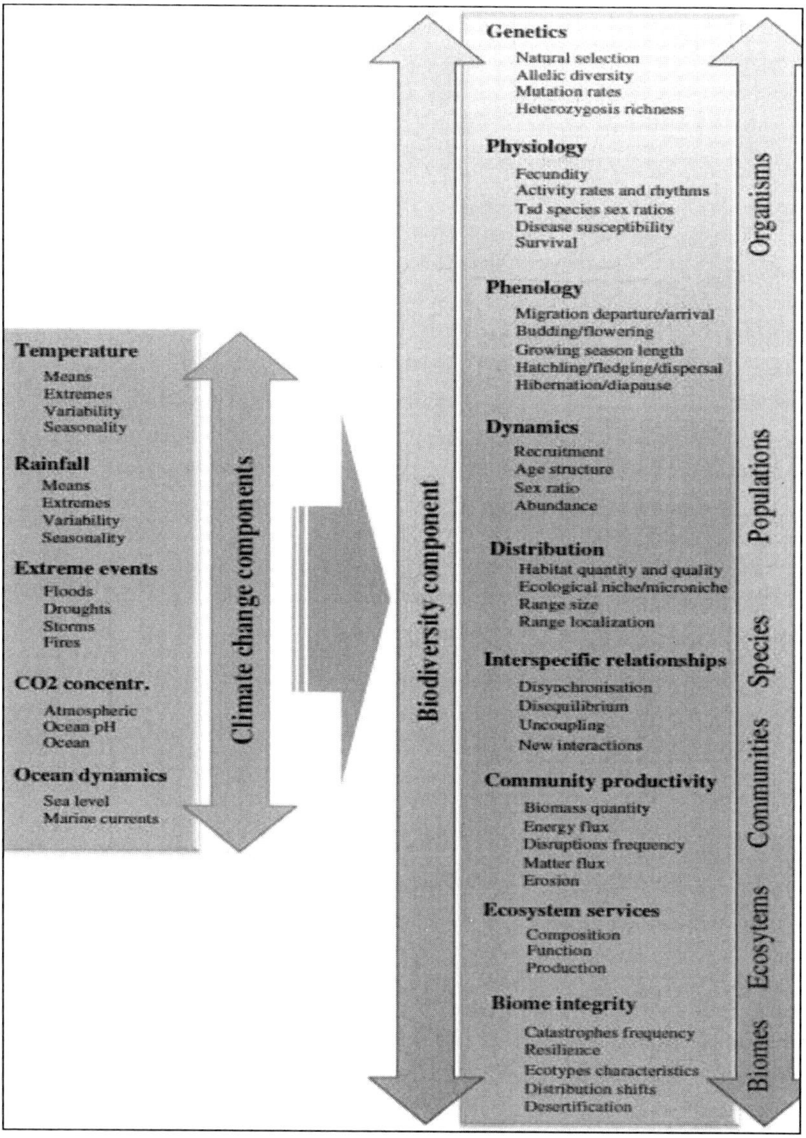

Fig 8.1: Summary of some of the predicted aspects of climate change and some examples of their likely effects on different levels of biodiversity.

At higher altitudes and latitudes, alpine and boreal forests are expected to expand northwards and shift their tree lines upwards at the expense of low stature tundra and alpine communities. Increased temperature and decreased rainfall mean that some lakes might dry out. Oceans are predicted to warm and become more acid, resulting in widespread degradation of tropical coral reefs. The implications of climate change for genetic and specific diversity have potentially strong implications for ecosystem services. The most extreme and irreversible form of fitness decrease is obviously species extinction. To avoid or mitigate these effects, biodiversity can respond in several ways, through several types of mechanisms.

Biodiversity Responses to Climate Change

Because of climate changes, species may no longer be adapted to the set of environmental conditions in a given region and could therefore fall outside its climatic niche. As other components of the ecological niche of species are not supposed to change directly, we will hereafter refer only to climatic niches of species (i.e. the climatic components of the n-dimensional hypervolume sensu Hutchinson). To persist, individuals, populations or species must produce adaptive responses, which can be of several types, and are provided by two categories of mechanisms.

Response Mechanisms: Plastic vs. Genetic

One of the crucial questions in the debate on ecological effects of climate change is whether or not species will be able to adapt fast enough to keep up with the rapid pace of changing climate. Whatever the types of adaptive responses, underlying mechanisms are either due to micro-evolution (i.e. species can genetically adapt to new conditions through mutations or selection of existing genotypes or plasticity, which provides a means of very short-term response (within Individual's lifetimes). It may involve intra-specific variation in morphological, physiological or behavioral traits, which can occur on different time scales within the population's spatial range (Chevin *et al.*, 2010).

Empirical evidence suggests that plastic contribution is often more important than genetic contribution, as observed in birds and marmots. On the other hand, there is increasing empirical evidence that evolution can be very rapid. This is the case for many introduced species, for which selection-driven phenotypic changes have enhanced the invasive potential. Recent experiments on evolutionary rescue also confirm that rapid evolution through mutation and selection could allow species with rapid life cycles to adapt very severe and rapid environmental changes.

Responses: Along Three Axes

Whatever the mechanisms involved in response to climate change, species can in theory change, and changes have already been observed, along three distinct but non-exclusive axes: spatial, temporal or self. The first two axes correspond to easily observable and well documented responses to global warming. "Self" corresponds to less visible physiological and behavioral changes that allow species to adapt to the new climatic conditions in the same spatial and temporal frame.

Spatial: First, species can track appropriate conditions in space and follow them. This is typically done through dispersion, but spatial changes are not limited to this: shifts to a different habitat at the local or micro-habitat levels are also relevant. One of the best-documented responses – from both paleontological records and recent observations – is a spatial shift of species tracking suitable climatic conditions at the regional scale. Latitudinal and altitudinal range shifts have already been observed in more than 1000 species – especially those with high dispersal capacities like birds, insects and marine invertebrates, leading to a reduction in range size particularly in polar and mountaintop species. However, individuals shift their distribution to stay in quasi equilibrium with the climatic conditions they are adapted to, but they may not be adapted to other abiotic variables such as photoperiod or novel biotic interactions. In these cases, micro-evolution may be needed for them to persist.

Temporal: To keep up with changing abiotic factors that show cyclic variation over time, such as temperature on a daily or yearly period, individuals can also respond to climate change through a shift in time (on a daily to seasonal basis). Phenology, i.e. the timing of life cycle events such as flowering, fruiting and seasonal migrations, is one of the most ubiquitous responses to 20th century climate warming. It has already been documented in many species. In a meta-analysis of a wide range of species including animals and plants, the mean response across all species responding to climate change was a shift in key phenological events of 5.1 days earlier per decade over the last 50 years. Flowering has advanced by more than 10 days per decade in some species. These phenological changes can help species keep synchrony with cyclical abiotic factors. Yet, they can also be disruptive, by increasing asynchrony in predator-prey and insect-plant systems which may lead to species extinction. Temporal shifts may also occur at a small temporal scale, e.g. with behaviors' or activity patterns adjusted in daily activity rhythms to match the energetic costs of a different climatic condition.

Self: Last, species can cope with changing climatic conditions by adapting themselves to the new conditions in their local range, rather than by tracking their current optimal conditions in space or time. For lack of a better term, we refer to these in situ changes that are not related to spatial or temporal changes, as changes in "self". Species can move along this third, 'self' axis by physiological alterations that allow tolerance to warmer or drier conditions or by behavioral modifications of their diet, activity and energy budget, for example.

Although they are often less obvious than changes in time or space, some physiological responses have already been reported during the 20th century climate change, especially from many ectoderms', as their locomotion, growth, reproduction and sex determination are temperature sensitive. However, for many traits, plastic phenotypic responses should reach a physiological limit and 'saturate' in extreme environments. For example, body size or metabolic rate cannot increase or decrease indefinitely under sustained environmental change. In this case, strong selection is needed to cope with climate change. As they remain in the same spatial and temporal frame, thereby limiting alterations of inter-specific relationships, changes in self also have different consequences for ecosystem responses than do changes in time and space.

Failing to adapt along one or several of these three axes, population or species will go extinct locally or globally. There is thus a multitude of possible responses for species to cope with climate change, and in fact relatively few taxa went extinct following climate change during the Quaternary period. This should help temper catastrophist predictions regarding the global effects of the current climate change on biodiversity. However, the responses of many populations are likely to be inadequate to counter the speed and magnitude of the current climate change. In addition, unlike in past periods of climate change, species have now to cope with additional threats, some of which may act in synergy with climate change. As we are already facing an irrefutable biodiversity crisis, the number of species that may go extinct following climate change has become a major concern over the last few years.

Assessing the Future of Global Biodiversity

The effects of global climate change on biodiversity and its different levels of response is still insufficiently well developed. Yet, it is enough to raise serious concern for the future of biodiversity. The most pressing issue is too quantitatively assessing the prospects for biological diversity in the face of global climate change. Although several methods exist to draw inferences, starting with existing palaeontological or recent data, experiments, observations and meta-analyses, ecological modeling is the most commonly used tool for predictive studies.

Progress in this field is characterized by both an extremely high pace and a plurality of approaches. In particular, there are three main approaches to projecting species loss, concentrating either on future changes in species range or species extinction or changes in species abundance. However, all three modeling approaches have so far largely focused on one axis of response (change in space), largely overlooking the importance of the other aspects. In addition, they seldom account for the mechanisms of these responses (plasticity and evolution).

Table 8.1: Advantages and disadvantages of the two components of major modeling approaches used to estimate loss of biodiversity due to climate change for illustrations of how the two components can be combined to estimate biodiversity loss

Biodiversity Models	Advantages	Disadvantages	Key References
Bioclimatic envelope models (BEM)	Can be applied to a large number of species and a variety of taxonomic groups. Implicitly capture many ecological processes in the relationship between occurrence data and spatial information.	Do not explicitly account for mechanisms that mediate species range. May handle novel climates poorly. Lack temporal dynamics Assume that the current distribution of a species is a good indicator of suitable climate	Alo & Wang (2008), Bradley *et al*., (2010), Brook *et al*., (2008)
Dynamic vegetation models (DVMs)	Include the dynamics of plant growth, competition and, in a few cases, migration Allow the identification of future trends in ecosystem function and structure Can be used to explore feedbacks between biosphere and atmospheric processes	Require detailed physiological data. Don't include plant interactions with other taxonomic groups Limit biodiversity to a very small number of plant functional types. Do not take into account fine scale spatial heterogeneity. Are not adapted for predicting species extinctions at local scales	Anderson *et al*., (2009), Brook *et al*., (2009), Campbell *et al*., (2009)
Species loss model component Species-area relationships (SAR)	Are easy to couple with distribution models because they are based on range or habitat loss. Can be applied to a variety of taxonomic groups.	Lack empirical evidence concerning applicability of SAR for climate change or at species range level. Lack temporal dynamics Don't account for processes influencing extinction rates.	Alo & Wang (2008), Campbell *et al*., (2009), Carnaval *et al*., (2009)
IUCN status methods	Use a widely accepted measure of threat Are simple to couple with distribution models because partly based on criteria	Depend on thresholds that are somewhat arbitrary. Rely on sole criteria of declining range size in most studies. Often don't respect time frame for declines (i.e. 10	Alo & Wang (2008), Chevin *et al*., (2010)

	of range or habitat loss	years or three generations in most cases)	
Dose–response Relationships	Are anchored in measured responses of biodiversity to global change drivers. Can assess the impact of a wide range of global change factors alone or their cumulative effects. Can include time lags	Undisturbed- ecosystems used as baseline are difficult to define. Inadequately account for interactions between global change drivers. Lack validation at large regional or global scales. Use metrics that are difficult to relate to common biodiversity indices	Bakkenes *et al.*, (2002), Barnosky *et al.*, (2011)

Biodiversity Range Models Components

Studies modeling species' range shifts are generally based on the assumption that species niches are defined by a small set of environmental variables, i.e. the species' "climate niche", defines the suitable climatic habitat for that particular species (Table 8.1). These Bioclimatic Envelope Models relate current species ranges to multiple climatic variables and thereby define the climatic niche (envelope) for each species. It is then possible to project this niche for different future climate scenarios to determine the potential redistribution of the suitable climate space of the species. The extinction risks can then be calculated in different ways using species–area relationships (Thomas *et al.*, 2004). Despite a number of limitations (lack of biological process as well as methodological and theoretical issues, species distribution models still constitute the bulk of studies in this domain.

Shifts in distinct vegetation types, often referred to as biome or habitat shifts are often simulated with Dynamic Vegetation Models (DVM). These models forecast shifts in vegetation and associated biogeochemical and hydrological cycles in response to climate change. DVMs use time series of climate data (e.g. temperature, precipitation, humidity, sunshine days) and take into account constraints of topography and soil characteristics to simulate monthly or daily dynamics of ecosystem processes. Plant species are represented as groups with similar physiological and structural properties, termed Plant Functional Types (PFTs), which are designed to represent all major types of plants (Sitch *et al.*, 2008). PFT distributions can then be used to estimate changes in biome or habitat ranges. Currently, DVMs are of limited use for projecting responses in biodiversity directly (i.e. the absence of animals and the limitation to c. 10 PFTs exclude direct utilization). However, coupled with extinction models, they allow extinction risk for species to be estimated at the regional or global scale.

Species Loss Models Components

The simplest method for calculating extinction risk is to assume that species go extinct when they no longer have any suitable habitat. This may underestimate extinctions because species often enter an 'extinction vortex' well before they lose their entire habitat. Yet, it could also substantially overestimate extinctions because many species have weak habitat specificity.

The species–area relationship (SAR), an empirical relationship between the number of species and the land area of a region, is often used to estimate extinction risk. Species extinctions are calculated as a direct function of habitat loss or climate-induced range contraction based on the observation that extinction risk increases with decreasing range and population size.

SAR methods may over- or underestimate species extinction risk depending on the capacity of species to persist in small populations or adapt to novel environments, and they also fail to provide a time frame in which extinctions are likely to occur because the extinction debt is not accounted for the relevance of using this relationship to project biodiversity loss is still hotly debated (He & Hubbell 2011).

The IUCN has developed criteria to assess species extinction risk for their Red List of the conservation status of plant and animal species. These include 90 biological traits identified as enhancing species_ vulnerability (e.g. habitat specialization, narrow environmental tolerance, dependence on specific environmental triggers, dependence on inter-specific interactions or poor ability to disperse to or colonize a new range. Recent attempts at projecting climate change impacts on biodiversity have used the IUCN status metric to obtain estimates of extinction rates based on projected range shifts although the use of IUCN status for this purpose has also been questioned.

In dose–response relationship (DRR) models, observational data and experiments can be employed to generate empirical relationships between the relative importance of global drivers (dose) and changes in species loss (response). Most models focus on species extinctions, which are only the last step of a decline in abundance and are a less immediate (although dramatic) impact of climate change. This fact has led to the development of new models that attempt to quantify the impact of human activities on species abundance (i.e. an equivalent of DRR for species abundance). Impacts on biodiversity are estimated using biodiversity indicators including mean species abundance and Biodiversity Intactness Index. For example, the GLOBIO model uses a matrix of changes in mean local species abundances following conversion between two land use or land cover categories, derived from empirical studies. These types of models use species abundances in pristine ecosystems as a baseline, and thereby provide a measure of the distance from "naturalness" of plant and animal communities following human disturbance. Species abundance models based on plant traits and abiotic characteristics can also provide evidence of changes in ecosystem services.

There are, however, several major limitations to these models, especially concerning the indices of biodiversity derived from DRR applied to species abundance. They depend largely on the quality of the input data (of which much more is needed than in most other models), uncertainty cannot be taken into account, and they are difficult to relate to commonly used biodiversity indices.

Model Combinations

Current modeling techniques usually incorporate a succession of one (or several) future socio-economic scenario(s), one (or several) extinction driver(s), and one biodiversity range or species loss models. They then express their projections in terms of various extinction metrics. This has generated a variety of 'model combinations' leading to a wide range of projections that are not straightforward to compare, as the underlying assumptions differ greatly. Despite important differences and some bias (e.g. marine biodiversity is still poorly represented), these models generally indicate that many species can be expected to decline rapidly at a global scale.

What Future Improvements Needed?

Climate change ecology is still in its infancy, and tremendous improvements are made rapidly in virtually all aspects of this emerging field. Critical requirements to be able to predict future trends include the need to study a much larger part of biodiversity, to overcome several major model limitations, to account for co-extinctions and other major drivers of biodiversity loss and to validate models by comparing projections with observations. These limitations are detailed below. Fig 8.2 provides a summary of the directions in which each of these weaknesses is likely to affect the current projections on biodiversity loss, as well as the major types of challenges that have to be faced to overcome these uncertainties.

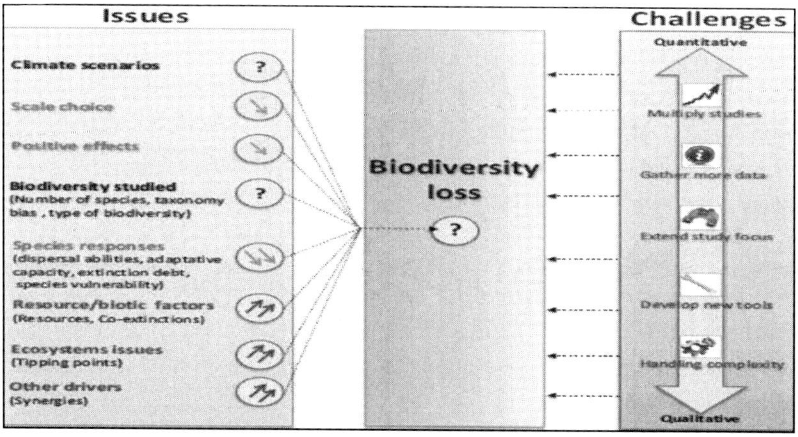

Fig 8.2: Issues and challenges for climate change driven biodiversity loss

Climate Scenarios

Climate scenarios depend on a wide range of socio-economic storylines for greenhouse gas emissions in the future, including the Special Report on Emissions Scenarios Millenium Assessment and Global Biodiversity Outlook scenarios and on a broad suite of General Circulation Models used to calculate climate change for given trajectories of greenhouse gas emissions. This means that the projections of species loss can yield highly contrasting results depending on the choice of combinations of emissions scenarios and climate models, independently of the model of biodiversity response that is used. In addition, internal climate model variability could result in greater differences in projected species_ distributions than variability between climate models. In addition, 4–39% of the world's landmass will experience combinations of climate variables that do not currently have equivalent values anywhere on the globe (so called no-analogue climates). One key challenge is to provide robust and credible uncertainty intervals for all model outcomes and, if possible, to reduce them.

Scale Choice

The choice of the spatial resolution scale is probably one of the most important factors generating variability. For example, a coarse, European-scale model (with 10' x 10' grid cells) predicted a loss of all suitable habitats during the 21st century, whereas a model using local-scale data (25 x 25 m grid cells) predicted persistence of suitable habitats in up to 100% of plant species. These differences are probably explained by the failure of coarser spatial scale models to capture both local topographic diversity and habitat heterogeneity. On one hand, global models can be used for a large number of species but focus on one type of specie's response and therefore lack biological realism. On the other hand, population or species models provide insight into a very limited range of species, typically at regional scales (i.e. adaptation phenology, dynamic population) but cannot provide global scale trends. This is a classical trade-off between precise small scale models and coarse large-scale models that lack biological realism.

Positive Effects

Climatic changes could also have positive effects on biodiversity. For instance, more clement temperatures and increased CO_2 are likely to be beneficial to many plants, resulting in an acceleration of biomass production. Milder winters might increase survival of many currently threatened species in temperate regions. Increased precipitation may also benefit some plant communities and species depending on them. Moreover, several studies reported detrimental effects of climate change on biological invasions. Although few studies report beneficial effects of global changes on biodiversity, they certainly exist and add to the difficulty of getting a clear overview of the effects of climate changes on the biodiversity of our planet.

Biodiversity Measures

Even in the most ambitious studies, the range of species studied always represents a small percentage of known biodiversity. All studies are taxonomically biased, as they generally concentrate on a few conspicuous taxonomic groups such as plants, mammals and birds with a particularly strong bias towards terrestrial vs. marine biodiversity. However, it is generally recognized that the vast majority of biodiversity in terms of species richness, evolutionary divergence, biomass and even ecosystem functioning is represented by "cryptic biodiversity", especially micro-organisms and insects. Similarly, there are important biases in data collection both across regions and ecosystems. Furthermore, most studies focus on species richness, because it is thought to influence the resilience and resistance of ecosystems to environmental change. However, a few studies have explored the impact of climate change on functional and phylogenetic diversity and the effects on genetic diversity are only beginning to be explored. Moreover, it is likely that different levels of diversity are affected differently by climate change, so these should be evaluated in parallel to provide a broad picture of biodiversity response to climate change.

In addition, there are different indicators of biodiversity change, such as the number of species 'committed' to extinction risk or change in abundances. The number of species 'committed' to extinction is probably not the most appropriate metric to forecast the future of biodiversity because the extinction debt could vary from decades to centuries. A complementary metric of biodiversity can be changes in MSA, which is an index defined as the mean abundance of original species relative to their abundance in undisturbed ecosystems. As extinction is only the last step in the process of species decline, the MSA can provide a metric to evaluate ecosystem degradation. In addition, we need new approaches to predict the impacts of climate change at the community level. Currently, neither the estimates of species losses nor of the qualitative impacts on ecosystem functioning and services are satisfactory to give a global overview of the impacts of climate change.

Limitations of Predictive Tools

Each of the modeling approaches reviewed has methodological, spatial and temporal limitations that constrain their predictive power. In general, models do not take into account the multiple responses of species to climate change, but rather focus mainly on one axis, spatial shifts of potential habitat. The temporal and the physiological responses are generally overlooked, as are the genetic and plastic capacity of species. In addition, species are commonly considered as static and independent entities, although their dynamics and their role in ecological networks are both known to be essential. However, a new generation of models is emerging that focuses on more realistic biological hypotheses and meets some of the challenges posed by each limitation.

Species Responses

Current global extinction models make very coarse assumptions about species responses. For example, the dispersal capability is a major issue for projections of future biodiversity. Until recently, models often addressed dispersal issues by using two extreme assumptions of either unlimited or no species dispersal (Thomas *et al.*, 2004). This is clearly convenient for practical purposes, but most species are between these extremes. In addition, exceptional occurrences of long distance dispersal are thought to have helped past species surmount prehistoric climate changes. Although rare, these events are of crucial significance in the current context, as they could in many cases make the difference between species survival and extinction, especially as human-mediated long-distance dispersion is now common for many organisms.

Inherent differences in vulnerability are not always taken into account when making large-scale correlative predictions of the future of biodiversity.

Indeed, most models work under the very strong hypotheses that species currently live in an optimal climatic niche and that they cannot survive if there is a change in the climatic conditions that contribute to define this niche. Such a hypothesis amounts to overlook the potential of adaptability of species. Numerous studies have shown that species are capable of fast responses through phenotypic plasticity or microevolution. Before being able to forecast species trajectories with some credibility, we therefore need to assess properly the vulnerability of species (including exposure, sensitivity, adaptive capacity and migration potential), habitats and regions to different components of climate change. Despite growing evidence for rapid adaptive evolution in response to climate change, the consequences of such evolution on species persistence remain to be explored. Currently, more flexible models of micro-evolutionary processes combine population-based information with phylogenetic comparative methods to provide an estimation of the evolutionary potential. Moreover, populations can be locally adapted to specific climatic conditions and therefore, models treating a species as a single homogeneous unit might be flawed. Consequently, studying under which circumstances losses in genetic and species diversity at local to regional scales occurred in the past could improve models outputs.

Population and Meta-Population Dynamics

An inherent property of current global extinction models is that they do not take into account population or meta-population dynamics that determine species distributions, population structure and extinction risk at a local scale. Models are often based on habitat reduction and do not offer a mechanism by which the species_ fitness is reduced and ultimately led to extinction. Recently, new mechanistic niche models have endeavored to estimate key fitness components such as survival, growth, development and reproduction as a function of species attributes (i.e. physiology, phenology, behaviour) that vary with climatic conditions.

Currently the probability of extinction is based on a projection of suitable habitats entirely lost (species will be committed to extinction) or partially lost (species will increase their extinction risk), although it is possible to combine niche models and abundance data. In fact, on one side some species are more likely to go extinct because they consist of small and disconnected populations and on the other side; some species have shown compensatory changes in demographic rates following climate change. The approach of population viability analysis (PVA) in this context could provide a good alternative. It is used to estimate the likelihood of a population's extinction within a given number of years through the species_ demographic characteristics and environmental variability, instead of species range shifts.

Only a few studies have used PVA to predict the effects of climate change on extinction risks in part because this approach requires data on population dynamics that are not available for most species. In addition, physio-demo genctic models could provide the most realistic method to forecast the future of key species, as they consider simultaneously demographic (e.g. growth, rate of survival and reproduction, migration processes, environmental stochasticity), physiological (e.g. metabolic rate) and genetics parameters (heritability and norm of reaction). However, studies using physio-demo genetic models to evaluate the impacts of climate changes remain scarce.

Co-Extinctions

Whether centered on a single species or taking into account large taxonomic groups, most studies and all models have disregarded inter-specific relationships such as competition, facilitation or mutualism.

Beyond single-species extinctions, both direct and indirect processes can lead to cascading and catastrophic co-extinctions, also called "chains of extinction". Despite the importance of inter-specific interactions, these relationships are exceptionally difficult to model; this is especially cogent in the context of lack of data on population dynamics and trophic webs (McMahon *et al.*, 2011). As each species comes with its cortege of specific parasites and symbionts, as well as many trophic relationships, the consequences of global change on biodiversity might be substantially underestimated when focusing on species-specific extinction rates. There is an urgent need not only to go beyond the single-species approach, but also to get past the species richness approach and to consider inter-specific interactions, trophic webs and ecological networks.

Synergies between Extrinsic Drivers of Extinction

Research on future species extinctions has so far mostly considered a limited set of the predicted changes. For example, sea level rise has seldom been considered, although the most recent scenarios projecting up to 2 m of rise by 2100 are raising new concerns for coastal and insular biodiversity. In addition, studies have mainly focused on the sole impact of climate change, sometimes considering a single additional driver (e.g. land-use change). These limitations may often lead to overly optimistic estimates. Indeed, other components of global change, such as habitat fragmentation, pollution, overexploitation and biological invasions have all been documented as major, additional threats for the future of biodiversity, with possible synergies, or reinforcing feedbacks between them. For instance climate induced facilitation of invasions can occur at all stages of the invasion process. In an experimental context, habitat fragmentation and overfishing combined with global warming have led to a decline in rotifer populations of up to 50 times faster than when either threat acts alone. In this case, dose-response relationship models could offer a powerful approach because they could simulate the impacts of many drivers of species extinctions.

Overall, despite important uncertainties and much conflicting imprecision, both under- and overestimating species loss, the very large underestimations due to co-extinctions, synergies and tipping points are extremely worrisome for the future of biodiversity (Fig 8.2). Given the attractiveness of quantitative projections from mathematical models, it is tempting to underestimate their limitations. To minimize this risk, it is crucial to use a variety of complementary approaches, ranging from observational studies (palaeontological data, phonological responses, adaptive capacity) to field and laboratory experiments (on genetic, species and ecosystem levels).

Biodiversity Management

The large variation of responses of different species necessitates the use and integration of multiple approaches to further our understanding of the impacts climate change can have on biodiversity. Similarly, our responses in terms of biodiversity management ought to transcend disciplines. Beyond this, global climate change prompts several methodological issues and the implications for conservation and management of biodiversity and ecosystem services.

Conservation of Species and Ecosystems

The large projected impacts of climate change on biodiversity at all levels mean that ecologists must quickly raise to the challenge of providing scientific guidance for the development of conservation strategies. A major role of conservation planning is to design reserve networks that protect biodiversity in situ. Currently few studies have attempted to use modeling for conservation purposes. It is increasingly important to protect the heterogeneity of habitats as well as genetic diversity within a species to sustain the capacity of a species to adapt. Also, the characteristics of protected areas, where planning has to be done decades in advance need to be reviewed under climate change.

Climate modeling can help to reevaluate the current set of protected areas, their places, size, layout and design. In particular, protection should be prioritized for places that minimize the effects of climate change-like forests, which contribute strongly to local climatic conditions-as well as for climate refuges for biodiversity.

Finally, model predictions of where, when and how future risks may affect species, biomes or ecosystems could assist in identifying the most appropriate conservation measures. For instance, species or ecosystem projected to be primarily affected by climate change may require adapted measures compared to species negatively affected by land-use change that could persist through protection of their remaining natural habitat. In many cases there may no longer be any overlap between a species- current range and its possible future range. As this situation will lead to extinction, climate change has been a major argument for the proponents of human assisted colonization. In this increasingly hot debate, case-by-case decisions have been advocated based on the balance between threatened status of a species and threat of that species for the recipient ecosystem, as well as the socioeconomic context in which conservation is taking place (Dawson *et al.*, 2011). A widespread view is that an important strategy is to enhance landscape connectivity to enable species to move through a matrix of interconnected habitats to favor escapes from unsuitable climatic conditions.

It is also essential to shift from a species centered focus to a holistic view encompassing species interaction networks, and other aspects of biodiversity such as functional and phylogenetic diversities. Beyond these various strategies, there is a growing call to go past the predictive focus and start aiming for an integrated and unified framework to identify species vulnerability and adapt biodiversity management interventions. In this regard, preventive actions are of foremost importance. For example, it should be remembered that the proportion of species extinction is a power function of the expected global warming (Hansen *et al.*, 2010).

Minimizing global warming could therefore have nonlinear effects in the preservation of species from extinction, with each tenth of degree avoided saving an increasing number of species. Acting to reduce global warming itself, and not only its effect on biodiversity, should remain a priority in conservation sciences. In addition, reducing other global change drivers could increase overall resilience of biodiversity in the face of climate change.

Ecosystem Services

Other aspects of biodiversity management will be affected by global change and will need adapting, including wildlife exploitation [e.g. forestry or fisheries], agronomy, pest and invasive species control or human and wildlife disease management. For example, major challenges in agronomy include the need to shift to species or varieties better adapted to particular components of climate change or to rethink strategies to control invasive and pest outbreaks, finding solutions in the increasing competition for water between the natural and the agricultural ecosystems, improving infrastructures and adapting cropping systems to meet future demands of a growing population living on poorer biodiversity resources.

Given the tremendous ecological impact of alien invasive species, and the expected exacerbation of invasion due to climate change, it is urgent to increase predictive power in this field. It is also crucial to move beyond predictions and to strengthen risk assessment, protocols of screening and of early detection, vector control and integrated management in area and/or of invasive species that will become at higher risk following climate change. Similar efforts are imperative for other drivers of biodiversity loss, such as overexploitation or habitat destruction.

The issues which represents the main factors of uncertainty and the direction of the likely current bias (green factors and arrow down is a likely overestimation, while red factors and arrow up is a likely underestimation; both can occur simultaneously, according to the approach taken in different studies; black factors and question marks are for unknown direction of misestimating; double arrow means very large expected effect. Challenge types range from quantitative (such as increasing the number of studies and the quantity of available data, or extend the focus of studies to include more factors to more qualitative ones (developing new tools and progressing in resolving complexity).

Adaptation Response Options

To increase the resilience of biodiversity and ecosystem service delivery under future climate conditions, synergies could be developed during adaptation planning and implementation between biodiversity, poverty reduction and development objectives. Specifically, the potential of biodiversity and ecological infrastructure to achieve sector-specific adaptation and development outcomes/benefits would need to be mainstreamed into adaptation, development and poverty reduction processes and strategies at national and local level. Coordination across sectors (including the land use planning sectors) would be essential in this mainstreaming process. The appropriate and specific types of local action required would need to be further defined in biome specific adaptation plans. Key elements of this would be to include strong stakeholder engagement and implementation focused at local level; prioritizing low-cost approaches with multiple benefits; integrating adaptation and mitigation responses; and making use of indigenous knowledge.Ecosystem-based Adaptation and expansion of protected areas using climate-resilient approaches offer two adaptation response options for the biodiversity sector that are appropriate for achieving increases in the climate resilience of biodiversity and maintaining and/or enhancing ecosystem service delivery. However, these two approaches should be adapted as necessary to build the resilience of ecological infrastructure to support economic sectors and livelihood activities. For example:

Agriculture and forestry hold great potential for gaining adaptation benefits while mitigating greenhouse gas emissions through reforestation and restoration, managing soil carbon and invasive species, implementing integrated crop and livestock management, and improving management of emissions from livestock and crop production. Specific related approaches include climate smart agriculture, conservation agriculture, agro-ecology and community-based adaptation.

Fisheries that are successfully managed to achieve resource sustainability will be better positioned in the long term to adapt to the effects of climate change. An Ecosystem Approach to Fisheries can contribute to resource recovery through protection of spawning and nursery areas and the maintenance of other essential fish habitats which are in turn dependent on water flows that are mediated by land-based ecosystems.

Human health and ecosystem impacts that are associated with increased rainfall intensities, flash floods and regional flooding include overflowing sewers and climate change impacts on water resources (quality and quantity). These can be addressed through collaboration between the health, water and biodiversity sectors on integrated water resource management, restoration of catchment areas and effective land-use planning.

Developing a climate-resilient network of protected areas with greater ecological connectivity, and designed specifically with future climate conditions in mind will work towards assisting with species migration and the maintenance and/or enhancement of ecosystem service delivery to support local communities and all sectors. Implementing efficient biodiversity stewardship programmers can assist in achieving connectivity between core biodiversity and ecosystem services areas.

References

1. Alo CA, Wang GL. Potential future changes of the terrestrial ecosystem based on climate projections by eight general circulation models. J Geophys. Res. Biogeosci. 2008; 113:16.

2. Anderson BJ, Akcakaya HR, Araujo MB, Fordham DA, *et al.*, Dynamics of range margins for metapopulations under climate change. Proc. R. Soc. B. Biol. Sci. 2009; 276:1415–1420.

3. Bakkenes M, Alkemade JMR, Ihle F, Leemans R, Latour JB. Assessing effects of forecasted climate change on the diversity and distribution of European higher plants for 2050. Global Change Biol. 2002; 8:390–407.

4. Barnosky AD, Matzke N, Tomiya S, Wogan GOU, *et al.*, Has the Earth s sixth mass extinction already arrived? Nature. 2011; 471:51–57.

5. Bradley BA, Blumenthal DM, Wilcove DS, Ziska LH. Predicting plant invasions in an era of global change. Trends Ecol. Evol. 2010; 25:310–318.

6. Brook BW, Sodhi NS, Bradshaw CJA. Synergies among extinction drivers under global change. Trends Ecol. Evol. 2008; 23:453–460.

7. Brook BW, Akcakaya HR, Keith DA, Mace GM *et al.*, Integrating bioclimate with population models to improve forecasts of species extinctions under climate change. Biol. Lett. 2008; 5:723–725.

8. Campbell A, Kapos V, Scharlemann JPW, Bubb P *et al.*, Review of the literature on the links between biodiversity and climate change: impacts, adaptation and mitigation. In: CBD Technical Series no. 42 (ed. Diversity Sot CoB). Secretariat of the Convention on Biological Diversity, Montreal, 2009, 124.

9. Carnaval AC, Hickerson MJ, Haddad CFB, Rodrigues MT, Moritz C. Stability predicts genetic diversity in the Brazilian Atlantic forest hotspot. Science. 2009; 323:785–789.

10. Chevin LM, Lande R, Mace GM. Adaptation, plasticity and extinction in a changing environment: towards a predictive theory. PLoS Biol. 2010; 8:e1000357.

11. Dawson TP, Jackson ST, House JI, Prentice IC, Mace GM. Beyond predictions: biodiversity conservation in a changing climate. Sci. 2011; 332:453–58.

12. Gilman SE, Urban MC, Tewksbury J, Gilchrist GW, Holt RD. A framework for community interactions under climate change. Trends Ecol. Evol, 25, 325–331.

13. Hansen L, Hoffman J, Drews C, Mielbrecht E. Designing climatesmart conservation: guidance and case studies. Conserv. Biol. 2010; 24:63–69.

14. He F. Hubbell SP. Species–area relationships always overestimate extinction rates from habitat loss. Nature. 2011; 473:368–371.

15. Leadley P, Pereira HM, Alkemade R, Fernandez-Manjarres JF *et al.*, Biodiversity scenarios: projections of 21st century change in biodiversity and associated ecosystem services. In: Secretariat of the Convention on Biological Diversity (ed. Diversity Sot CoB). Published by the Secretariat of the Convention on Biological Diversity, Montreal, Technical Series no. 50. 2010, 1–132.

16. McMahon SM, Harrison SP, Armbruster WS, Bartlein PJ, *et al.*, Improving assessment and modelling of climate change impacts on global terrestrial biodiversity. Trends Ecol. Evol. 2011; 26:249–259.

17. Thomas CD, Cameron A, Green RE, Bakkenes M, *et al.*, Extinction risk from climate change. Nature. 2004; 427:145– 148.

18. Sitch S, Huntingford C, Gedney N, Levy PE *et al.*, Evaluation of the terrestrial carbon cycle, future plant geography and climate-carbon cycle feedbacks using five Dynamic Global Vegetation Models (DGVMs). Global Change Biol. 2008; 14:2015–2039.

Chapter - 9
Climate Change and Insect Pest Dynamics

Authors

Dr. Umesh Chandra

[1]Narendra Dev University of Agriculture &Technology, Kumarganj, Faizabad, Uttar Pradesh, India

M.P Gautam

Sardar Vallabhbhai Patel University of Agriculture & Technology, Meerut, Uttar Pradesh, India

Chapter - 9

Climate Change and Insect Pest Dynamics

Dr. Umesh Chandra and M.P Gautam

Climate change and global warming will have serious consequences for the diversity and abundance of arthropods, and the extent of losses due to insect pests, which will impact both crop production and food security. Prediction of changes in geographical distribution and population dynamics of insect pests will be useful for adapting pest management strategies to mitigate the adverse effects of climate change on crop production. Pest outbreaks might occur more frequently, particularly during extended periods of drought, followed by heavy rainfall. Some of the components of pest management such as host-plant resistance, bio-pesticides, natural enemies, and synthetic chemicals will be rendered less effective as a result of the increase in temperatures and UV radiation, and decrease in relative humidity. The rate of insect multiplication might increase with an increase in CO_2 and temperature. Thus climate change might change the population dynamics of insect pests differently in different agro-ecosystem and ecological zones. Climate change related factors like rise in temperature, changes in precipitation patterns, milder and shorter winters, rise of sea levels and increased incidence of extreme weather events can directly influence insects by affecting their rate of development, reproduction, distribution, migration and adaptation. In addition, indirect effects can occur through the influence of climate on the insect's host plants, natural enemies and inter-specific interactions with other insects. The impacts include changes in phenology, distribution and community composition of ecosystem that finally leads to extinction of species. As insects represent huge numbers of taxa and individuals, with their short generation times, high mobility and high reproductive rates, they will respond more quickly to climate changes than long lived organisms, like higher plants and mammals. In fact, insects may be the first predictors of climate change.

One of the major constraints globally to increased food production and food security is the action of crop pests, whether native or introduced species. For example, reported global losses in several crops, varying from 50% in wheat to over 80% in cotton; for maize, rice and potatoes losses were

31, 37 and 40%, respectively. Further, according to this report, weeds generally caused the highest potential loss (34%), with animal pests and pathogens responsible for losses of 18 and 16 percent respectively. Today, one of the major factors facilitating the movement and establishment of pests into new environments is that of climate change. While extreme weather changes and events now make front page headlines globally because of the increasing incidence of flooding, drought and heat waves in major cities where such events had only rarely occurred in the past, there is also increasing concern of the effects of changes in climate on agriculture and food security. One such concern is the impact of climate change driven or, more particularly, global warming driven movement of pest species into areas from which they had previously been unable to colonize because of adverse climatic and/or other environmental factors. Such adverse factors which population ecologists have classified as abiotic and biotic, impact on pest population growth, limiting the ability of any species to achieve full reproductive potential or maximum population levels. While climatic or abiotic factors are primarily important in this regard, they have been shown to sometimes have the opposite effect, producing more favorable conditions for the distribution, survival and establishment of pest species into new environments.

The population dynamics is the aspect of population ecology dealing with factors affecting changes in population densities. The seasonal effects of weather and ongoing changes in climatic conditions will dispersal and development of insect species. The changes in surrounding temperature regimes certainly involve alterations in development rates, voltinism and survival of insects and subsequently act upon size, density and genetic composition of populations, as well as on the extent of host plant exploitation Bale *et al.*, 2002. The change in the environment affects the pest population dynamics in two ways either directly or indirectly by altering the host physiology. Host availability and the probability of pest outbreaks are further determined by the incidence and character of abiotic disturbances. The developmental success of insect herbivores also indirectly depends on climate, as environmental parameters impact on plant physiology. Insects and plants are exposed to complex interactions among changes in temperature, precipitation and, increased levels of CO_2 and variations in nutrient availability.

Insect Distribution

Environmental factors influenced by global climate change determine the distribution ranges of organisms. It plays a major role in defining the distribution limits of a species. It is predicted that the distribution of most insect species will shift towards the poles and to higher elevations with predicted temperature increase due to climate change and temperate regions will bear the main burden of these shifts. With changes in climate, these limits are shifting as species expand into higher latitudes and altitudes and disappear from areas that have become climatically unsuitable (Parmesan,2006; Menéndez, 2007). Climate change will alter the distribution of many species in different taxa (Hickling *et al.*, 2005) especially, ectothermic animals are expected to shift their distribution ranges northwards in the next hundred years or so (Vanhanen *et al.*, 2007). It has been recognized that global warming affects the individual species and communities in a form of range shifts and extinctions (Walther *et al.*, 2002, Battisti *et al.*, 2005). The increasing winter temperatures have been proposed to be the key factor affecting range shifts in insects by reducing winter mortality (Battisti *et al.*, 2005). As a result of temperature increase, the ranges of species could expand poleward and in the mountainous areas also, an upward elevation takes place because the number of insect species is inversely related to latitude and elevation from the sea level (Hickling *et al.*, 2005). According to a study conducted on 1100 insect species, climate changes due to global warming may cause 15-37% of those species to extinct by 2050 (Hance *et al.*, 2007).

It is now clear that poleward and upward shifts of species ranges have occurred across many taxonomic groups and in a large diversity of geographical locations during the 20th century. According to one survey of about 1600 species, about 940 of them showed the effects of climate change. For instance, in Europe, 35 species of butterflies have already shifted their ranges 35-240 km northward. In California, 70% of 23 butterfly species now start their first flight about 24 days earlier than they used to do 31 years ago (Parmesan, 2007). Parmesan and Yohe (2003) reported that more than 1700 Northern Hemisphere species have exhibited significant range shifts averaging 6.1 km per decade towards the poles (or 6.1 m per decade upward). At the same time effects of defoliators, wood borers and bark beetles could become more detrimental due to prolonged growing season leading to multivoltinism, absence of extreme temperatures in winter that diminish population levels and possible shifts to novel host plants (Battisti *et al.*, 2006 and Stastny *et al.*, 2006).

Climate change will alter the ranges and abundances of insects and therefore have profound impacts on agriculture by the movement of existing crop pests into new areas and potentially, by rising currently disregarded insect species to pest status. The range and abundance of insect changes under global warming would not necessarily be derived from the physiology of individual insect species. Species may increase in abundance at physiologically non-optimum temperatures raising the concern that rare species currently not regarded as pests may become economically important with global warming (Jenkinson *et al.*, 1996).

Insect Phenology

Phenology is the timing of biological events such as seasonal activities of plants and animals like flowering or breeding. These events are dictated by photoperiod, temperature or other stimuli. Many insects which are in synchronization with such events are affected if global warming initiates changes in phenology. It is one of the easiest impacts of climate change to monitor and is by far the most documented in this regard for a wide range of organisms (Root *et al.*, 2003). With increased temperatures, it is expected that insects will pass through their larval stages faster and become adults earlier. Therefore, expected responses in insects could include an advance in the timing of larval and adult emergence and an increase in the length of the flight period (Menéndez, 2007). Members of the order Lepidoptera are the best examples of such phenological changes.

Changes in butterfly phenology have been reported in UK, where 26 of 35 species have advanced their first appearance (Roy and Sparks, 2000). First appearance for 17 species in Spain has advanced by 1-7 weeks in just 15 years (Stefanescu *et al.*, 2003). Seventy percent of 23 butterfly species in California, USA have shown an advancement of first flight date of approximately eight days per decade (Forister and Shapiro, 2003). Early adult emergence and an early arrival of migratory species have also been reported for aphids in the UK (Harrington *et al.*, 2007). Gordo and Sanz (2005) investigated climate impacts on four Mediterranean insect species viz. butterfly, bee, fly and beetle and indicated that all species exhibited changes in their first appearance date over the last 50 years, which was correlated with increases in spring temperature. Parmesan and Yohe (2003) estimated that more than half (59%) of 1598 species investigated exhibited measurable changes in their phenologies and/or distributions over the past 20 to 140 years. They also estimated a mean advancement of spring events by 2.3 days/decade based on the quantitative analyses of phenological responses for these species. In a similar quantitative study, Root *et al.*, (2003) estimated an advancement of 5.1 days per decade. Degreeday or phenology based models are often used to predict the emergence of insects like cabbage maggot, onion maggot, European corn borer, Colorado potato beetle etc. and their potential to damage crops. Increased temperatures will accelerate the development of these types of insects resulting in more generations and possibly more crop damage per year.

PEST Dynamics and Ecology

The establishment and survival of any introduced species depend on a range of factors, both biotic and abiotic, in its new habitat and the establishment of any such species to pest status is equally dependent on these factors. Simmonds and Greathead (1977) argue that while introduced species are often the more important pests in many areas, this will depend, inter alia, on their taxonomic grouping (mites, aphids and other sucking insects are more likely to achieve pest status), geographical area (islands as opposed to continents), crop, and means of dispersal (small species can be transported undetected on living plant material, fruits, tubers etc. in trade or dispersed naturally, by air currents for example).

Density Dependent and Density Independent Factors and Pest Population Size

Population ecologists distinguish between density dependent and density independent effects which can regulate the size of the population (Fig 9.1). Density dependent effects alter the birth rate or death rate of a population as a function of the population density or size, i.e. they become more effective at high population density and less effective at low population density. Such density dependent factors may result from intra- or inter-specific competition for food, living space, etc. or as a result of the impact of natural enemies (predators and parasites). These factors change in effectiveness as the population density changes and may positively or negatively affect final population size. For example, as the pest population size grows, the effect of natural enemies will increase (more available food) to the point where there are not sufficient hosts to support the increasing natural enemy population itself. As the natural enemy population itself now decreases, the pressure on the host population is removed, so allowing its build-up once again. Studies have shown that such density dependent factors are usually the most critical factor in the regulation of populations. Density independent factors, on the other hand, impact the population irrespective of the population density or size; for example, extreme weather conditions or application of a pesticide could decimate a population whatever its size.

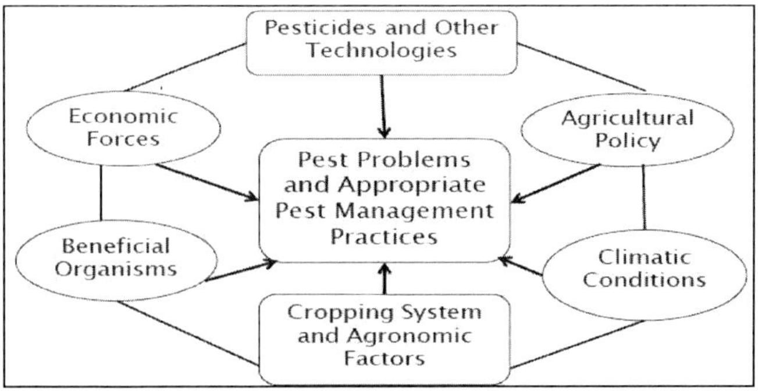

Source: Norton and Mumford, 1993

Fig 9.1: The many factors that affect pest status, damage and pest management

Beneficial Organisms

All organisms have their natural enemies (predators and parasites and which may be referred to as beneficial organisms) which help to keep their populations in check (density dependent effects). However, when a new species with the potential to achieve pest status is introduced into a new environment, this introduced species is usually unaccompanied by its natural enemies or, any existing natural enemies in that environment are so inefficient in effecting a decline in population numbers, that a rapid build-up of the pest population results leading to the establishment of the pest and a subsequent a pest outbreak situation. However, natural enemies may be manipulated to have a greater negative impact on their pest host populations. Usually, these natural enemies are sourced from the home of origin of the introduced pest species in sufficient numbers for their release into the new ecosystem of the pest (classical biological control) or alternatively, they may be artificially reared in a laboratory and released back into the environment as a controlling agent (applied biological control). Whatever the approach, the objective is to establish the density dependent interaction between pest host and natural enemy to the detriment of the pest population.

Climatic Conditions

Climate or weather on pest populations has traditionally been with regard to their role as density independent factors. However, climate variability is now the focus, having major impacts on pest biology, pest movement and distribution, survival and adaptation to new environments and on crop production.

Climate Change and Crop Pests

Climate and weather influence the functioning of any agro-ecosystem, affecting the growth, development and survival of crops, pest organisms and their natural enemies. Additionally, climate change and climate variability are major determining factors for the spread and distribution of crops, pests and natural enemies. Global warming together with changing rainfall patterns and other weather variables which result from climate change are most important in affecting the movement and survival of these organisms. The population dynamics of insect crop pests, for example, may be severely affected by these climate variables which may directly impact their survival, development, distribution and abundance; also, these climate variables affect host crop plants and natural enemies of the pest as well.

Temperature, particularly, will affect the distribution and establishment of insect pest populations in new geographic areas which may have been previously unsuitable because of unfavorable climatic conditions, especially in higher and lower latitudinal regions as historical climate barriers are breached; the same holds for high altitudes as well. However, successful establishment of such pests will also be constrained by the presence or absence of their host plants and what control any natural enemies could have on their populations bearing in mind that the later species may themselves be subject to climate constraints. However, Gregory *et al.*, (2009), for example, stated that "the extent to which crop pests and pathogens have altered their latitudinal ranges in response to global warming is largely unknown".

At an FAO conference on climate change and transboundary pests (FAO, 2008), several reasons were suggested for climate change resulting in a higher probability of the establishment and spread of such pests. Climate change will result in a higher probability of entry, establishment and spread of vector-borne diseases of animals, parasites of animals with free-living life stages, and pests of plants, diseases of fish and invasive alien aquatic species for the following reasons: Climate change will create winners and losers. For some animal and plant pests and diseases and invasive alien aquatic species, the climate will become more conducive and for others the meteorological conditions will become less favorable. This will result in unstable situations with a high probability of entry and establishment in areas that are presently protected by unsuitable conditions. Meteorological and related environmental circumstances may change the geographical distribution of host species, putting them in contact with animal and plant pests and diseases of related hosts to which they do not possess resistance. New animal and plant pests and diseases may emerge due to evolving selection and adaptation to new situations.

Any consideration of the impact of meteorological conditions on crop/pest interactions must be viewed as being more of a combination of several separate weather parameters rather than the effect of any one single factor. IPCC (2007), for example, quote several reports which suggest that CO_2-temperature and CO_2-precipitation interactions are "key factors" in determining plant pest damage. Of course extreme weather events like drought or flooding will have a one-off major impact on both crops and pest populations at that particular time of occurrence but the effects may persist over a long period. Also, it must be emphasized that while meteorological factors affect both crop and pest, they do not do so in the same fashion. Whether crops or pests, each requires certain conditions of water (rainfall), light and temperature for optimal growth and development, hence any change in these will have an adverse effects on their survival. Insect pests tend to favors higher temperatures and some pest species such as, thrips and mites, proliferate under drier conditions but high levels of precipitation may cause a reduction in their populations. Good soil moisture enhances nematode populations while flooding is a deterrent to soil pests like the banana weevil (Cosmopolites) or burrowing rodents. Temperature, high humidity, dew, wind and air circulation patterns all influence the growth, survival and spread of crop pathogens; e.g. fungal spores generally require high humidity and moisture to germinate.

Effects of Climatic Parameters on Population Dynamics

Climate change is likely to involve a higher frequency of abiotic disturbance. Depending on the dimension of the disturbance, local to regional dynamics of insect populations and species composition will be strongly affected insects find optimum conditions for development within a certain range of temperature limited by species specific lower and upper developmental thresholds. These alteration and gradual changes might affect the population dynamics parameters.

Development and Reproduction

As long as species optima for development are not exceeded, there might be positive direct responses of insects to increasing temperature conditions, such as enhanced reproductive potential, are to be expected. Polyvoltine species, e.g. some economically important bark beetles like Ips typographus, will profit from accelerated development rates allowing for an earlier completion of life cycles and the establishment of additional generations within a season (Lange *et al.*, 2006; Jönsson *et al.*, 2009). Temperatures above the specific optimum range lead to decreased growth rates, reduced fecundity and increased rates of mortality for a multitude of species Promoted by milder winters and decreasing frequencies of temperature extremes, enhanced reproductive capacity and changes in distribution (if host plants will be available) are to be expected for a variety of pest species. For instance, increased survival of hibernating eggs in fringe areas of occurrence is prognoses for the winter moth, Operophthera brumata and the autumnal moth, Epirrita autumnata and the gypsy moth, Lymantria dispar in North America. Temperature increase coupled with reduced snowfall and earlier snow melt may on the one hand lead to decreased overwinter survival of species hibernating under the protection of snow.

Diapauses and Winter Mortality

A phase of dormancy is essential for many species of temperate regions to complete their life cycles and to endure the low temperatures of the winter season. Increased temperature conditions may be beneficial for species with low frost resistance (like individuals 2 of the anholocyclic development of Elatobium abietinum) and those, which are actively feeding during winter (like the pine processionary moth, Thaumetopoea pityocampa). They are not beneficial for species which do need low temperatures to manifest diapauses or to increase their frost resistance Bale *et al.*, (2002).

Migration and Movement

The onset of flight as well as the extent and endurance of dispersal are essential parameters in the phenology of insect herbivores, impacting the timing and intensity of mating, host finding and colonization and brood establishment. Temperature thresholds for insect flight vary both among and within species, with season and also with region. Different thresholds have also been described for different phases of flight activity. In the aphid Aphis fabae, 17°C is required for take-off Johnson *et al.*, (1957), 15°C for sustained upward flight, 13°C for horizontal flight and 6.5°C for wing beating Cockbain, (1961). The regional spread of mountain pine beetle, Dendroctonus ponderosae outbreaks in British Columbia might for instance primarily be due to the dispersal of epidemic populations Aukema *et al.*, (2008). Besides their physiological constitution, insects are influenced in their flight activities by environmental conditions, such as the actual weather situation. Furthermore, insect emergence and first appearance after hibernation may depend on a combination of day length and temperature thresholds. E.g., the spring swarming of the European spruce bark beetle, Ips typographus after its termination of reproductive diapause can be predicted accurately in consideration of a specific thermal threshold for flight activity and a certain accumulated effective temperature sum above a photoperiod threshold Baier *et al.*, (2007). The geographic range of many insect species of the temperate zones is determined by temperature thresholds of frost resistance. Decreasing snowfall might for instance promote the recent expansion of the pine processionary moth, Thaumetopoea pityocampa into high elevation stands of mountain pine. More than 50 species of butterflies showed northward range expansions and ten species of previously migrant butterflies established on Nansei Islands during 1966–1987. Global warming may be responsible for the recent decline in abundance of Plutella xylostella and the increase in Helicoverpa armigera and Trichoplusia Ni Kiritani, (2006).

Effect Temperature on Survival Rate of Insects

Warm temperature halved the time required to reproduce (Spruce beetle, Dentroctonus rufipennis). The population of pine beetle is reduced drastically at -16°C (lethal) but better survival and distribution at warm winter In case of aphids a 2°C temperatures increase causes one to five additional life cycles per season [17]. The adult survival of the brown plant hopper, Nilapavata lugens remained almost unchanged between 25 and 35°C, but was drastically reduced at 40°C. The oviposition of females at 35 and 40°C was relatively higher than at 25 and 30°C, but egg survival was markedly reduced at 35°C. At the higher temperatures, durations of pre-oviposition periods were also reduced.

The global warming is likely to increase BPH abundance in areas with temperatures below 30°C. Survival of the different stages of the rice leaf folder Cnaphalocrocis medinalis was greatly affected at 35°C. Adults emerging from pupae reared at 35°C were unable to lay eggs. The upper temperature threshold for survival of this species appears to lie between 30 and 35°C. Shifts in climate can differentially affect the development rates of pest and predator species. In addition, temperatures can affect predator search. The egg predator (Cyrtorhinus lividipennis) had increased instantaneous attack rates and decreased handling times with increasing temperatures until 32°C. At 35°C the attack rate and handling time decreased drastically. This implies that predator activity is likely to increase with increasing temperatures up to a critical temperature of about 35°C. In addition to affecting biological characters, global warming may cause temporal asynchrony between interacting populations Heong *et al.*, (1995).

Insect may also evade climate change stress through changes in life cycle. Insect populations from environments with higher temperatures may have higher fecundity and shorter growth stage durations to increase fitness.

The egg duration was 10.4 days at 25°C and 7.9 days at 27-28°C (Dhillon and Sharma, 2007). Helicoverpa armigera and day degrees required for egg hatching decreased with an increase in temperature from 10 to 27°C and egg age from 0 to 3 days. The DD Degree Days requirements were highest for 0-day-old eggs at 10°C and lowest at 27°C. At 27°C with a hatchability is >75.0%. (Dhillon and Sharma, 2007). In Rice ear head bug, Leptocorisa acuta, the 0.5 and 2°C rise in daily average caused no affect on the generation time but 3°C rise caused 1 to 3 days increase in generation time. 0.5 to 1°C rise reduced the population of non wing-padded nymphs, wing padded nymphs and adults (Reji and Chander, 2008).

Effect of Temperature on Growth Rate of Insects

The rate of development of pests will enable a more rapid response to a change in temperature. The population of Nephotettix cincticeps will increase by a ratio of from 3 to 4. Global warming has seemingly similar influences on the abundance of C. suppressalis and N. cincticeps. An increase in winter temperature enhances the abundance of both species. However, the increment is much larger for N. cincticeps than for C. suppressalis. Such a difference may be related to the difference in the number of generations per year. N. cincticeps has twice the number of generations, which may cause a sensitive response to the change in temperature Yumamura *et al*., (2006).

Winter mortality of adults of Nezara viridula and Halyomorpha halys is predicted to be reduced by 15% by each rise of 1°C and in C. suppressalis two generation per year after 2°C warming Kiritani, (2006). The infestation of mustard aphid will change with the variation in the degree days. Thermal time calculations for mustard aphid Accumulation of 250 or more degree days in January at New Delhi caused less aphid infestation. If the degree day is 200 or less it favours the higher attack. Slower accumulation of heat units promotes aphid infestation Chakravarty *et al*., (2002). The rise in temperature beyond 3°C, affects the population growth of rice ear head bug Leptocorisa acuta in (Reji and Chander, 2008).

Global Warming vs Species Distributions

Distributions of spruce budworm Choristoneura fumiferana under climate change they will shift towards the poles, distribution shifts may be good or bad, depending on the species and the regions concerned.

Precipitation and Drought

There are few scientific evidences on the effect of precipitation on insect pest population and their growth. Some insects are sensitive to precipitation and are killed or removed from crops by heavy rains. However, some insects that over-winter in soil, such as cranberry fruit worm and other cranberry insect pests, flooding the soil has been used as a control measure (Vincent *et al.*, 2003). It is predicted that more frequent and intense precipitation events forecasted during climate may change have negative impacts on these insect pest population. Similar to temperature, precipitation changes can impact insect pest predators, parasites and diseases resulting in a complex dynamic manner. Fungal pathogens of insects are favored by high humidity and their incidence would increase by climate changes that lengthen periods of high humidity and reduced by those that result in drier conditions.

Enhanced summer rainfall and drought conditions on soil dwelling Agriotes lineatus (wireworms) promote rapid increase in the population of wireworms in the upper soil Staley *et al.*, (2007). The deviation of rainfall during monsoon and November and its relationship with level of Helicoverpa armigera damage severity showed higher November rainfall favored higher infestation. Average rainfall is predicted to decrease in several regions and the occurrence of summer droughts is likely to increase. Root herbivore species responded differently to the summer rainfall manipulations. The larvae of the dominant root-chewing species, Agriotes lineatus, were more numerous under enhanced rainfall, in contrast, abundance of the Coccoidea lecanopsis formicarum was unaffected by the rainfall manipulations. The responses of root herbivores to an increased incidence of summer droughts are therefore likely to vary, depending on their feeding strategy and life history Staley *et al.*, (2007). Drought conditions severely affected egg viability of Scopelosaurus Lepidus eggs and did not hatch at all under very dry conditions Johnson *et al.*, (2010).

Effect of Temperature on Voltinism

Climate change can cause major changes to the dynamics of individual species and to those communities in which they interact. One effect of increasing temperatures is on insect voltinism, with the logical assumption that increases in surface temperatures would permit multivoltine species to increase the number of generations per year. The relationship between climate change and voltinism could be complex. Effect of increasing temperatures is on insect voltinism, with the logical assumption that increases in surface temperatures would permit multivoltine species to increase the number of generations per year. The phenology based model for grape berry moth, Paralobesia viteana (Clemens), predicted that the increases in mean surface temperatures > 2°C can have dramatic effects on insect voltinism by causing a shift in the ovipositional period that currently is subject to diapause-inducing photoperiods.

Effects of Climate Change on Insect Pests via Host Plants

The nutritional quality of plant tissue for insect herbivores generally increases with nitrogen content and decreases with lower water content and rising concentrations of secondary metabolites. The atmospheric CO_2 is likely to stimulate plant defense and resistance to the colonization of phytophagous insects. Under moderate water stress, turgor pressure and water content of plant tissue decrease, leading to tougher foliage. Under increased concentrations of tropospheric CO_2, plants react by enhanced productivity due to direct fertilization effects Lindroth *et al*., (1993).The larvae fed on 700 and 550 ppm CO_2 foliage exhibited greater consumption. The final larval dry weights were also higher on elevated CO_2 foliage. Higher significant approximate digestibility (AD) foliage and of relative consumption rate (RCR) were observed on elevated CO_2 foliage. Lower efficiency of conversion digested food (ECD), efficiency of conversion ingested food (ECI) and relative growth rate (RGR) of larvae were noticed on elevated CO foliage than chamber and open ambient foliages. Under elevated CO_2 lower N concentration, higher C, C/N ratio and ployphenols was observed Rao *et al*., (2009). Exposure to elevated CO_2 increases significantly western corn rootworm (Diabrotica virgifera) adults (foliage chewer) and soybean aphids (Aphis glycines). Higher concentrations of CO_2 in the atmosphere may increase herbivory in the soybean agro-ecosystem Dermody *et al*., (2008).

Effect of Climate on Pest Population via Natural Enemies

Shifts in climate can differentially affect the development rates of pest and predator species. In addition, temperatures can affect predator search. This way it affects the insect population dynamics indirectly. Normal increases in metabolic rate with increasing temperature are accompanied with an increase in developmental rate. Hence, in biological control systems of insect pests utilizing parasitoids, the temperature response of the parasitoids determines their success in controlling the pest population. The egg predator Cyrtorhinus lividipennis of BPH Brown Plant Hopper had increased instantaneous attack rates and decreased handling times with increasing temperatures until 32°C. At 35°C the attack rate and handling time decreased drastically. This implies that predator activity is likely to increase with increasing temperatures up to a critical temperature of about 35°C. In addition to affecting biological characters, global warming may cause temporal asynchrony between interacting populations. Although natural selection will tend to increase synchrony between hosts and parasitoids, asynchrony may occur if host and parasitoid respond differentially to changes in weather patterns.

Asynchrony may introduce a partial refuge effect that can reduce parasitism. However, the lack of temporal coincidence between searching parasitoids and hosts can contribute to stability. Insect may also evade climate change stress through changes in life cycle. Insect populations from environments with higher temperatures may have higher fecundity and shorter growth stage durations to increase fitness. For example, in BPH, egg duration was 10.4 days at 25°C and 7.9 days at 27 to 28°C (Heong *et al.*, 1995). The parasitoid Campoletis chlorideae developed successfully over the temperature range of 12-37°C. However, the developmental period was found to be inversely correlated with temperature in the range of 12–37°C. The percentage pupal mortality of C. chlorideae increased above and below 22°C, with the highest mortality rate occurring at 37°C. As temperature increased, longevity decreased accordingly. The highest longevity of parental females (17.2 ± 3.6 days) was recorded at 12°C and longevity decreased continuously with rising temperature, with the longevity of adult female parasitoids being less than 4 days at 37°C. The pattern of daily reproduction and the estimates of other life-table statistics of C. chlorideae demonstrate that its performance decreased below and above 22°C.

Climate Change Brings Additional Complexity to Crop-Pest Systems

Complexity is intrinsically high in agricultural systems, and remains the main barrier to their study and management, with climate change and invasive species being additional factors that complicate management issues further point out that understanding and managing insects under climate change is vexing as contradictory results likely arise in field studies, large-scale experiments are challenging, and driving factors include interactions with other species such as host plants, competitors and natural enemies that each respond differently to climate change. This latter point is illustrated in Fig 9.2 which shows that control by a natural enemy might be compromised by a 2^{0}C increase in temperature that decreases its growth rates relative to that of the pest herbivore. As a result management strategies for critical pest problems often fail because holistic analyses that underpin sound decision making at the field level are unavailable.

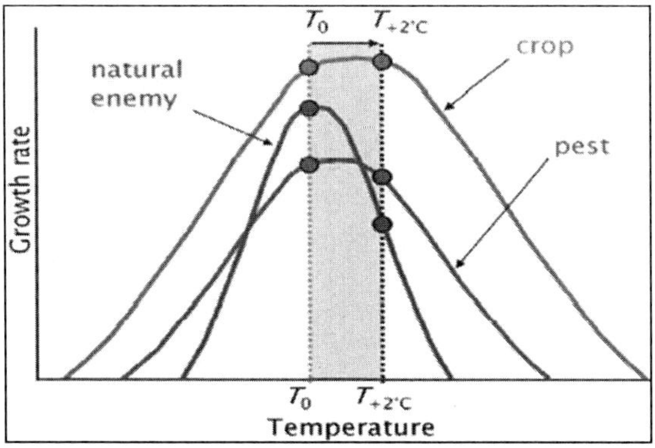

Fig 9.2: Plots of the growth rates on temperature of a crop, pest and natural enemy to intuitively illustrate how an increase in average temperature from T_0 to $T_{+2°C}$ may affect the physiological basis of trophic interactions. With temperature warming, biological control of the pest may decrease due to the narrower temperature tolerance of the natural enemy compared to the pest. Source: Gutierrez *et al.*, (2010).

The issue of climate change and its impacts is one of the most challenging global issues today with the most visible effects being manifest in the many extreme weather events that are now experienced so regularly in nearly all geographic regions of the world and which are responsible either directly or indirectly in several tragedies in so many countries. Less newsworthy, perhaps, but of extreme concern to its many stakeholders are the impacts of changing climatic factors on agriculture. Also of concern to the global agriculture community is the problem of invasive pest species or new pest introductions and their establishment in areas where they had not been historically present. While this phenomenon of pest movement is not a new occurrence, however, with the increased movement of peoples and trade between countries, this has provided an artificial pathway mechanism for the movement of pests, either inadvertently or deliberately. Additionally, pests could also move naturally into new regions. However, what is very worrisome today is the facilitation and exacerbation of this problem of the distribution of new and invasive pest introductions as a result of changing climatic conditions. Such climate variability may enhance the establishment of pests into areas which had been previously unfavorable for survival and establishment because of unsuitable climatic conditions.

One of the constraints to better management of these pests has been insufficient knowledge and understanding of pest-crop dynamics and the roles, for example, that beneficial organisms, climate changes and even agricultural policies may have on the actual pest populations and on the selection of appropriate pest management solutions. The role of changing climatic conditions on pest distribution and establishment is difficult to correlate. The more long term effect on climate change on future pest distribution is even more difficult to predict and several reports have emphasized the complex interactions between changes in climate, crops and pests. However, there is much scientific evidence to support the view that climate change will result in a higher probability for the entry, establishment and spread of crop pests.

Implementation of certain government policies may also impact on the introduction and establishment of new pests in a country. Such policy decisions as establishing new trade links for agricultural commodities, increased movement of peoples through the encouragement of tourism, changes in pesticide usage or promotion of agricultural diversification could all lead to new pest risks.

Governments need to define new policy measures or implement those relevant policies that have already been identified to deal with the critical problems on introduced crop pests and the impacts of changing climatic factors on invasive pest species. Both national action and regional collaboration are critical in the mitigation of the impacts of the growing number of actual introductions and potential threats of new invasive pest species into the Uttar Pradesh or even wider country region. This is even more vital, especially given the fact that any new pest introduction in one country will invariably and rapidly spread throughout the region. However, Asian countries face many constraints to effectively deal with the prevention, early warning and early control eradication of invasive pest introductions.

References

1. Aukema BH, AL Carroll Y, Zheng Zhu J *et al.*, Movement of outbreak populations of mountain pine beetle: influences of spatiotemporal patterns and climate, Ecography. 2008; 31(3):348-358.
2. Baier J, Pennerstorfer A, Schopf P. A comprehensive phenology model ofIps typographus (L.) (Col. Scolytidae) as a tool for hazard rating of bark beetle infestation.Forest Ecol. and Mgt, 2007, 171-186.
3. Bale J, Masters G, Hodkinson I, Awmack C *et al.*, Herbivory in global climate change research: direct effects of rising temperature on insect herbivores. J Global Change Biol. 2002; 8:1-16.
4. Battisti A, Stastny M, Netherer S, Robinet C *et al.*, Expansion of geographic range in the pine processionary moth caused by increased

winter temperatures-Ecological Appli- Ecological Applications. 2002; 15:2084–2096.

5. Chakravarty NVK, RD Gautam, Forewarning mustard aphid. Division of Agricultural Physics. Indian Agricultural Research Institute, New Delhi, 2002, 49.
6. Cockbain AJ, Low temperature thresholds for flight in Aphis fabae Scop. Entom. Expt. Et Appl. 1961; 4:211-219.
7. Dermody O, BFO'Neill, AR Zangrel, MR Berenbaum, EH Delucia, Effects of elevated CO_2 and O_3 on leaf damage and insect abundance in soybean ecosystem. Arthropod -Plant Interaction. 2008; 2:125-135.
8. Dhillon MK, HC Sharma. Effect of storage temperature and duration on viability of eggs of Helicoverpa armigera (Lepidoptera: Noctuidae). Bull. Entomol. Res. 2007; 97:55-59.
9. FAO. Climate-related transboundary pests and diseases: Technical background document from the Expert Consultation held on 25 to 27 February 2008. Rome: FAO, 2008.
10. Gregory PJ, Ingram JSI. Climate change and the current 'food crises. CAB Reviews: Perspectives in Agriculture, Veterinary Science, Nutrition and Natural Resources 3, No. 099, 2008.
11. Gregory PJ, Johnson SN, Newton AC, Ingram JS. Integrating pests and pathogens into the climate change/food security debate. J Experimental Botany. 2009; 60:2827-2838.
12. Gutierrez AP, L Ponti, G Gilioli. Climate change effects on plant-pest-natural enemy interactions, In Handbook of climate change and agro-ecosystems: impacts, adaptation, and mitigation, D. Hillel and C. Rosenzweig (eds.). Imperial College Press, London, UK, 2010, 209-237.
13. Gutierrez AP, L Ponti. Eradication of invasive species: why the biology matters, Environmental Entomology. 2013; 42:395–411.
14. Hassall C, DJ Thompson, GC French, IF Harvey. Historical changes in the phenology of British Odonata are related to climate – Global Change Biology. 2007; 13:933–941.
15. Heong KL, YH Song, S Pimsamarn, R Zhang D, SD Bae. Global Warming and Rice Arthropod Communities In: Climate Change and Rice. (Eds. Peng S, Ingram KT, Neue HU, Ziska LH), Springer publications, Berlin, 1995, 327-335.

16. Hickling R, DB Roy, JK Hill, R Fox, CD Thomas. The distributions of a wide range of taxonomic groups are expanding polewards. – Global Change Biology. 2006; 12:450–455.

17. IPPC. Climate Change impacts, adaptation and vulnerability. Contribution of Working Group II to the Fourth Assessment Report of the Intergovernmental Panel on Climate Change. ML Parry, Canziani OF, Palutikof JP, van der Linden PJ Hanson CE (eds). Cambridge, UK: Cambridge University Press, 2007.

18. Jönsson AM, Appelberg G, Harding S, Bärring L. Spatio-temporal impact of climate change on the activity and voltinism of the spruce bark beetle, Ips typographus. Glob Chang Biol. 2009; 15:486-499.

19. Johnson CG, LR Taylor. Periodism and energy summation with special reference to light rhythms in aphids. J Experimental Biol. 1957; 34:209-221.

20. Johnson SN, PJ Gregory, JW McNicol, Y Oodally *et al.*, Effects of soil conditions and drought on egg hatching and larval survival of the clover root weevil (Sitona lepidus). Appld. Soil Ecol. 1957; 44:75-79.

21. Karuppaiah V, Sujayanad GK. Impact of Climate Change on Population Dynamics of Insect Pests. World J Agric. Sci. 2012; 8(3):240-246.

22. Kiritani K. Predicting impacts of global warming on population dynamics and distribution of arthropods in Japan. Popln. Ecol. 2006; 48:5-12.

23. Lange H, Okland B, Krokene P. Thresholds in the life cycle of the spruce bark beetle under climate change. Int. J for Complex Systems. 2006; 1648:1-10.

24. Lindroth RL, Kinney KK, Platz CL. Responses of deciduous trees to elevated atmospheric CO_2 productivity, phytochemistry and insect performance. Ecology. 1993; 74:763-777.

25. Menéndez R, González-Megías AG, Collingham Y, Fox R *et al.*, Direct and indirect effects of climate and habitat factors on butterfly diversity. – Ecology. 2007; 88:605–611.

26. Parmesan C. Ecological and Evolutionary responses to recent climate change – Annual Review of Ecology, Evolution and Systematics. 2006; 37:637–669.

27. Parmesan C. Influences of species, latitudes and methodologies on estimates of phenological responses to global warming – Global Change Biology. 2007; 13:1860-1872.

28. Parmesan Yohe CG. A globally coherent fingerprint of climate change impacts across natural systems– Nature. 2003; 421:37–42.

29. Rao MS, Srinivas K, Vanaja M, GGSN Rao *et al.*, Host plant (*Ricinus communis* Linn.) mediated effects of elevated CO_2 on growth performance of two insect folivores. Current Sci. 2009; 97:7-10.

30. Reji G, Chander S. A degree-daysimulation model for the population dynamics of the rice bug, Leptocorisa acuta (Thunb.). J Appld. Entomol. 2008; 132:646-653.

31. Root TL, JT Price, KR Hall, SH Schneider *et al.*, Fingerprints of global warming on wild animals and plants – Nature. 2008; 416:626–629.

32. Roy DB, TH Sparks. Phenology of British butterflies and climate change – Global Change Biology. 2000; 6:407–416.

33. Simmonds FJ, Greathead DJ. Introductions and pest and weed problems. In: Cherrett JM, Sagar GR (Eds). Origins of pest, parasite, disease and weed problems, the 18th Symposium of The British Ecological Society, Bangor, 12-14 April. UK: Blackwell Scientific Publications, 1976, 1977, 109-124.

34. Staley JT, Hodgson CJ, Mortimer SR, Morecroft MD *et al.*, Effects of summer rainfall manipulations on the abundance and vertical distribution of herbivorous soil macro-invertebrates. Eur. J Soil Biol. 2007; 43:189-198.

35. Stefanescu C, Peñuelas J, Filella I. Effects of climatic change on the phenology of butterflies in the northwest Mediterranean Basin–Global Change Biology. 2003; 9:1494–1506.

36. Tobin PC, Nagarkatti S, Loeab G, Saunders MC. Historical and projected interactions between climate change and insect voltinism in a multivoltine species. Global Change Biol. 2008; 14:951-957.

37. Walter GR, Post E, Convery P, Menzel A *et al.*, Ecological responses to recent climate change-Nature. 2002; 416:389-395.

38. Yamamura K, Kiritani K. A simple method to estimate the potential increase in the number of generations under global warming in temperate zones. Appld. Entomol. Zool. 1998; 33:289-298.

39. Yumamura K, Yokazawa M, Nishimori M, Ueda Y, Yokosuka T. How to analyse long- term insect population dynamics under climate change: 50 year data of three insect pests in paddy fields. Poplnulation Ecol. 2006; 48:38-48.

Chapter - 10
Climate Change and Weed Dynamics

Authors

Dr. Vivek

Sardar Vallabhbhai Patel University of Agriculture & Technology, Meerut, Uttar Pradesh, India

Dr. S.S. Tomar

Rajamata Vijayaraje Scindia Krishi Vishwa Vidyalala, Gwalior ZARS, A B, Road Morena, Madhya Pradesh, India

Chapter - 10

Climate Change and Weed Dynamics

Dr. Vivek and Dr. S.S. Tomar

Climate change is expected to affect agriculture very differently in different parts of the world (Parry *et al.*, 2004). The resulting effects depend on current climatic and soil conditions, the direction of change, and the availability of resources and infrastructure to cope with change. There is a large variation in climatic conditions, soils, land use, infrastructure, political and economic conditions (Bouma *et al.*, 1998). These differences are also expected to influence the responsiveness to climatic change (Olesen and Bindi, 2002). Changes in atmospheric CO_2 levels, rainfall, temperature and other growing conditions will affect weed species distribution and their competitiveness within a weed population and within a rice crop. This may necessitate adaptations in crop management practices, which in turn will affect weed growth or the proliferation of certain species.

Environmental conditions also have a large impact on the effectiveness of weed management operations such as chemical and mechanical control. The magnitude of these effects will largely depend on the extent to which environmental conditions change locally and regionally. Major global changes will comprise further increases in atmospheric greenhouse gases and likely changes in temperature (> 0.2°C per decade), soil degradation, and competing claims for land and water (IPCC, 1996).

Weed Floral Composition and Distribution

The impacts of climate change on single species and ecosystems are likely to be complex. In the case of aggressive weed species of tropical and sub-tropical origins, which are currently restricted to Mediterranean environments, future climatic conditions may lead to an expansion of their potential range into temperate regions. However, in contrast to crops, weeds are troublesome invaders, ecological opportunists and resilient plants with far more genetic diversity. Weed populations include individuals with the ability to adapt and flourish in different types of habitats. Any factor which increases environmental stress on crops may make them more vulnerable to attack by insects and plant pathogens and less competitive with weeds. The geographical and seasonal distribution of pests likely will change as the climate changes.

The physiological plasticity of weeds and their greater intra specific genetic variation compared with most crops could provide weeds with a competitive advantage in a changing environment events such as cyclones, flooding, drought and fires (from high temperature) will become more common and weeds will be the first to gain a stronghold after these events Remigijus *et al.*, (2008). There are also mitigation and adaptation mechanism to climate change which might affect the composition and distribution of weeds (sleeper weeds will come to flourish).

Table 10.1: Potential impacts of climate change on weeds significant to agriculture

Common Name	Scientific Name	Expect Impact
Blackberry	*Rubus fruticosus* L. agg.	Expected to retreat south wards and to higher altitudes because it is sensitive to higher temperature and drought
Chilean needle grass	*Nassella neesiana* (Trin. & Rupr.) Barkworth	Expected to increase its ranges because it highly invasive (long lived, seed dispersal by wind and water) and drought tolerant.
Gorse	*Ulex europaeus* L.	Expected to retreated southwards because it is drought sensitive
Lantana	Lantana camara L.	Expected to continue its moves southwards into high rainfall zone of northern new south wales.
Mesquite	Prosopis glandulosa Torr.	Some risk that it may move into lower rainfall areas because it is very drought tolerant
Parthenium	Parthenium hysterophorus L.	Not suited to winter dominant rainfall areas, may move into summer dominant high rainfall (>500mm) region
Serrated tussock	Nassella trichotoma (Nees.) Hack. ex Arechav.	Expected to retreat southwards and to higher altitude because it is sensitive to temperature As drought tolerant plant, it should become more invasive in an areas where temperature allows
Prickly acacia	Acacia nilotica	Expected to move southwards and into arid areas

Weed-Crop Competition under Climatic Change

There is an ever-growing consensus that the earth's climate is changing, and despite the efforts made to reduce CO_2 emissions, there is an increasing pressure to identify adaptive mechanisms in agro-ecosystems. The projected increase in atmospheric CO_2 is known to favour net photosynthesis in C_3 plants (three quarters of global agriculture is represented by C_3 crops by limiting the loss of CO_2 via photorespiration and increasing the CO_2 concentration gradient from air to the leaf interior (Ziska, 2000). By contrast, plants with a C_4 photosynthetic pathway manifest little response to elevated CO_2 as they have an internal mechanism to concentrate CO_2 at the site of CO_2 carboxylation. Thus, from this perspective, ongoing increase in the atmospheric CO_2 concentration will have crucial implications for weed-crop competition and crop yield losses. Numerous studies have addressed weed-crop interactions by evaluating the comparative growth and physiology of C_3 crops and C_4 weeds, and concluded that an elevated CO_2 concentration generally favours the vegetative growth of C_3 plant species over those with C_4 pathways (Patterson, 1995). However, not all crops are based on C_3 pathways, and not all weeds are C_4 based (Ziska *et al*., 2010). Hence, while the above concept is relevant for C_3 cereals such as rice, which compete, in the main, with C_4 grassy and broad-leaved weeds, this situation is not universal.

The third important grain crop, maize, is a C_4 species, and as noted, although there were some variations in the experimental conditions, C_3 species generally responded more favourably than C_4 species to an increased concentration of CO_2. Of further interest is that Ziska and McClung (2008) indicated a greater physiological plasticity and genetic diversity among weedy (red) rice relative to cultivated rice, which may impact weed-crop competition with increased atmospheric CO_2.

It is becoming clear that predicting competitive outcomes based on species grown in isolation, may not adequately quantify crop-weed competition as a function of increasing CO_2; as weeds usually occur in a mixture. Therefore, there is a need to evaluate the effects of weed competition on crops in an environment of mixed of weeds and crops since most of the studies on the effect of CO_2 on crops and weeds have included weed and crop species in isolation. In a study by Ziska *et al*., (2010), an increase in CO_2 concentration resulted in a significant decrease in plant relative seed yield of a cultivated rice (C_3 crop species) variety, but the reverse for a weedy rice biotype, also a C_3 species. It is thought that this was probably due to a greater physiological plasticity and genetic diversity between wild and cultivated lines. Presumably, an increase in CO_2 concentration stimulated growth and development of many invasive plant species, for example Cirsium arvense L., an invasive perennial C_3 weed species, had registered a 70% increase in growth with elevated CO_2 (Ziska *et al*., 2011). The authors suggested that increasing CO_2 levels may also increase wind dispersal of weed seeds by either increasing the height of the weed plant or by increasing the plant size. Some of these wind dispersed invasive weed species are Cirsium arvense, Sonchus arvensis L., Sonchus oleraceus L., and Carduus nutans L.

Changes in weed communities in response to crop establishment methods, wet or dry moisture conditions, and tillage regimes, and other management practices are well established in the literature. Variations in the weed competitiveness response to elevated CO_2 among diverse lines of rice may necessitate screening of vulnerable and resistant cultivars for wider adoption. Although rice could benefit from CO_2 fertilization, the greater response of its wild relatives, particularly the weedy biotypes of rice, could offset the associated benefits and competitive outcomes as crop yield losses could increase.

The differential responses of C_3 and C_4 plants to increasing CO_2 are especially relevant to weed-crop competition in agro-ecosystems. However, studies on competition outcomes between C_3 crops and C_4 weeds, or vice versa, are limited in the literature. In general, elevated CO_2 levels would stimulate the growth of major C_3 crops of the world; the same effect is also likely to increase the growth of both C_3 and C_4 weeds. In all probability, this would lead to increased weed-crop competition, negating some of the otherwise beneficial effects of CO_2 'fertilization' of the C_3 crops and their yields. Some examples of relevant crop/weed competition studies are discussed below:

Carter and Peterson (1983) found that Festuca elatior L., a C_3 grass, out-competed Sorghum halepense (L.) Pers., a C_4 grass, in mixed cultures, under both ambient CO_2 levels and elevated CO_2, even under temperatures unfavorable to C_3 photosynthesis (between 25 and 40°C). The authors predicted that global CO_2 enrichment would alter the competitive balance between C_3 and C_4 plants and this may affect seasonal niche separation, species distribution patterns, and net primary production within mixed communities.

Ziska (2000) evaluated the outcome of competition between 'Round-up Ready' Soybean (Glycine max L.) and a C_3 weed (lambsquarters, Chenopodium album L.) and a C_4 weed (redroot pigweed, Amaranthus retroflexus), grown at ambient and enhanced CO_2 (ambient+250 $\mu L\ L^{-1}$). In a weed-free environment, elevated CO_2 resulted in increased soybean growth and yield, compared to the ambient CO_2 conditions.

However, soybean growth and yield were significantly reduced by both weed species at both levels of CO_2. With lamb's quarters, at elevated CO_2, the reduction in soybean seed yield relative to the weed-free control increased from 28 to 39%. Concomitantly, the dry weight of lambsquarters increased by 65%. Conversely, for pigweed, soybean seed yield losses diminished with increasing CO_2 from 45 to 30%, with no change in weed dry weight. This study suggests that rising CO_2 could alter yield losses due to competition from weeds, and that weed control will be crucial in realizing any potential increase in the yield of crops, such as soybean, as climate change occurs.

Alberto *et al.*, (1996), studied competition outcomes between rice and Echinochloa glabrescens L., which is a C_4 weed, using replacement series mixtures at two different CO_2 concentrations (393 and 594 μL L^{-1}) under day/night temperatures of 27/21°C and 37/29°C. Increasing the CO_2 concentration, at 27/21°C resulted in a significant increase in above ground biomass (+47%) and seed yield (+55%) of rice, averaged over all mixtures. For the C_4 weed, higher CO_2 concentration did not produce a significant effect on biomass or yield. When grown in mixture, the proportion of rice biomass increased significantly relative to that of the C_4 weed in all mixtures at elevated CO_2 indicating the increased competitiveness of rice.

However, under elevated CO_2 levels and the higher temperature regime, competitiveness and reproductive stimulation of rice was reduced compared to the lower growth temperature, suggesting that while a C_3 crop like rice may compete better against a C_4 weed at elevated CO_2 alone, simultaneous increases in CO_2 and temperature could still favor a C_4 species.

Herbicide efficacy may also decrease as CO_2 concentrations increase (Ziska *et al.*, 2004). An increase in CO_2 induces morpho-physiological and anatomical changes in plants affecting the rate of uptake and translocation of herbicides (Manea *et al.*, 2011). C_3 plants showed a decrease in stomata number and conductance and increased leaf thickness, which might interfere with foliar uptake of herbicides, as well as an increase in starch accumulation on the leaf surface. Moreover, perennial weeds may become even more noxious, if vegetative growth is stimulated as a result of increased photosynthesis in response to elevated CO_2. These changes are expected to reduce the efficacy of the most commonly used herbicides, such as glyphosate, due to a dilution effect, although the precise mechanism conducive to increased tolerance to glyphosate remains elusive. This could be due to less translocation as the root system becomes vigorous. In addition, an increase in the root-shoot ratio may play a critical role in herbicide efficacy. CO_2-induced increase in root biomass could make perennial weeds harder to control in a higher CO_2 environment. Thus, research is needed to assess the comparative growth and physiological response of C_3 and C_4 weeds of different age groups (seedlings or mature plants), and their molecular and biochemical bases to herbicide tolerance under ambient and elevated CO_2 concentrations. Allocation of resources to below ground parts, source-sink relationships, and mitochondrial respiration also need to be reassessed in the wake of climate change scenarios.

Weed Response to Elevated Temperature

Atmospheric temperature is regarded as an important indicator of weed species distribution in a geographical area. Its rise could alter weed proliferation and competitive behavior in weedy vegetation as well as in crop stands. The indicated likely climate change may favor C_4 over C_3 weeds, since under conditions of elevated CO_2, reduced CO_2 solubility, and decreased affinity of RUBISCO for CO_2 would deter C_3 photosynthesis. A variation in weed distribution will affect the world's most important cropping systems, for example, rice based cropping systems through weed shifts to high latitudes and altitudes. In addition, Striga spp. might extend their range to moderate climatic zones. However, *Striga asiatica* (L.) Kuntze is relatively insensitive to temperature, and changes in the geographical range of the host plants seem to play a critical role in its distribution rather than the direct effects of temperature.These weeds could pose a serious threat to global crop production, especially in fodders, in the near future. Many C_4 weeds, such as *Amaranthus retroflexus* L., Setaria sp., Digitaria sp., *Sorghum halepense* (L.) Pers. and *Paspalum dichotomiflorum* (L.) Michx., are expected to expand which would have a more pronounced effect in the northern Europe, where the number of weed species is lower than in the south (Fried *et al.*, 2010). Milder and wetter winters would tend to increase the survival of winter annual weeds, while thermophile summer annuals will grow more profusely in areas with warmer summers under prolonged growing seasons, enabling them to grow further north (Hanzlik and Gerowitt, 2012).

Patterson (1995) predicted that climate change would spread arable weed species. For example, *Datura stramonium* L., which needs high temperature for profuse growth, would become a more competitive candidate under the climate change scenarios. Warm temperature regimes augmented the abundance of *Hieracium aurantiacum* L. through accelerated growth and reproduction. Under warmer conditions, *Setaria viridis* (L.) P. Beauv. Germinated later in the (August) season. This was a beneficial temporal non-synchrony with emergence of a maize crop, avoiding crop-weed competition. In contrast, that this species would be a problematic weed in maize-based cropping systems elsewhere, through synchrony with maize emergence, which is probably due to stimulation by, increased temperature (Peters and Gerowitt, 2014). Therefore, S. viridis would become a competitor of maize at enhanced temperatures at the time of emergence.

Lee (2011) opined that increased temperature had a greater effect on plant phenological development than elevated CO_2. The author observed that increasing temperature by 4^0C advanced the emergence timing of C. album and S. viridis by 26 and 35 days, respectively, and flowering time by 50 and 31.5 days. Increased temperatures strongly affected the biomass accumulation by annual grass species during their reproductive phase as compared with the vegetative phase, and such effects are more pronounced in C_3 than C_4 plant species. However, the increased temperature was believed to offset potential benefits of elevated CO_2 by reducing the reproductive output. The uptake and translocation of herbicides in plants and their persistence in soil will also be affected by rising temperatures. In addition to these effects, temperature will also affect the rate of water absorption and movement, which affects the rate of leaf development, cuticle thickness, and stomata number and their aperture, thus indirectly affecting herbicide selectivity and efficacy.

Weed Response to Variation in Rainfall and Drought Spells

A variation in rainfall pattern and increased aridity consistent with a warming climate could alter weed distribution and their impact on crop production. In the near future, aridity is expected to increase in many agronomically important areas, since an anticipated increase in temperature (1–5^0C) is expected with each doubling of the atmospheric CO_2 level. Rising temperatures also causes greater evaporation, and global trends in rainfall variability suggest that the monsoon regions will become drier. Trends in future rainfall prediction are difficult to predict, except to forecast more erratic rainfall and consequently, drought and flood would become recurrent phenomena. Under such a scenario, the distribution and prevalence of weeds will be problematic in crop ecosystems, and in particular summer droughts will affect weed management in spring-sown crop. Rodenburg *et al*., (2010) postulated that under prolonged drought spells, C_4 and parasitic weeds like S. hermonthica will thrive better. Under excess water environments, weeds such as Rhamphicarpa fistulosa (Hochst.) Benth. Will be favored.

A change in rainfall patterns would favor hydromorphic weeds while prolonged drought spells will benefit C_4 over C_3 weeds. Under rainfed or dryland environments, little or no rainfall will hamper adequate land preparation for wet season rice as a result of limited water availability for flooding, especially early in the season when the rice is most susceptible to weed competition. This will limit traditional weed management in flooded rice and necessitate the use of herbicides. Rice yield losses are expected to be higher under such circumstances.

Asch *et al*., (2005) emphasized that drought-tolerant rice cultivars would be required to prevent water stress-induced yield losses and to increase rice competitiveness against weeds under rainfed conditions. A change from transplanting to direct seeding of rice, in relation to water saving in South Asia, has already resulted in increased weed competition and changed weed dynamics. Competition of cotton with Abutilon theophrasti Medic and *Anoda cristata* (L.) Schltdl, increased under drought conditions. A yield reduction due to *Xanthium strumarium* L. was more pronounced in well-watered soybeans compared with water-stressed soybeans. An increase in rainfall provided greater competition to wheat growth and yield against C. arvense. According to Patterson (1995), weed competition had little effect on crops under water deficit conditions, as the potential crop yield was already reduced by water stress. By contrast, Amaranthus spinosus and *Leptochloa chinensis* (L) Nees survived under water stress conditions and produced a significant number of tillers/branches and leave even at the lowest soil water content.

Increased rainfall frequency and intensity will have an adverse effect on uptake, retention, and activity of soil-applied herbicides. Increased cuticle thickness and leaf pubescence in response to drought, will reduce herbicide entry into leaves. These attributes can also affect growth and recovery of crops and weeds following herbicide application. Increasing aridity and drought will increase herbicide volatilization, and, moreover, frequent rain showers will reduce the "rain safe periods" available for herbicide application in a given cropping system posing multidimensional challenges for weed management. An unprecedented increase in rainfall (either as a single rain even tor annually) may promote leaching of soil-applied herbicides, and subsequent ground water contamination. A general conclusion that can be drawn from the above discussion is that an increase in rainfall would lead to additional weed pressure, thus increasing the herbicide costs and overall cost of production of major crops.

Herbicide-Climate Interactions

Herbicide effectiveness is dependent on the local climate/microclimate, and herbicides are no exceptions, particularly foliage applied post-emergence. A rise in temperature will increase volatility of certain herbicides such as trifluralin, rendering it less efficient. In the 1970s, rapid volatilization of surface-applied alachlor, butachlor, and propachlor occurred from continuously moist soils exposed to a constant 21^0C. Temperatures (day/night $32/22^0$C and $26/16^0$C) were not as critical as that of relative humidity in influencing acifluorfen (diphenylether group) phytotoxicity on Xanthium strumarium L. and Ambrosia artemisiifolia L. However, temperature had a significant effect on the degradation of imazapyr (imidazolinone group), flumetsulam (sulfonanilide family), and thifensulfuron (sulfonylurea group). Glyphosate absorption is dependent on the atmospheric temperature, as evident from Desmodium tortuosum (Sw.) DC. A C_3 weed.

An increase in temperature or relative humidity increased the efficacy of mesotrione on X. strumarium and A. theophrastii three-fold. The efficacy of the herbicide pyrithiobac (pyrimidinylthiobenzoic acid group) on

Amaranthus palmeri L. was reduced at temperatures outside the range of 20-34^{0}C. Anderson *et al*., (1993) found that relative humidity had the most significant effect on the phytotoxic action of glufosinate-ammonium, since this is attributed to changes in cuticle hydration and droplet drying. Studies under controlled environmental chambers in Australia using varying night/day temperatures of 5/10, 15/20, and 20/25^{0}C showed that Raphanus raphanistrum L., grown under cooler temperatures of 5/10^{0}C, was poorly controlled with 1,200 g ai ha^{-1} of glufosinate. By comparision, 100% mortality was achieved under 15/20 and 20/25^{0}C for the same dose (Kumaratilake and Preston, 2005), suggesting enhanced efficacy of glufosinate under enhanced atmospheric temperature.

Weed Management under Climate Change

Effectiveness of weed management is also hypothesized to change along with environmental conditions. Weeds that are under moisture stress can respond by thickening their leaf cuticles, slowing down vegetative growth and flowering rapidly. Drought stressed weeds are more difficult to control with post-emergent herbicides than plants that are actively growing for example, systemic herbicides that are translocate within the weed need active plant growth stage to be effective. Pre-emergent herbicides or herbicides absorbed by plant roots need soil moisture and actively growing roots to reach their target sites. Occurrence of drought has the potential to reduce the effectiveness of pre-emergent herbicides [5]. Overall; there are strong empirical reasons for expecting climate and/or rising CO_2 to alter weed management. Adaptation strategies are available, but the cost of implementing such strategies (e.g. new herbicides, higher chemical concentrations, new bio control agents) is unclear.

Chemical Weed Control

Clearly, any direct or indirect impacts from a changing climate will have a significant effect on chemical management. Changes in temperature, wind speed, soil moisture and atmospheric humidity can influence the effectiveness of applications. For example, drought can result in thicker cuticle development or increased leaf pubescence, with subsequent reductions in herbicide entry into the leaf. These same variables can also interfere with crop growth and recovery following pesticide application. Overall, pesticides are most effective when applied to plants that are rapidly growing and metabolizing, i.e. those free from environmental stress. Environmental factors such as temperature, precipitation, wind and relative humidity influence the efficacy of herbicides Ziska *et al*., 2004.

Most of the herbicides required optimum temperature for their activity (function). But prolonged high temperature after application (pre plant pre and post emergent herbicides) the efficiency and selectivity may affected. That means at high temperature, selective herbicides may become nonselective. There are an increasing number of studies Ziska and Teasdale, (2000) that demonstrate a decline in chemical efficacy with rising CO_2. But some scholar suggest that as CO_2 increases the growth and development of some weeds (C_3) will increase and plant can develop immunity that can tolerate or detoxifying mechanism (high tissue volume). Recent work with Canada thistle grown in monoculture under field conditions suggested a greater root to shoot ratio and subsequent dilution effect of glyphosate when grown at elevated CO_2. The scholar pointed out CO_2 does reduce efficacy of the herbicides. However, then additional work is needed to determine herbicide specificity, concentration and application rates as possible means of adaptation.

Furthermore, there is the possibility that increased CO_2 and temperature will reduce the effectiveness of herbicides, requiring an increase in frequency of application and concentration of herbicide to obtain adequate control. Similarly although the effects of elevated atmospheric CO_2 concentrations on weed/crop/herbicide interactions are not well known, a study showed that elevated CO_2 levels diminished the efficacy of the widely used herbicide glyphosate. If these effects are common and widespread, they will have a significant impact on agriculture. Moreover, according to Bale *et al*., (2002) weeds that are under moisture stress can respond by thickening their leaf cuticles, slowing down vegetative growth and may flower rapidly. Drought stressed weeds are more difficult to control with post-emergent herbicides than plants that are actively growing for example, systemic herbicides that are translocate within the weed need active plant growth stage to b e effective. Pre-emergent herbicides or herbicides absorbed by plant roots need soil moisture and actively growing roots to reach their target sites. Occurrence of drought has the potential to reduce the effectiveness of pre-emergent herbicides.

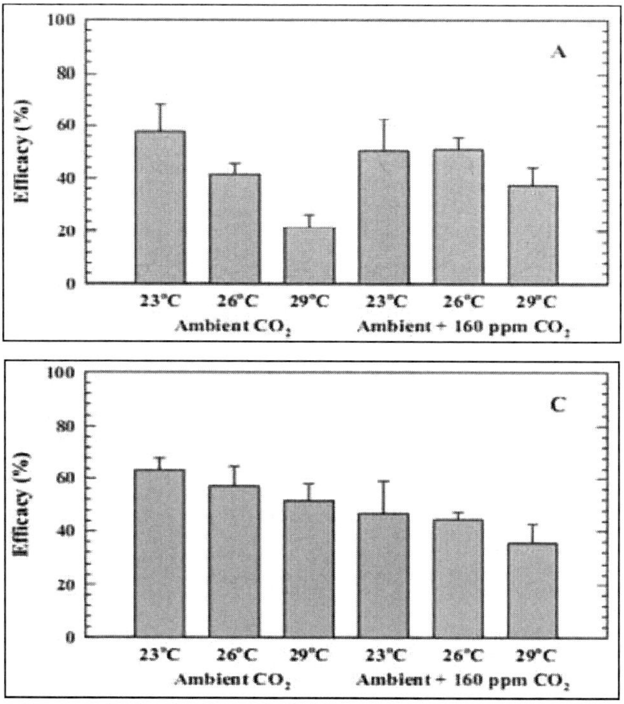

Source: Archambault *et al.*, 2001.

Fig 10.1: Herbicide efficacy of Fusion (A), and Liberty (C) applied to wild oats grown at different temperatures at ambient or ambient + 160 ppm CO_2

Effects of Climate Change on Chemical Weed Control Methods

These are Greater root to shoot ratio leading to subsequent dilution effect on glyphosate when grown at elevated CO_2. Decreased stomatal aperture or number, cuticle thickening, development or increased leaf pubescence with subsequent reductions in herbicide entry into the leaf. (Altered transpiration leading to reduced uptake of soil herbicides. Changes to the wetting and drying of soil profiles affecting microbial activity and herbicide degradation. Increased temperatures affecting the efficacy and selectivity of herbicides. Increased growth rate implying narrower temporal windows in which to apply herbicides before crop yield losses from weed competition and before weed set (Fig 10.1). High temperatures, low relative humidity and unsuitable wind conditions reduce the window of spraying opportunity.

Mechanical Weed Control

A standard means of controlling weed populations, and the one most widely used in developing countries is mechanical removal. Tillage (by animal or mechanical means) is regarded as a global method of weed control in agronomic systems. Under heavy frequent rainfall carrying out of tillage and other mechanical weed control methods will very difficult due to logging of soil and under condition undertaking of cultivation will help to reduce soil moisture which may decrease the growth of crop plant and make less competitive with weeds.

Elevated CO_2 could lead to further below ground carbon storage with subsequent increases in the growth of roots or rhizomes particularly in perennial weeds (Fig 10.2). Consequently, mechanical tillage may lead to additional plant propagation in a higher CO_2 environment, with increased asexual reproduction from below ground structures and negative effects on weed control Ziska *et al.*, 2004.

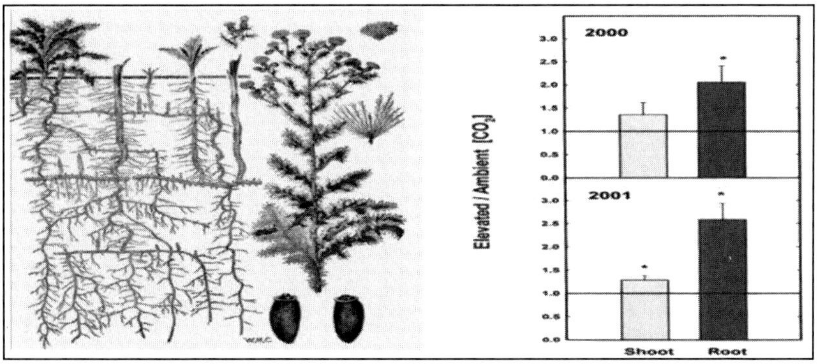

Fig 10.2: Shoot and root biomass at elevated CO_2 relative to ambient CO_2 (**Source:** Ziska *et al.*, 2004

Biological Control

Biological control involves the introduction of host-specific agents, usually insects or fungi, for the control of a weed species. The biological control agents have evolved to feed only on the target plant (Darren, 2010). Most bio-control systems perform best under a stable environment. However, in addition to the forecast increase in mean temperatures, a CO_2 and shift in rainfall distribution climate variability is expected to increase.

Extreme weather events such as droughts, flood and even unseasonal frosts are predicted to occur more frequently. While many species have mechanisms to cope with extremes, they require time to acclimatize and/or enter the resistant state. The relative vulnerability of the host plants, and bio-control agents to extremes of temperature, desiccation or flooding will determine whether the drought or flood will be followed by a weed or pest outbreak (Gerard *et al.*, 2010). Moreover, according to [30] increases in CO_2, changes in water availability and increases in temperature will alter plant phenology, growth and distribution, all of which will have flow on effects on the plant herbivores and those that prey on them. Aspects of plant life cycle events controlled by temperature, such as the timing and duration of seed germination, bud burst, and flowering, are likely to change with warmer temperatures, subject to photoperiod and water availability. Similarly, the degree that individual species benefit will depend on how well the life cycle events are synchronized to ensure optimum survival. Changes in host plant quality in response to elevated CO_2, drought or flooding may also cause shifts in herbivore and natural enemy fitness. In general, elevated CO_2 leads to a decrease in the nutritional value of plants. There is a tendency for foliage feeders to consume more plant material (greater damage), but to grow and develop more slowly, and have reduced fecundity and survival.

The direct impacts of climate change will be either on the biology of the biological control agent and/or on the ability of the host plant to tolerates or compensate for the presence of the herbivore or plant pathogen. Increased temperature would be expected to increase the rate of life cycles of both the biological control agents and the weeds. Increased water stress will affect the host plant's development, and through this, the development of biological control agents, so they might be less effective in drier situations. Increase in greenhouse gases will also affect the herbivore plant relationship so impacts at both the spatial and temporal scales are possible Darren *et al.*, 2010. Changes in temperature can influence production of plant defense compounds against plant herbivores. It is known that levels of many insect-resistance allele-chemicals increase during drought Gerard *et al.*, 2010.

At the spatial scale, the effects of moisture and temperature are likely to play out through differential changes in the preferred distribution of the agents and target weed. For example, an increase in temperature may enable biological control agents from subtropical areas to survive on weeds in more temperate regions. Some examples are the agents of aquatic weeds such as salvinia (Salvinia molesta) and alligator weed (Alternanther aphiloxeroides), and the agents of terrestrial weeds such as prickly pear cactus (Opuntia spp.) which may be favored in southern regions of their distribution. Understanding of the agent's and the weed's physiology, capacity for evolution, and dispersal is needed to determine the potential for a weed to migrate under climate change and "escape" their control agents Darren *et al.*, 2010. CO_2 could alter the efficacy of weed bio-control agents by potentially altering the development, morphology and reproduction of the target pest.

Direct effects of CO_2 would also be related to changes in the ratio of C: N and alterations in the feeding habits and growth rate of herbivores. As pointed out by Patterson, 1995 warming could also result in increased overwintering of insect populations and changes in their potential range. Although this could increase both the biological control of some weeds, it could also increase the incidence of specific crop pests, with subsequent indirect effects on crop-weed competition. Overall, synchrony between development and reproduction of bio-control agents and their selected targets is unlikely to be maintained in periods of rapid climatic change or climatic extremes.

Range of possible impacts (based on generally polyphagous invertebrate herbivores) identified in a meta-analysis of CO_2 induced changes on plant chemistry and herbivore performance. The impacts included:

Significantly decreased herbivore abundance (-21.6%)

Increased: relative consumption rates by herbivores (+16.5%) development time (+3.87%); and total consumption (+9.2%).

Significantly decreased: relative growth rate (-8.3%); conversion efficiency (-19.9%); and pupal weight (-5.03%).

No difference between herbivore associations.

(Plants showed significantly larger biomass (+34.4%), increased C: N (+26.57%), decreased N concentration (-16.4%), increased tannins (+29.9%), and other phenolics.

Adapting Mechanisms to Climatic Change

It is clear that both crops and weeds will respond to climate change, but the overall winners of their competition in the field will be the colonizing species, because of their superior adaptations and wide ecological amplitudes (i.e., the limits of environmental conditions within which an organism can live and function). Although it is not possible to be specific, under climate change, weed management will become more important in the future at every scale, from farmlands to regional landscapes. As colonizing species become abundant, and possibly more aggressive in many regions, humans will have to adapt to manage weed populations more effectively, to maintain productive landscapes, and achieve food security.

Control of weeds, pests, and diseases are all likely to be more difficult and more expensive under climate change, and there will have to be more emphasis on regional cooperation for preventing the spread of certain weeds, pests, and diseases (as in the case of control of diseases, such as HIV). Given that some well-known invasive species are likely increase their bio-geographical ranges, and other, relatively mild species may become aggressive invaders, all countries need to be able to conduct risk assessments at the appropriate level, for national planning to reduce the new threats posed by weeds.

Global and regional co-operation is essential to establish new networks and the capacity to implement early detection and rapid response systems. Increased gathering of information, through local and regional surveys of distribution and abundance of potential invaders, sharing of such information and increased border protection of countries through quarantine, are likely to be of greater importance in the future. More effective integration of on-ground control methods (manual, mechanical, chemical, and biological control) with broader pest control at the farm level will be part of the solutions. What this means is that natural resource managers need to cooperate more with each other, and weed managers and researchers need to be even more effective than before. A new paradigm in weed management might include the view: '**Do what you have been always doing... better...**' because the stakes are much higher now.

Projected changes in climate and crop yields in the latter part of the 21st century suggest that there will be yield increases in mid and high latitudes. These regions are recognized as having sufficient technology-based adaptive capacity to face the changing global climate. In contrast, yield decreases are predicted for tropical and sub-tropical regions of lower latitudes, mainly developing countries, including the Indian sub-continent, Middle East and Southeast Asia, with important regional differences (Parry, 1998). In the latter regions, presently characterized by persistent poverty and food insecurity, temperature maxima are already near the optimum under the current climatic conditions.

The agricultural systems in many developing countries are more vulnerable to climate change, because they are dependent on declining natural resource bases, are labor intensive and less capital and technology dependent. The increasing population pressure on natural resources in developing countries is well known; it has already led to pronounced degradation of land and water resources and increased the risk of hunger. Under this scenario, in Africa, predictions are that by 2080, cereal production will decrease by 10% and the consequent risk of hunger will increase by 20%, although such effects can be partly offset by various farmer adaptations, technological changes, and CO_2 fertilization effects (Rosenzweig and Hillel, 1998).

Nevertheless, it is also predicted that the aggregate agricultural production in developing countries may not change much, as climate change occurs. Despite this prediction, there are specific regions within some countries that would be disproportionately affected by climate change, leading to increased poverty. Most experts agree that the future of global agriculture will be shaped by the: (a) dynamics of change and developments in science and technology; (b) sharing of knowledge and transfer of technology to developing countries; (c) expected production gains in developed countries (mainly Europe); and (d) impacts of trade liberalization.

Technically, adapting to climate change will require significant transformation of agriculture production across the globe, by tapping three main sources for growth: (a) expanding the land area, (b) increasing the land cropping intensity (mostly through irrigation), and (c) boosting yields. The view that we may be approaching the ceiling for all three sources is not supported at the global level, although severe problems exist in specific countries and even whole regions. There will be major changes of land use, probably involving changes in farming locations. For instance, in tropical and sub-tropical countries, flood-prone areas will be less attractive to cropping, because of increased rainfall and flooding frequency. On the other hand, areas previously not farmed, due to varying degrees of aridity, salinity or low productive potential, may become important, also due to modified rainfall patterns.

In temperate countries, global warming will reduce climatic constraints on agriculture, which is likely to expand and extend into uplands. In Europe, a 10°C warming may raise climatic limits to cultivation by approximately 150m (IPCC 2001). Changes in the types of crops grown are also likely in regions where there are substantial increases in the temperature of the growing seasons, and in areas where agricultural productivity is currently limited by temperature.

In many situations, tropical and sub-tropical crops with higher thermal requirements would become more attractive. In all areas of the world, there will be a need to have stress tolerant and hardy crop cultivars, including more drought-tolerant cultivars, to face the uncertainties of climate change. As rainfall patterns change and areas become prone to drought, irrigation will be crucial to maintain world food supplies and its role is expected to increase under climate change. One in five developing countries will face water shortages and water availability is already critical in west Asia and North Africa and will be so also in south Asia in 2030 (IPCC, 1996, 2001). Greater efficiency in water use needs to be achieved, and a new irrigation infrastructure will have to be installed, to substitute for moisture losses due to increased transpiration.

Maintaining soil fertility will be challenging, because in some areas, increased rainfall will cause increased leaching, while in other areas, warming may increase productive potential, so that yields can be maintained without additional fertilizer. Adopting farming methods that reduce the costs of production and minimize environmental damage, while maintaining or even increasing production will be crucial. In this regard, no-till or conservation agriculture, which can raise crop yields by 20-50%, will have a major role under climate change. Experts agree that 80% of increased crop production in developing countries still has to come from intensification of agriculture, which involves: (a) increased cultivable land; (b) higher yield crops; (c) increased crop diversification and multiple cropping; and (d) shorter fallow periods. In the past, environmental policies for agriculture have traditionally focused largely on practices of soil conservation, reducing land and water quality and reducing the impacts of excessive use of herbicides and pesticides in farming landscapes. More recently, agriculture has turned attention to conserving biological diversity on rural landscapes.

Given that agriculture is a major contributor of the greenhouse gases methane and nitrous oxide, it seems prudent to expand these policies to limit the emissions of CO_2, CH_4, and N_2O from agricultural practices. It is also necessary to encourage agriculture to more aggressively adopt and expand on agro-forestry opportunities for carbon sequestration benefits. On a farm level, this will require revitalizing well established conservation farming practices, including avenue cropping, minimum tillage, allelopathic crop residues, and similar ecological approaches to holistic management of populations of weeds, pests, and pathogens.

Humans must act to reduce the primary root cause: the high rate of CO_2 emissions, by a variety of approaches, such as decreased burning of fossil fuels, eradicating large-scale deforestation, and reclamation of large wilderness areas for agricultural or other human uses. Among the most feasible actions to mitigate the CO_2 buildup involve some combination of conserving energy, substituting alternative energy sources (e.g., solar, wind and hydropower) for fossil fuels, and reducing the deforestation occurring in the tropics.

The trend of increasing concentrations of greenhouse gases and an enhanced greenhouse effect is likely to continue in the coming decades, presenting serious threats to both agricultural systems and natural ecosystems. Climate change is therefore the biggest challenge faced by humanity. The response of crops, weeds, or natural vegetation communities is inexorably linked to the climate modifications that humans have exacerbated. This review has provided an overview of some key issues and the complex and multiple-driver nature of global change.

Overall, climate change can be expected to favor invasive plants over established, and slow growing native vegetation, especially if accompanied by an increase in extreme conditions, such as droughts alternating with very wet years. Pioneering species with various physiological adaptations and wide ecological amplitudes are better equipped to adapt to new climatic conditions. Weeds generally have excellent propagule dispersal mechanisms, often by human activities or by birds, and are likely to spread rapidly into new areas, quickly exploiting changing climatic conditions that favor their establishment.

More effective management solutions will therefore be required to reduce the threat posed by aggressive colonizers, which can make production of food and management of land and water resources much more difficult. Global change is a somewhat deceptively simple expression for what is actually an exceedingly complex array of dynamic processes and specific interactions and manifestations in different regions (Rosenzweig and Hillel, 1998). Climate change, sea level increases, higher CO_2 concentrations, UV radiation, and tropospheric ozone are but a few of the potentially fateful factors involved. In dealing with an issue as complex as climate change, there are many uncertainties, including the disordered behavior of the physical climate and our inadequate understanding of that system, especially in regard to the interactions of oceans, clouds, and ice. Still other uncertainties are the fast pace and unknown directions of future social, political, and technological changes. Such uncertainties and unpredictable developments will affect how the Earth's ecosystems and our agricultural landscapes respond to climate change, and ultimately, how humans will respond.

However, climate is not the only factor that will be changing as the 21st century unfolds. Population growth and varying economic and technological changes are likely to affect the environment no less than will climate change per se. Furthermore, the socio-economic and technological conditions will seriously interact with agriculture as well, and our ability to sustain effective production, whilst ensuring sustainable land use. To define how and what we may realistically achieve is a problem in itself, but taking no action is not an option.

The inherent size differential of transplanted rice seedlings in conjunction with flooded environments provided a distinct competitive advantage, i.e., an earlier growth and germination over a wide range of weed species that otherwise are quite problematic in aerobic rice. With a dwindling water supply and more severe drought spells, flooding will not be available as a potential weed management tool in the near future. Hand weeding was 35% higher when the flooding regime was altered from permanent to temporary flooding (Latif *et al.*, 2005). This means that farmers lacking alternate means and resources to combat weeds will suffer significant yield losses. Moreover, dry tillage practices, alternate wetting and drying regimes, and extended periods during which soil is not flooded, will result in the insurgence of non-native and difficult-to-control weeds (Chauhan *et al.*, 2014). Under drought conditions, rice (C_3) is already a poor weed competitor and will be under greater pressure due to increased competition from C_4 weeds, which comprise the majority of weed flora infesting rice fields (Caton *et al.*, 2010).

In rainfed rice, a lack of rainfall early in wet seasons may compel farmers to adjust their timing of land preparation and subsequent planting. This might affect synchronization of rice sensitive growth periods with emergence and active growth period of troublesome weeds. Hence, it seems that strategies aimed at mitigating climate change effects on crop production like drought-tolerant rice germplasm and water saving rice cultivation, will also have implications for weed management.

References

1. Alberto AMP, Ziska LH, Cervancia CR, Manalo PA. The influence of increasing carbon dioxide and temperature on competitive interactions between a C_3 crop, rice (Oryza sativa) and a C_4 weed (Echinochloa glabrescens). Aust. J Plant Physiol. 1996; 23(6):795-802.

2. Anderson DM, Swanton CJ, Hall JC, Mersey BG. The influence of temperature and relative humidity on the efficacy of glufosinate-ammonium. Weed Res. 1993; 33:139–147.

3. Archambault DJX Li, Robinson D, O'Donovan Kurt JT, Klein K. The Effects of Elevated CO_2 and Temperature on Herbicide Efficacy and Weed/Crop Competition, Agriculture and Agric-Food Canada, 2001.

4. Asch F, Dingkuhn M, Sow A, Audebert A. Drought-induced changes in rooting patterns and assimilate partitioning between root and shoot in upland rice. Field Crops Res. 2005; 93:223–236.

5. Bale JS, Masters GJ, Hodkinson ID. Herbivory in global climate change research: direct effects of rising temperatures on insect herbivores, Global Change Biology. 2002; 8:1–16.

6. Bouma J, Varallyay G, Batjes NH. Principal land use changes anticipated in Europe. Agr. Ecosyst. Environ. 1998; 67(2):103-119.

7. Caton BP, Mortimer M, Hill JE, Johnson DE. A Practical Field Guide to Weeds of Rice in Asia, 2nd Edn. Los Baños: International Rice Research Institute, 2010, 118.

8. Cater DR, Peterson KM. Effects of a CO_2-enriched atmosphere on the growth and competitive interaction of a C_3 and a C_4 grass. Oecologia. 1983; 58(2):188-93.

9. Chauhan BS, Prabhjyot-Kaur, Mahajan G, Randhawa RJ, *et al.*, Global warming and its possible impact on agriculture in India. Adv. Agron. 2014; 123:65-121.

10. Darren Kriticos J, Neville Crossman D, Noboru Ota, John Scott K. Climate change and invasive plants in South Australia. National Research Flagship Climatic Adaptation- Australia, 2010.

11. Fried G, Petit S, Reboud X. A specialist-generalist classification of the arable flora and its response to changes in agricultural practices. BMC Ecol. 2010; 10:1–11.

12. Forister ML, AM Shapiro. Climatic trends and advancing spring flight of butterflies in lowland California-Global Change Biology. 2003; 9:1130–1135.

13. Gerard PJ, JM Kean, CB Phillips, SV Fowler *et al.*, Possible impacts of climate change on bio-control systems in New Zealand Agric research, 2010.

14. Gordo O, Sanz JJ. Phenology and climate change: a long-term study in a Mediterranean locality-Oecologia. 2005; 146:484-495.

15. Hanzlik K, Gerowitt B. Occurrence and distribution of important weed species in German winter oilseed rape fields. J Plant Dis. Prot. 2012; 119:107–120.

16. Harrington R, Clark SJ, Weltham SJ, Virrier PJ *et al.*, Environmental change and the phenology of European aphids-G-lobal Change Biology. 2007; 13:1556–1565.

17. IPCC. Climate Change 1995: The Science of Climate Change. (Eds.)

Houghton J, Meira Filho LG, Callander BA, Harris N, Kattenberg A, Maskell K, Intergovernmental Panel on Climate Change. Cambridge University Press, Cambridge, U.K, 1996.

18. IPCC. Climate Change: The Scientific Basis. Contribution of Working Group I to the Third Assessment Report of the Intergovernmental Panel on Climate Change, (Eds.) Houghton JT, Ding Y, Griggs DJ, Noguer M, van der Linden PJ, Dai X, Maskell K, Johnson CA, Cambridge University Press, Cambridge, UK, 2001, 881.

19. Kumaratilake AR, Preston C. Low temperature reduces glufosinate activity and translocation in wild radish (Raphanus raphanistrum). Weed Sci. 2005; 53:10–16.

20. Latif MA, Islam MR, Ali MY, Saeque MA. Validation of the system of rice intensification (SRI) in Bangladesh. Field Crops Res. 2005; 93:281–292.

21. Lee J. Combined effect of elevated CO_2 and temperature on the growth and phenology of two annual C_3 and C_4 weedy species. Agric. Ecosys. Environ. 2011; 140:484-491.

22. Md. Shahidul Haque Bir, Min Yong Eom, Md. Romij Uddin, Tae Seon Park *et al.*, Weed Population Dynamics under Climatic Change. Weed Turf. Sci. 2014; 3(3):174-182.

23. Menéndez R, González-Megías AG, Collingham Y, Fox R *et al.*, Direct and indirect effects of climate and habitat factors on butterfly diversity-Ecology. 2007; 88:605–611.

24. Olesen JE, Bindi M. Consequences of climate change for European agricultural productivity, land use, and policy. Eur. J Agron. 2002; 16:239-262.

25. Parmesan C. Ecological and Evolutionary responses to recent climate change-Annual Review of Ecology, Evolution and Systematics. 2006; 37:637–669.

26. Parry ML. The impact of climate change on European agriculture. In: T. Lewis (Ed.). The Bawden Memorial Lectures 1973-1998, Silver Jubilee Edition, British Crop Protection Council, Surrey, UK. 1998, 325-38.

27. Parry ML, Rosenzweig C, Iglesias A, Livermore M, Fischer G. Effects of climate change on global food production under SRES emissions and socio-economic scenarios. Global Environ. Chang. 2004; 14(1):53-67.

28. Patterson DT. Weeds in a changing climate. Weed Sci. 43, 685–701.

29. Peters K, Gerowitt B. Important maize weeds profit in growth and reproduction from climate change conditions represented by higher temperatures and reduced humidity. J Appl. Bot. Food Qual. 2014; 1995; 87:234–242.

30. Ramesh K, Matloob A, Aslam F, Florentine SK, Chauhan BS. Weeds in a Changing Climate: Vulnerabilities, Consequences, and Implications for Future Weed Management. Front. Plant Sci. 2017; 8:95. Doi: 10.3389/fpls.2017.00095.

31. Remigijus Šmatas, Roma Semaškiene, Sigitaslazauskas. Effect of Climate Changes on Plant Pests and Weeds, Zemdirbyste-Agriculture. 2008; 95(3):235–241.

32. Rodenburg J, Riches CR, Kayeke JM. Addressing current and future problems of parasitic weeds in rice. Crop Prot. 2010; 29:210–221.

33. Rosenzweig CR, Hillel D, Climate Change, Global Harvest. Oxford University Press, Oxford Stern N. Stern Review: The Economics of Climate Change. London: Treasury Office of the Government of the United Kingdom and Northern Ireland, UK. 1998, 2006.

34. Tesfay Amare. Review on Impact of Climate Change on Weed and Their Management. Am J Bio Environ Statis. 2016; 2(3):21-27.

35. Ziska LH. Rising atmospheric carbon dioxide and plant biology: The overlooked paradigm. DNA. Cell Biol. 2008; 27(4):165-172.

36. Ziska LH. The impact of elevated CO_2 on yield loss from a C_3 and C_4 weed in field grown soybean. Global Change Biol. 2000; 6(8):899-905.

37. Ziska LH, Blumenthal DM, Runion GB, Hunt ER Jr, Diaz-Soltero H. Invasive species and climate change: an agronomic perspective. Clim. Change. 2011; 105:13-42.

38. Ziska LH, Faulkner SS, Lydon J. Changes in biomass and root: shoot ratio of field-grown Canada thistle (Cirsium arvense), a noxious, invasive weed, with elevated CO_2: implications for control with glyphosate. Weed Sci. 2004; 52:584-588.

39. Ziska LH, McClung A. Differential response of cultivated and weedy (red) rice to recent and projected increases in atmospheric carbon dioxide. Agron. J, 2008; 100:1259-1263.

40. Ziska LH, McConnell LL. Climate change, carbon dioxide, and pest biology: monitor, mitigate, management. J Agric. Food Chem. 2015; 64:6–12.

41. Ziska LH, Teasdale JR. Sustained growth and increased tolerance to glyphosate observed in a C_3 perennial weed, quack grass (Elytrigia repens), grown at elevated carbon dioxide. Aust. J Plant Physiol. 2000; 27:159–164.

42. Ziska LH, Tomecek MB, Gealy DR. Competitive interactions between cultivated and red rice as a function of recent and projected increases in atmospheric carbon dioxide. Agron. J, 2010; 102:118–123.

Chapter - 11
Climate Change and Carbon; Water Footprints

Authors

Dr. R.K. Naresh
Sardar Vallabhbhai Patel University of Agriculture & Technology, Meerut, Uttar Pradesh, India

Dr. Saurabh Tyagi
Sardar Vallabhbhai Patel University of Agriculture & Technology, Meerut, Uttar Pradesh, India

Chapter - 11

Climate Change and Carbon; Water Footprints

Dr. R.K. Naresh and Dr. Saurabh Tyagi

The carbon and water footprint concepts were introduced about a decade ago, simultaneously, but independently from one another. The CF arose out of the debate on climate change, as a tool to measure GHG emissions. The WF was introduced in the field of water resources management, as a tool to measure water use in relation to consumption patterns. In both cases, the terminology chosen was inspired by the ecological footprint (EF), which had been introduced in the 1990s (Rees, 1992).

All footprints measure, in different ways, human appropriation of the planet's natural resources and carrying capacity (Galli *et al.*, 2012; Hoekstra, 2009) (Fig 11.1). The EF measures the use of bio-productive space in hectares; the WF measures the consumption and contamination of freshwater resources in cubic metres per year; and the CF measures the emission of gases that contribute to heating the planet in carbon dioxide (CO_2)-equivalents per unit of time or product. A common property of all footprints is that they can be related to specific activities, products and consumption patterns. Recently, the nitrogen footprint was introduced as a tool to measure the amount of nitrogen released into the environment in relation to consumption (Leach *et al.*, 2012).

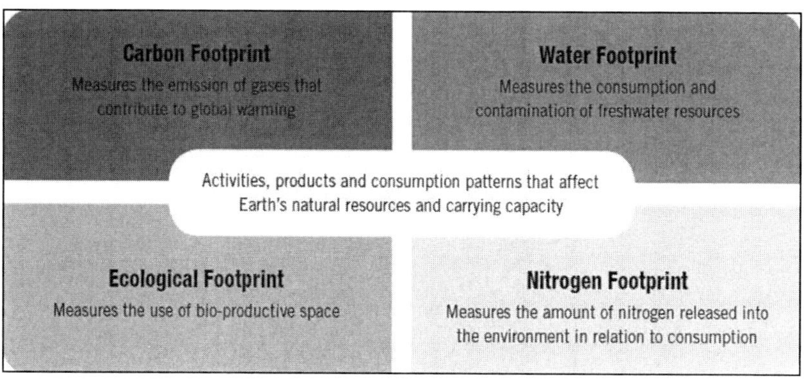

Fig 11.1: Footprint concepts

The Carbon Footprint

A 'carbon footprint' is a measure of the greenhouse gas emissions associated with an activity, group of activities or a product. Nearly everything that we do produces greenhouse gas (GHG) emissions either directly or indirectly usually expressed in equivalent tone of carbon dioxide (CO_2). The most important greenhouse gas produced by human activities is carbon dioxide. Direct GHG emissions sources are often easy to identify – for example burning fossil fuels for electricity generation, heating and transport. It is sometimes less obvious that products and services also cause indirect emissions throughout their life-cycles. Energy is required for production and transport of products, and greenhouse gases are also released when products are disposed of at the end of their useful lives.

Greenhouse gases affect the ability of the earth's atmosphere to retain heat.

Higher greenhouse gas concentrations in the earth's atmosphere cause global warming through this 'green- house effect'. The Kyoto protocol, which originated at the 3rd Conference of the Parties to the United Nations Convention on Climate Change in 1997, has identified six greenhouse gases whose atmospheric concentrations are strongly influenced by human activity. The most important of these is carbon dioxide (CO_2). The global warming potential (GWP) of each greenhouse gas can be expressed in CO_2 equivalents. For gases with a high global warming potential, a relatively small emission can have a considerable impact.

Kyoto Gas	GWP*	Example Sources
Carbon dioxide (CO_2)	1	Burning fossil fuels
Methane (CH_4)	23	Cattle, landfill sites, leaks from disused mines, burning fossil fuels
Nitrous oxide (N_2O)	296	Emissions from fertilised soils, burning fossil fuels.
Sulphur Hexafluoride (SF_6)	22,200	Leaks from electrical and electronics industries.
Per fluorocarbons (PFCs)	4,800 – 9,200	Electronics industries, fire extinguishers
Hydro fluorocarbons (HFCs)	12-12,000	Leaks from air conditioning and refrigeration systems. LPG storage.

*Note: the 'global warming potential' of a gas is its relative potential contribution to climate change over a 100 year period, where $CO_2 = 1$ Source: IPCC (2001)

Basic Calculation

Fuel	Unit	CO_2 Emitted Per Unit
Petrol	I litre	2.3 kg
Gasoline	I litre	2.3 kg
Diesel	I litre	2.7 kg
Oil (heating)	I litre	3.0 kg

If your car consumes 7.5 litre diesel per 100 km, then a drive of 300 km distance consumes 3x7.5 = 22.5 litre diesel, which adds 22.5x2.7 kg = 60.75 kg CO_2 to your personal carbon footprint.

Clean Development Mechanism (CDM): An industrialized country with an emission-reduction or emission limitation commitment can implement emission reduction projects in developing countries. In this way, the country earns saleable certified emission reduction credits (CERs).

Joint Implementation (JI): An industrialized or in transition country with an emission-reduction or emission-limitation commitment can earn emission reduction units (ERUs) from an emission-reduction or emission-removal project in another industrialized country or a country in transition.

A mechanism that allows countries to earn removal units (RMUs) through projects that sequester CO_2, such as reforestation.

Table 1: Comparison of carbon and water footprints

	Carbon Footprint (CF)	**Water Footprint (WF)**
What is measured	The anthropogenic emission of greenhouse gases (GHG).	The human appropriation of freshwater resources in terms of volumes of water consumed and polluted.
Unit of measurement	Mass of carbon dioxide (CO_2)-equivalents per unit of time or per unit of product.	Water volume per unit of time or per unit of product.
Spatiotemporal dimension	Timing within the year and place of emissions are not specified. It does not matter where and when carbon emissions occur; carbon emission units are inter changeable.	WFs are specified in time and by location. It matters where and when a WF occurs; WF units are not inter-changeable. For some uses, total/ average WFs are shown, thus leaving out spatiotemporal specifications.
Footprint components	CF per type of GHG: CO_2, CH_4, N_2O, HFC, PFC and SF_6. Emissions per type of gas are weighted by their global warming potential before adding.	Blue, green and grey WF. If added, the three components are added without weighting.
Entities for which the footprint can be calculated	Processes, products, companies, industry sectors, individual consumers, groups of consumers, geographically	Processes, products, companies, industry sectors, individual consumers, groups of consumers, geographically delineated areas.

	delineated areas.	
Calculation methods	Bottom-up approach: • For processes, products and small Entities • The method of Life Cycle Assessment (LCA) Top-down approach: • For sector, national and global studies The method of Environmentally Extended • Input-Output Analysis (EE-IOA) Hybrid approach: • LCA and EE-IOA for products, nations, organizations	Bottom-up approach: • For processes, products and businesses, but also for sector, national and global studies • The method of bottom-up accounting in Water Footprint Assessment (WFA) • For products, the accounting along supply chains in WFA is similar to the accounting in the Life Cycle Inventory stage of LCA studies Top-down approach: • For sector, national and global studies • The method of top-down accounting in WFA, which is based on drawing national virtual water trade balances • The method of EE-IOA is used as an alternative
Scope	1. Direct emissions 2. Indirect emissions from electricity used 3. Other indirect emissions	Always includes direct and indirect WF.
Sustainability of the footprint	Additional information is required to assess the sustainability of the CF. For the planet as a whole, a maximum allowable GHG concentration needs to be estimated, which needs to be translated to a CF cap. For specific processes and products, CF bench- marks can be used.	Additional information is required to assess the sustainability of the WF. Per catchment area, freshwater availability and waste assimilation capacity need to be estimated, which form a WF cap for the catchment? For specific processes and products, WF benchmarks can be used.

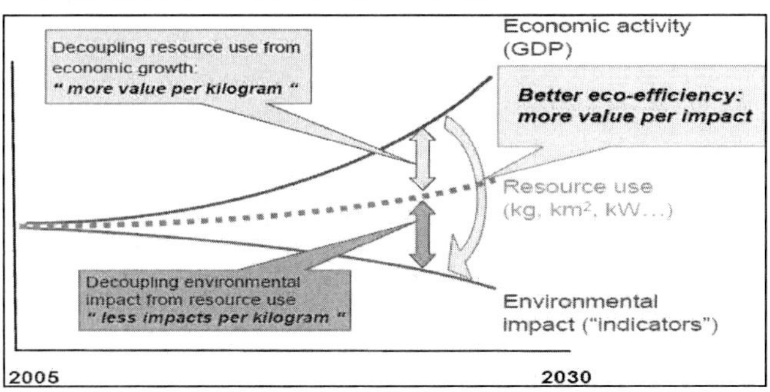

Fig 11.2: The double decoupling objective from the Thematic Strategy on the Use of Natural Resources

To achieve this through so-called double decoupling: the decoupling of resource use from economic growth and environmental impacts from resource use (Fig 11.2). The Resource Strategy requires resource-specific indicators to evaluate the environmental impact of resource use. As part of the implementation of the Resource Strategy, a basket of indicators is being developed, which aims to illustrate the different types of impacts (Best *et al.*, 2008).

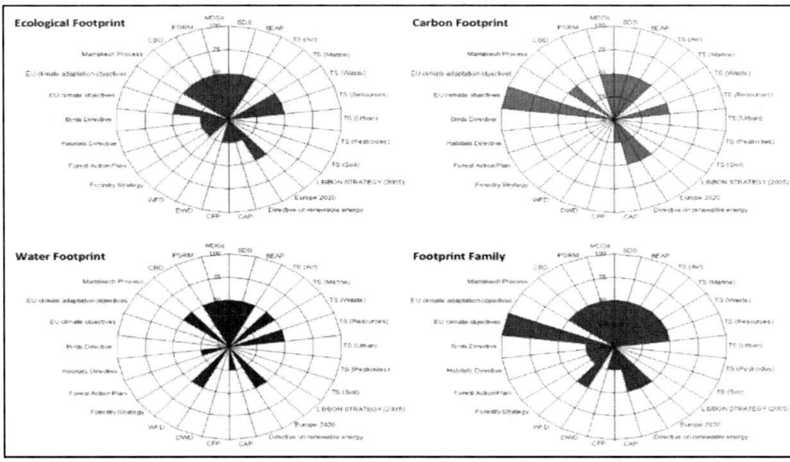

Fig 11.3: Indicator-Policy Radar. It summarizes the range of applicability and the depth of the assessment for each of the Footprint indicators as well as the whole Footprint Family. For any given policy, the radar highlights whether the indicator is able to address such policy fully (100), sufficiently (75), partly (50), marginally (25) or not at all (0).

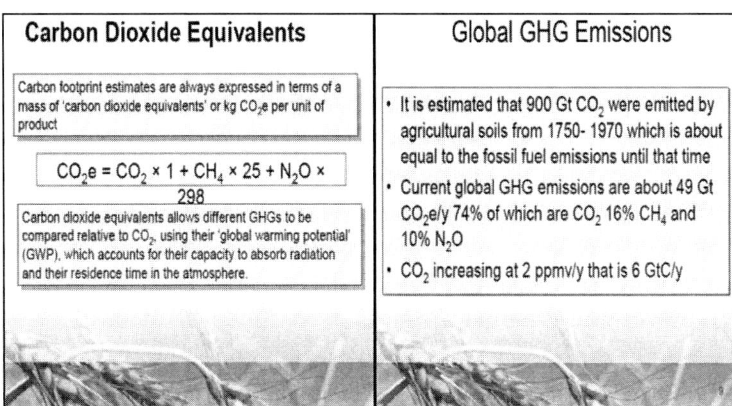

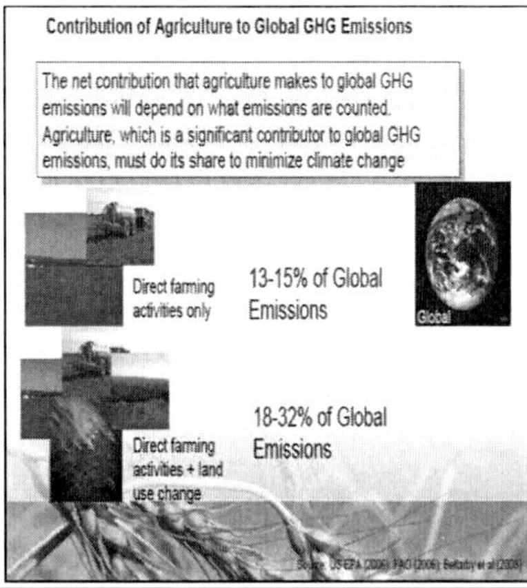

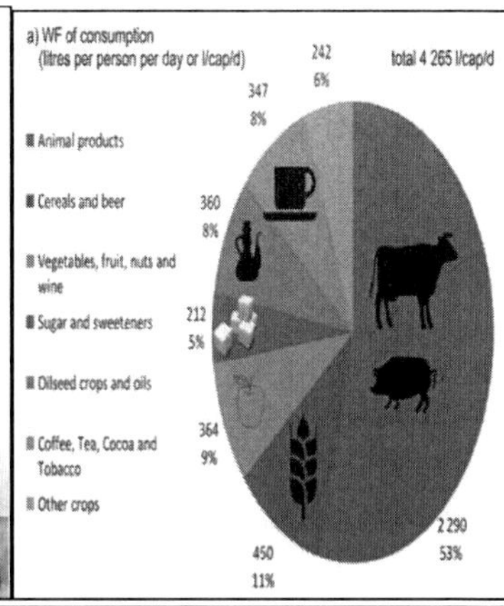

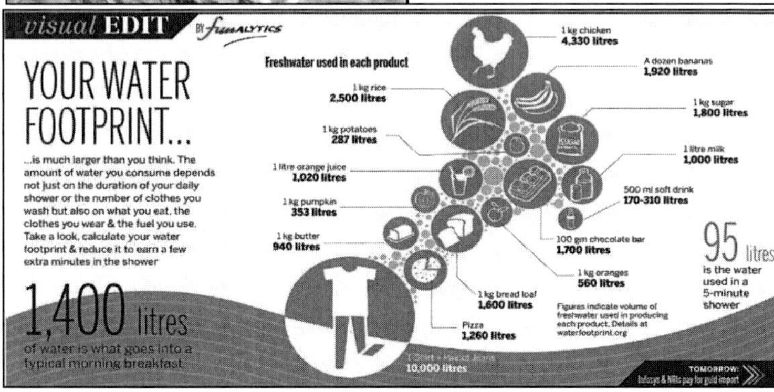

The water footprint of an individual, community or business is defined as the total volume of freshwater that is used to produce the goods and services consumed by the individual or community or produced by the business.

Water use is measured in terms of water volumes consumed and/ or polluted per unit of time. The water footprint is a geographically explicit indicator that not only shows volumes of water use and pollution, but also the locations.

Types of Water Footprint

Water footprint for any products or processes consists of three components: Blue, Green and Grey.

1. Blue Water Footprint (WF_{blue}) is the volume of freshwater that evaporated/uses from global blue water resources (surface water and ground water) to produce the goods and service consumed by the individual or community (either loss through evapo-transpiration, incorporated in products or transferred to non-blue

catchments). or Blue Water Footprint (WF_{blue}) refers to the ratio of volume of blue water consumed (m^3 ha^{-1}) during the life cycle of a crop to the quantity of economic crop yield (tha^{-1}) produced.

$$WF_{blue}\,(m^3\,t^{-1}) = \frac{Volume\ of\ Blue\ Water\ Use(m^3\,ha^{-1})}{Grain\ yield\ of\ the\ crop\ (t\,ha^{-1})} \text{ .... } (1)$$

2. Green Water Footprient (WF_{green}) refers to the ratio of loss of green water resources (profile stored soil moisture or rainwater in so far as it does not become runoff) due to evaporation or evapo-transpiration during the crop growth period to the quantity of economic crop yield (tha^{-1}) produced. Thus,

$$WF_{green}\,(m^3\,t^{-1}) = \frac{Volume\ of\ Green\ Water\ Use(m^3\,ha^{-1})}{Grain\ yield\ of\ the\ crop\ (t\,ha^{-1})} \text{ .... } (2)$$

When no rainfall is received during crop growth period, effective rainfall components is zero but stored profile residual soil moisture of rainy season (PSMC) may serve as source of green water footprients.

3. Grey Water Footprient (WF_{grey}) is defined as the volume of polluted water that associates with the production of all goods and services for the individual or community. The latter can be estimated as the volume of water that is required to dilute pollutants to such an extent that the quality of the water remains at or above agreed water quality standards.

$$WF_{grey}\,(m^3\,t^{-1}) = \frac{Volume\ of\ Grey\ Water\ Use(m^3\,ha^{-1})}{Grain\ yield\ of\ the\ crop\ (t\,ha^{-1})} \text{ .... } (3)$$

The water footprint of a product is always expressed as volume of green, blue and grey water consumed per product unit Examples:

- Water volume per unit mass (for products where weight is a good indicator of quantity).
- Water volume per unit of money (for prodicts where value tells more than weight).
- Water volume per piece (for products that are counted per piece rather than weight).
- Water volume per unit of energy (per kcal for food products, or per joule foe electricity or fules).
- The water footprints (WFs) of a product can be used to provide

information to consumers about the water-related impacts of products they use or to give policy makers an idea of how much water is being 'traded' through imports and exports of the product. As for example to produce one kilogram of rice approximately 3000 litres of water or for production of one kilogram of beef requires 15 thousand litres of water (Dourte and Fraisse, 2012). The actual water footprint of a product depends upon the type of production system, the composition and origin of the raw materials etc.

- WFs of a consumer will provide insight about his or her direct and indirect freshwater use. The direct water use is the water used at home, while the indirect water use relates to the total volume of freshwater that is used to produce the goods and services consumed. The water footprint of consumers can be expressed in terms of water volume per unit of time per capita. Example, the average resident in China and India has a water footprint of 1,071 and 1,089 m^3/ person/year, respectively. (Dourte and Fraisse, 2012).

- WFs of a geographical or land area (country, watershed, or nation) provide information on water used to produce the goods and services consumed by the inhabitants of the nation. The internal water footprint is the appropriation of domestic water resources of that country; the external water footprint is the appropriation of water resources in other countries. The water footprint within a geographically delineated area is expressed as water volume per unit of time. For example, the United States has the highest per capita total water footprint of any nation 2,480 m^3 water/person/year compared to 700 m^3 water/person/year for China. However, the U.S. is by far the leading exporter of water because of the large amount of agricultural exports (Dourte and Fraisse, 2012).

- WFs of an agricultural crop can be used to compare consumptive water use among different agricultural systems in different regions, or it can be used at a farm level to compare water use among different management practices.The water footprints of crops are expressed as volume of water consumed per unit quantity of produce (m^3 ton^{-1} or litre Kg^{-1}). Many products contain ingredients from agriculture of forestry. Crops are used for food, feed, fibre, fuel, oil, soaps, cosmetics, etc. Wood from trees and shrubs is used for timber, paper and fuel as well. Providing a water footprint label on agricultural based products could give consumers more information about the products water footprint.

- WFs can be used for making comparisons of consumptive water use among different agricultural systems: For example, converting a rain-fed system to conservation agriculture may decrease the water footprint as there may be an increase in infiltration of rainfall and a reduction in non beneficial soil evaporation.
- WFs in agriculture can be an important tool for reflecting water conservation impacts from various farm management options like changes in irrigation management, conservation tillage, crop selection, and rotations can all have meaningful impacts on farm-level water footprints.
- When comparing different management strategies, a lower water footprint in a low-rainfall or a highly variable rainfall situation suggests higher water use efficiency.
- Regional comparisons of water footprints will suggest that production should be shifted to an area where production would have a lower water footprint by comparing management systems in different regions/climates.

Accounting Farm Level Crop Water Footprints

The water footprint of a crop is a special case of a process water footprint and has three components viz., green water footprint (soil evaporation or crop evapo-transpiration of water supplied from the rainfall or contribution from the stored soil moisture of the profile for crop production), blue water footprint (soil evaporation or crop evapo-transpiration of the irrigation water supplied from surface and renewable ground water sources) and the grey water footprint (volume of water required to dilute pollutants to such an extent that concentrations are reduced to agreed maximum acceptable limits). (Hoekstra *et al.*, 2011).

Thus, total water footprints (WF_{total})

$$WF_{total} = (WF_{green}) + (WF_{blue}) + (WF_{grey}) \left({volume}/{mass} \right) \quad \text{..................} (4)$$

The unit of crop water footprints is thus volume per unit mass (often litres Kg^{-1} or, equivalently, m^3 ton^{-1}; 'ton'refers to a matric ton of 1000kg). The yield in the denominator of the water footprint components is the yield at standard marketable moisture content. Therefore, if a yield is measured in a field based on grains that were harvested above marketable moisture content, the yield value should be adjusted downward to account for the grain drying needed prior to marketing the crop (Hoekstra and Chapagain, 2008).

Thus, crop water footprints

$$\frac{WU_{green} + WU_{blue} + WU_{grey}\ (m^3\ ha^{-1})}{Economic\ yield\ of\ the\ crop\ (t\ ha^{-1})} \quad \ldots\ldots\ldots\ldots\ldots\ldots\ldots\ldots \quad (5)$$

WU_{green}= 'green' crop water use, WU_{blue}= 'blue' crop water use and WR_{grey}= 'grey' crop water use.

Water Footprint Concept

The concept of the water footprint has been introduced to create a consumption based indicator of water use (Hoekstra & Chapagain, 2007). This is contrast to the traditional production-sector based indicators of water use, that are useful in Water management but do not indicate the water that is actually needed by the inhabitants of a country in relation to their consumption pattern.

The concept of the water footprint has been introduced to create a consumption based indicator of water use (Hoekstra & Chapagain, 2007). This is contrast to the traditional production-sector based indicators of water use, that are useful in Water management but do not indicate the water that is actually needed by the inhabitants of a country in relation to their consumption pattern. This concept is developed in analogy to the concept of the ecological footprint (Wackernagel & Rees, 1996).

The water footprint can be divided into an internal and an external water footprint. The internal component covers the use of domestic water resources and the external component covers the use of water resources elsewhere. Furthermore, an agricultural, an industrial and a domestic component of water footprint can be assessed. Here, the agricultural component corresponds with the water use in the agricultural sector (i.e. in the form of crop evapo-transpiration or water pollution), the industrial component corresponds with the water use in the industrial sector and the domestic component with the water use in the domestic sector.

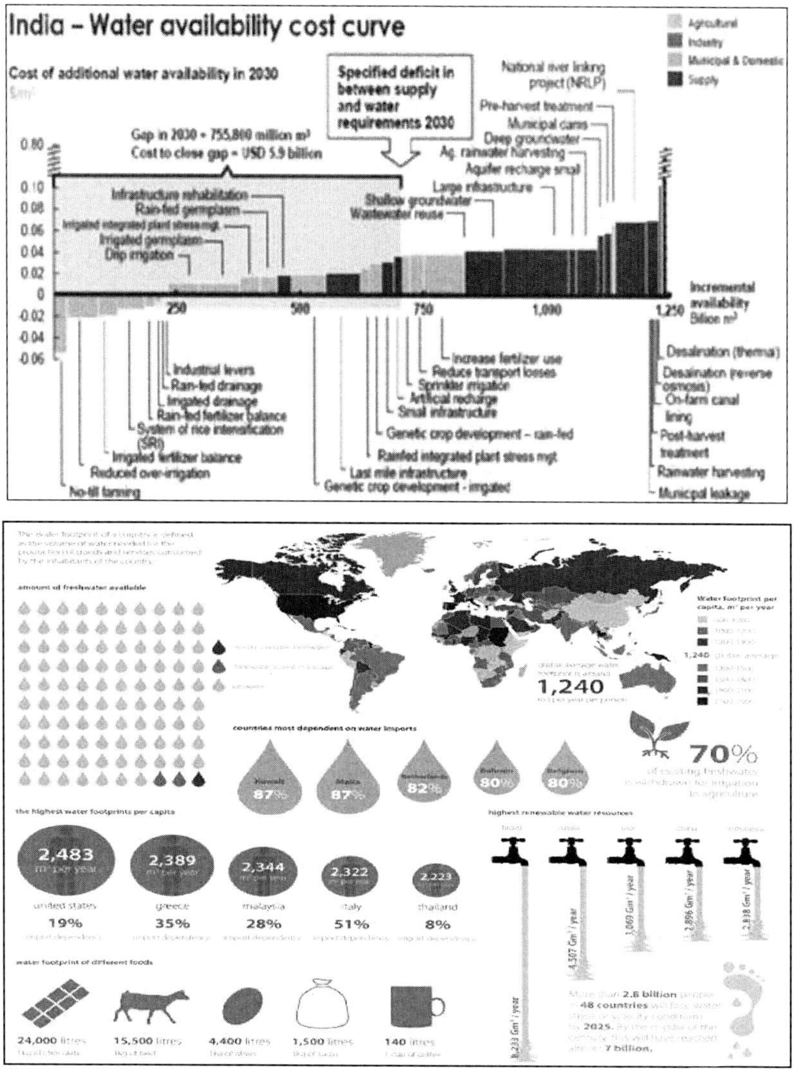

Source: 2030 Water Resource Group

Fig 11.4: India-water availability cost curve

Water Saving as a Result of Comparative Advantage in Water Productivity

The interstate trade in crops that leads to a global water loss can still be justified from a water management perspective with the theory of comparative advantage. The state of Rajasthan has an absolute disadvantage in the water productivity of most of its crops. This is because of a dry hot climate in combination with a low development of agricultural practice.

When only absolute advantages in water productivity are observed, the conclusion would be that the export of most crops leads to global water loss and that water will be saved when this export of virtual water from Rajasthan is stopped.

The theory of comparative advantage shows that the elimination of export of virtual water from Rajasthan does not necessarily lead to water saving. This is because the theory of comparative advantage focuses on relative advantages in water productivity instead of absolute advantages in water productivity. The logic behind this theory is that when a state has an absolute disadvantage in water productivity of all crops, the maximum global benefit from the state water resources is created when these water resources are used to produce the crop, of which the water productivity has the least absolute disadvantages. Eight countries- India, china, the USA, the Russian Federation, Indonesia, Nigeria, Brazil and Pakistan-together contribute fifty percent to the total global water footprient. India (13%), China (12%) and the USA (9%) are the largest consumers of the global water resources (Fig 11.5).

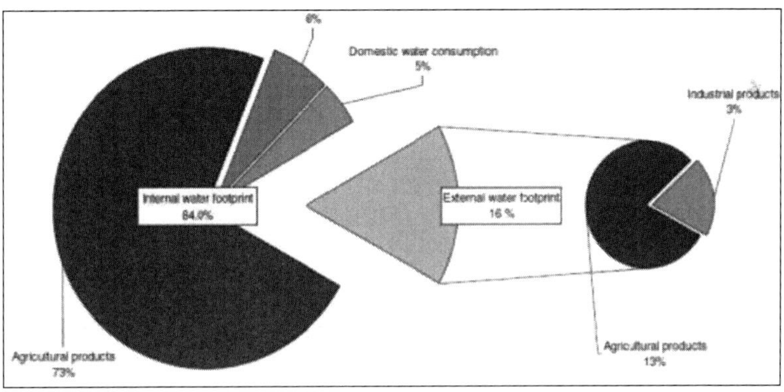

Fig 11.5: Contribution of different consumption categories to the global water footprient, with a distinction between the internal and external footprient

Moving Toward a Holistic Approach to Water and Carbon Footprints

In recent years, there has been an increase in the number of activities considering water related business risk and incorporating water into their sustainability strategies. To support these efforts, stakeholders have developed tools10 (e.g., Water Footprint Assessment Tool, Water Risk Filter, Aqueduct, Global Water Tool) and guidance11 (e.g., Water Footprint Assessment Manual, Draft Standard ISO 14046 Water Footprint, CEO Water Mandate, Alliance for Water Stewardship) to assist companies on the water stewardship journey.

A useful framework has been proposed by the CEO Water Mandate in August 2012. This has been called the Water Stewardship Maturity Progression (WSMP) and it has a similarity with the approach to Water Footprint Assessment and carbon footprint by firstly considering impacts at an operational level before venturing further afield. This five-step process comprises:

I. Improve operational water performance
II. Understand how the company interacts with river basins
III. Develop a comprehensive water strategy
IV. Leverage improved performance in the value chain
V. Advance sustainable water management and collective action

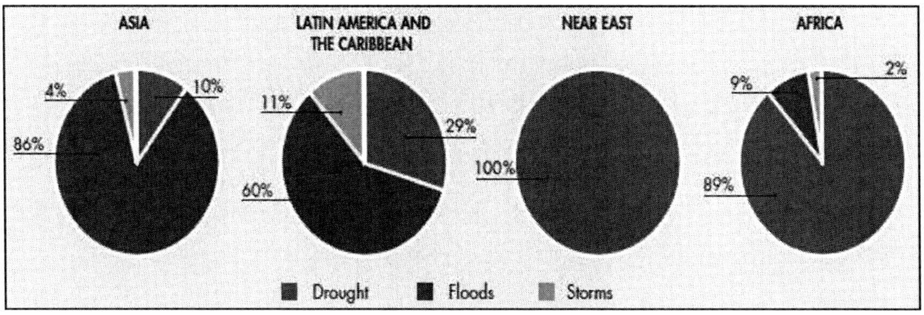

Fig 11.6: Crop and livestock production losses after medium to large scale climate related disasters, by type of hazard, 2003-13.

Extreme weather events are of major significance to agriculture. An FAO study estimated that, between 2003 and 2013, some 25 percent of the total economic impact of climate-related disasters in developing countries was felt in agriculture; when only drought is considered, the share rises to 84 percent (FAO, 2015). The types of hazard vary widely by region (Fig 11.6 and Fig 11.7). Assessing climate change impacts on agriculture requires integrated use of climate, crop, and economic models to take into account the reaction to changing conditions in the sector, including management decisions, land-use choices, international trade and prices, as well as consumers. For this reason, the climate research community has developed over the past two decades sets of scenarios that describe plausible future trajectories and represent many of the major driving forces that are important for informing climate change policy.

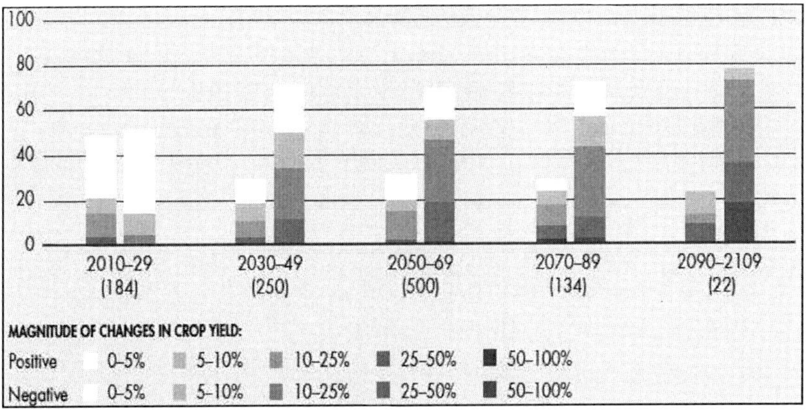

Sources: Porter *et al*., (2014) and Challinor *et al*., (2014).

Note: Number of estimates of change in crop yield is shown in parentheses.

Fig 11.7: Projected changes in crop yields for all locations worldwide owing to climate change

Efficient Water Management

As climate change alters rainfall and water availability patterns, the capacit y to deal with water scarcity or water excess will be crucial in efforts to sustainably improve productivity. Increasing water use efficiency in agricultural systems under climate change may require action in the fields of policy, investment and water management, and institutional and technical changes applied at different scales: on fields and farms, in irrigation schemes, in watersheds or aquifers, in river basins and at the national level (FAO, 2013a). Model projections indicate that the benefits to food security are greater – up to three times greater than that which can be obtained from the improved use of nitrogen alone – when a combined set of improved practices is adopted, compared with the benefits of a single practice (Rosegrant et. al., 2014). However, some technologies that improve water use efficiency, such as drip irrigation, require energy. More broadly, there are often trade-offs and possible synergies in the use of water, energy and land for food production (Fig 11.8). The "water-energy-food nexus" approach is a useful concept when planning the use of these resources in agrifood chains (FAO, 2014b).

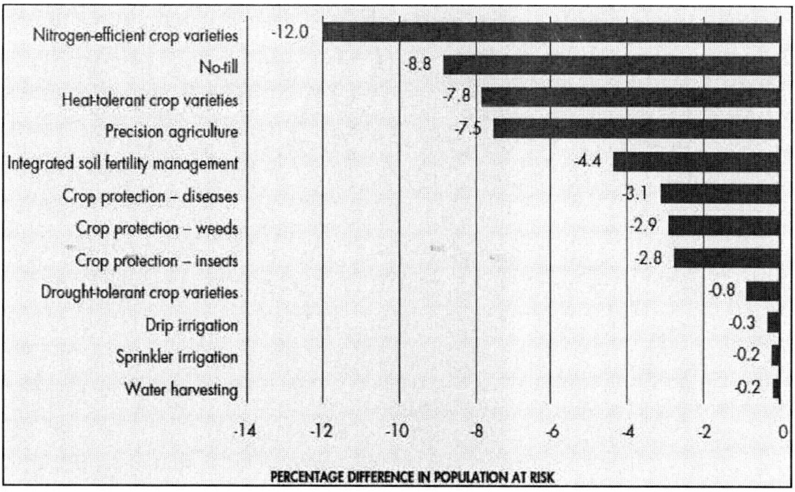

Source: Rosegrant *et al.*, (2014)

Fig 11.8: Change in 2050 in the number of people at risk of hunger, relative to the baseline scenario after adoption of improved agricultural technologies

Water Footprint of Primary Crops and Derived Crop Products per Ton

The average water footprint per ton of primary crop differs significantly among crops and across production regions. Crops with a high yield or large fraction of crop biomass that is harvested generally have a smaller water footprint per ton than crops with a low yield or small fraction of crop biomass harvested. When considered per ton of product, commodities with relatively large water footprints are: coffee, tea, cocoa, tobacco, spices, nuts, rubber and fibers (Fig 11.9). For food crops, the global average water footprint per ton of crop increases from sugar crops (roughly 200 m^3/ton), vegetables (~300 m^3/ton), roots and tubers (~400 m^3ton^{-1}), fruits (~1000 m^3ton^{-1}), cereals (~1600 m^3ton^{-1}), oil crops (~2400 m^3ton^{-1}), pulses (~4000 m^3ton^{-1}), spices (~7000 m^3ton^{-1}) to nuts (~9000 m^3ton^{-1}). The water footprint varies, however, across different crops per crop category. Besides, if one considers the water footprint per kcal, the picture changes as well. Vegetables and fruits, which have a relatively small water footprint per kg but a low caloric content, have a relatively large water footprint per kcal.

Global average water footprints of selected primary crops and their derived products are comparing the water footprints of different products:

The average water footprint for cereal crops is 1644 m^3ton^{-1}, but the footprint for wheat is relatively large (1827 m^3ton^{-1}),while for maize it is relatively small (1222 m^3ton^{-1}). The average water footprint of rice is close to the average for all cereals together.

Sugar obtained from sugar beet has a smaller water footprint than sugar from sugar cane. Besides, the blue component in the total water footprint of beet sugar (20%) is smaller than for cane sugar (27%).

For vegetable oils we find a large variation in water footprints: maize oil 2600 m^3ton^{-1}; cotton-seed oil 3800 m^3ton^{-1}; soybean oil 4200 m^3ton^{-1}; rapeseed oil 4300 m^3ton^{-1}; palm oil 5000 m^3ton^{-1}; sunflower oil 6800 m^3ton^{-1}; groundnut oil 7500 m^3ton^{-1}; linseed oil 9400 m^3ton^{-1}; olive oil 14500 m^3ton^{-1}; castor oil 24700 m^3ton^{-1}.

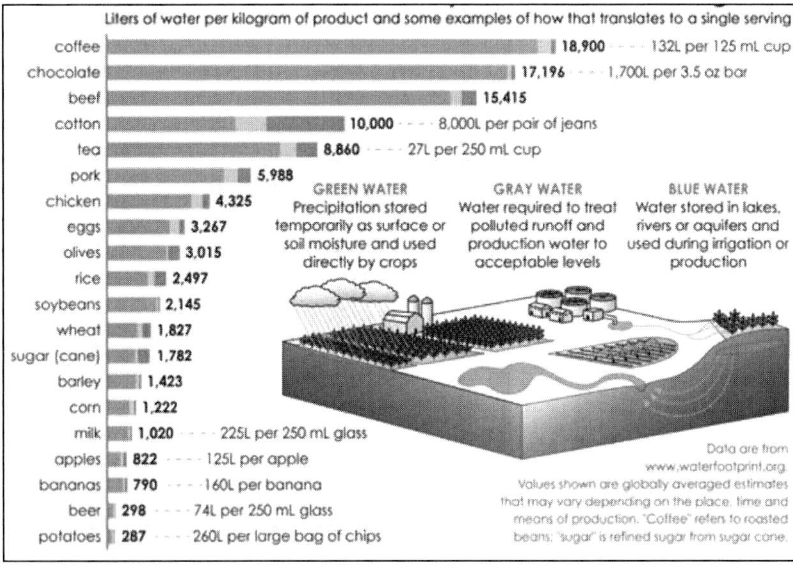

Source: Mekonnen & Hoekstra (2010)

Fig 11.9: How much water does it take to produce these common goods?

For fruits we find a similar variation in water footprints: water melon 235 m^3ton^{-1}; pineapple 255 m^3ton^{-1}; papaya 460 m^3ton^{-1}; orange 560 m^3ton^{-1}; banana 790 m^3ton^{-1}; apple 820 m^3ton^{-1}; peach 910 m^3ton^{-1}; pear 920m^3ton^{-1}; apricot 1300 m^3ton^{-1}; plums 2200 m^3ton^{-1}; dates 2300 m^3ton^{-1}; grapes 2400 m^3ton^{-1}; figs 3350 m^3ton^{-1}.

For alcoholic beverages we find: a water footprint of 300m^3ton^{-1} for beer and 870m^3ton^{-1} for wine.The water footprints of juices vary from tomato juice (270m^3ton^{-1}), grapefruit juice (675m^3 ton^{-1}), orange juice (1000 m^3ton^{-1}) and apple juice (1100 m^3ton^{-1}) to pineapple juice (1300m^3ton^{-1}).The water footprint of coffee (130 liter/cup, based on use of 7 gram of roasted coffee per cup) is much larger than the water footprint of tea (27 liter/cup, based on use of 3 gram of black tea per cup). The water footprint of cotton fibers is substantially larger than the water footprints of sisal and flax fibers, which are again larger than the water footprints of jute and hemp fibers.

Total Water Footprint of Crop Production at National and Sub-National Level

At the country level, the largest total water footprints were estimated for India (1047 Gm^3yr^{-1}), China (967 Gm^3yr^{-1}), the USA (826 Gm^3yr^{-1}), Brazil (329 Gm^3yr^{-1}), Russia (327 Gm^3yr^{-1}) and Indonesia (318 Gm^3yr^{-1}). These six countries together account for about half of the global total water footprint related to crop production. The largest green water footprints are also found in these six countries: India (716 Gm^3yr^{-1}), China (624 Gm^3yr^{-1}), the USA (612 Gm^3yr^{-1}), Russia (305 Gm^3yr^{-1}), Brazil (304 Gm^3yr^{-1}) and Indonesia (286 Gm^3yr^{-1}). Data per country are shown in (Table 11.2) for the largest producers. The water footprint of crop production per crop at sub-national level (state or province level), the largest green water footprints can be found in Uttar Pradesh (88 Gm^3yr^{-1}), Maharashtra (86 Gm^3yr^{-1}), Karnataka (65 Gm^3yr^{-1}), Andhra Pradesh (61 Gm^3yr^{-1}), and Madhya Pradesh (60 Gm^3yr^{-1}), all in India. The largest blue water footprints were calculated for India (231 Gm^3yr^{-1}), China (119 Gm^3yr^{-1}), the USA (96Gm^3yr^{-1}) and Pakistan (74 Gm^3yr^{-1}). These four countries together account for 58% of the total blue water footprint related to crop production. At sub-national level, the largest blue water footprints were found in: Uttar Pradesh (59 Gm^3yr^{-1}) and Madhya Pradesh (24 Gm^3yr^{-1}) in India; Punjab (50 Gm^3yr^{-1}) in Pakistan; and California (20 Gm^3yr^{-1}) in the USA. Large grey water footprints were estimated for China (224 Gm^3yr^{-1}), the USA (118 Gm^3yr^{-1}) and India (99 Gm^3yr^{-1}).

Table 11.2: The water footprint of crop production in selected countries

Country	Water Footprint of Crop Production (Gm^3yr^{-1})			
	Green	Blue	Grey	Total
India	716.0	231.4	99.4	1047
China	623.9	118.9	223.8	967
USA	612.0	95.9	118.2	826
Brazil	303.7	8.9	16.0	329
Russia	304.8	10.4	11.6	327
Indonesia	285.5	11.5	20.9	318
Nigeria	190.6	1.1	0.6	192
Argentina	157.6	4.3	5.0	167
Canada	120.3	1.6	18.2	140
Pakistan	40.6	74.3	21.8	137
World	5771	899	733	7404

Total Water Footprint of Crop Production at River Basin Level

At the river basin level, large water footprints were calculated for the Mississippi, Ganges, Yangtze, Indus and Parana River basins (Table 11.3). These five river basins together account for 23% of the global water footprint related to crop production. The largest green water footprint was calculated for the Mississippi river basin (424 Gm^3yr^{-1}). The largest blue water footprints were found in the basins of the Indus (117 Gm^3yr^{-1}) and Ganges (108 Gm^3yr^{-1}).These two river basins together account for 25% of the global blue water footprint. Both basins are under severe water stress (Alcamo *et al.*, 2007).

Table 11.3: The water footprint of crop production in selected river basins

River basin*	Water footprint of crop production ($Gm^3\ yr^{-1}$)			
	Green	**Blue**	**Grey**	**Total**
Mississippi	424	40 7	70	534
Ganges	260	108	39	408
Yangtze (Chang Jiang)	177	18	61	256
Indus	102	117	34	253
Parana	237	3.2	9.4	250
Niger	186	1.7	0.5	188
Nile	131	29	6.9	167
Huang He (Yellow River)	80	21	31	132
Nelson	108	1.5	18	128
Danube	106	1.8	11	119
Krishna	89	21	8.7	118
Volga	101	3.4	3.9	108
Ob	92	1.8	1.8	95
World	5771	899	733	7404

* River basins grid data from Global Runoff Data Centre (2007).

References

1. Challinor AJ, Watson J, Lobell DB, Howden SM *et al.*, A meta-analysis of crop yield under climate change and adaptation. Nature Climate Change. 2014; 4:287–291.

2. FAO. FAOSTAT. Online Statistical Database (retrieved 30 July 2016), 2016d.

3. Galli A, Wiedmann T, Ercin E, Knoblauch D *et al.*, Integrating

ecological, carbon and water footprint into a 'footprint family' of indicators: Definition and role in tracking human pressure on the planet. Ecological Indicators. 2012; 16:100–112.

4. Hoekstra AY. Human appropriation of natural capital: A comparison of ecological footprint and water footprint analysis. Ecological Economics. 2009; 68(7):1963–74.
5. Leach AM, Galloway JN, Bleeker A, Erisman JW *et al.*, A nitrogen footprint model to help consumers understand their role in nitrogen losses to the environment. Environmental Development. 2012; 1(1):40–66.
6. Porter JR, Xie L, Challinor AJ, Cochrane K *et al.*, Food security and food production systems. In: Field CB, Barros VR, Dokken DJ, Mach KJ, Mastrandrea MD, Bilir TE, Chatterjee M, Ebi KL, Estrada YO, Genova RC, Girma B, Kissel ES, Levy AN, MacCracken S, Mastrandrea PR, White LL, eds. Climate change: impacts, adaptation, and vulnerability. Part A: global and sectoral aspects. Contribution of Working Group II to the Fifth Assessment Report of the Intergovernmental Panel on Climate Change. Cambridge, UK, and New York, USA, Cambridge University Press, 2014.
7. Rees WE. Ecological footprints and appropriated carrying capacity:
8. What urban economics leaves out? Environment and Urbanization. 1992; 4(2):121– 30.

Chapter - 12
Climate Smart Agriculture (CSA): An Option for Changing Climatic Situation

<u>**Authors**</u>

Dr. R.K.Gupta
Borlaug Institute for South Asia (BISA), New Delhi, India

Dr. R.K. Naresh
Sardar Vallabhbhai Patel University of Agriculture & Technology, Meerut, Uttar Pradesh, India

Chapter - 12

Climate Smart Agriculture (CSA): An Option for Changing Climatic Situation

Dr. R.K. Gupta and Dr. R.K. Naresh

World population is increasing day by day and at the same time agriculture is threatened due to natural resource degradation and climate change. Production stability, agricultural productivity, income and food security is negatively affected by changing climate. Therefore, agriculture must change according to present situation for meeting the need of food security and also withstanding under changing climatic situation. Projected estimates based on food consumption pattern and population growth show that agriculture production will require enhancing by 65% to meet the need of burgeoning population by 2050. Agriculture is a prominent source as well as a sink of greenhouse gases (GHGs). So there is a need to modify agricultural practices in a more sustainable way to overcome these problems. Developing climate-resilient agriculture is thus crucial to achieving future food security and climate change goals. It helps the agricultural system to resist damage and recover quickly by adaptation and mitigation strategies. Mitigation strategies reduce the contribution of agriculture system to greenhouse gas emission, and adaptation strategies provide agriculture production under changing scenarios. This chapter explains different mitigation and adaptation strategies, including farming practices and engineering approaches.

Worldwide, more than a billion of farmers and their families face a great challenge of climate change because agriculture sector is the most vulnerable sector for climate change. Thus, their lives and livelihoods are directly affected by climate change, so there is an urgent need to implement many of the solutions to overcome this problem. Despite the attention paid to agricultural development and food security over the past decades, there are still about 800 million undernourished and 1 billion malnourished people in the world. At the same time, more than 1.4 billion adults are overweight and one-third of all food produced is wasted. Before 2050, the global population is expected to swell to more than 9.7 billion people (United Nations. Deptt. Eco & Social Affairs, 2015). At the same time, global food consumption trends are drastically changing, for example, increasing demand for meat-rich diets. If the current trends in consumption patterns and food waste continue, it is estimated that we will require 60% more food production by 2050 (Alexandratos and Bruninsma, 2012). The International Food Policy Research Institute (IFPRI) estimated that by 2050 about 50 million more people could be at a risk of undernourishment because of climate change.

Climate Smart Agriculture (CSA) as 'Agriculture that sustainably increases productivity, enhances resilience (adaptation), reduces/removes GHGs (mitigation) where possible, and enhances achievement of national food security and development goals'. CSA addresses the challenges which climate change (CC) poses to agricultural production. It is a pathway towards sustainable development and food security and is built on three pillars:

- Increasing agricultural productivity (crops, livestock and fisheries) and income.
- Enhancing resilience or adaptation of livelihoods and ecosystems towards climate extremes.
- Reducing and removing GHG emissions from the atmosphere (FAO 2016).

An agricultural technique or practice that contributes to the achievement of these pillars can be considered climate smart. But often, different techniques perform differently over the three pillars, and therefore have to be combined in an integrated CSA approach to complement each other and maximize their benefits (World Bank 2015, FAO 2015).

I. Sustainability increases productivity and income.

II. Strengthens resilience to climate change and variability.

III. Reduces agriculture's contribution to climate change [greenhouse gas emissions + carbon storage on farmlands].

Climate-Smartness Categories

In the farm is only counted as climate smart if it applies practices that are strong in all climate-smartness categories:

- Soil and water conservation structures.
- Integrate perennial and annual crops.
- Improved livestock enterprises.
- Diversification of enterprises.
- Readiness of a farm plan.

Sometimes it is difficult to assess how climate smart a specific agricultural technology is in a certain context. Climate-smartness indicators, divided in three categories,

- CSA-Technology indicators evaluate beforehand how well technologies will achieve CSA goals.
- CSA-Policy indicators asses to which extent the enabling environment (e.g. policies) support the implementation of CSA.
- CSA-Result indicators monitor the short term impacts of CSA interventions (Rawlins 2015).

How do You Implement CSA?

CSA requires site-specific assessments to identify suitable agricultural production technologies and practices (FAO 2015).

i. Soil and Water conservation measures increase ground cover and use little water.
ii. Manure and compost can decrease use of chemical fertilizers and adequate manure management for biogas production can reduce methane release.
iii. In agro-forestry systems trees and crops coexist and benefit from each other.

Activities that Amplify Climate Change Effects Include

- Inadequate tillage practices that expose the soil release carbon stored in the soil.
- Indiscriminative use and wrong timing of agrochemicals harm the ecosystem.
- Clearing land and burning plant biomass for farming releases carbon stored in the soil.

Why CSA?

Therefore CSA is a basket of agricultural practices and techniques that not only aims at increasing profits and resilience for farmers but does so without harming, often even bettering, environmental parameters. It improves input efficiency, soil quality and benefit-cost returns for farmers while limiting the expected negative effects of climate change on Kenyan agriculture for producers and consumers (World Bank 2015, FAO 2016).

Why is CSA Needed?

According to the FAO there are several reasons that call for the rapid transition of present agricultural production system to a more climate-smart and resilient production system in the backdrop of increasing risks from Climate Change and climate associated disasters. The following are the six important reasons to make the Indian agriculture production system Climate-Smart.

1. The demand for food is increasing and more food has to be produced with the same amount of resources such as the land, water and capital.
2. There is an overall depletion and degradation of natural resource that sustains agriculture production.
3. Subsistence farmers are highly vulnerable to the impacts of Climate Change and there is urgency for a more sustainable approach for adaptation to Climate Change.
4. There is a need for enhancing food security while mitigating Climate Change and preserving the natural resource base.
5. In the face of Climate Change the agricultural production systems have to be more productive, efficient, less variable, greater stability in their outputs and more resilient to risks, shocks and long-term climate variability.
6. The awareness and understanding of the farming communities on the potential impacts of Climate Change on agriculture is limited which calls for the urgent need to create awareness and build their capacity for adaptation.

What is Climate Change?

Climate change (CC) is the long-term or permanent shift of average climatic conditions (FAO 2015). They result in changes of weather patterns and directly affect agricultural production. Some of the most visible changes are:

- Increase in mean temperature;
- Shifts in the onset and end of the rainy seasons;
- Changes in duration, amounts and intensity of rainfall;
- Higher frequency of droughts and floods;
- Changing strength and direction of winds;

- Higher temperatures and stronger solar radiation;
- Occurrence of more and new pests and diseases (FAO 2015, World Bank 2015).

Impact of Climate Change on Agriculture

Climate change is very likely to affect food security at the global, regional and local level. Climate change can disrupt food availability, reduce access to food and affect food quality (Brown *et al.*, 2015). Many advance technologies, such as

- Improved crop varieties,
- Pest control methods,
- Genetically modified organisms and irrigation systems, are widely adopted for crop improvement.
- **By 1880, 1.5°C of Earth's temperature had been increased** (IPCC, 2016). It is projected that throughout the twenty-first century, India will experience warming above global level. It is also expected that India will experience more seasonal variation in temperature with more warming in the winters than in the summers.
- In recent years, extreme heat waves across India have extended together with warmer night temperatures and hotter days, and this trend is expected to continue. This is predicted that the average temperature will rise up to 2.33–4.78°C with a doubling in CO_2 concentrations at the end of this century. These heat waves will lead to increased variability in summer monsoon precipitation, which will result in drastic effects on the agriculture sector in India (Kumar and Gautam, 2014).
- Due to rainfall variability, changes in the frequency and severity of droughts and floods could pose challenges for farmers and ranchers and will ultimately threaten global food security. Meanwhile, the rise in sea level and warming of sea water are likely to cause the habitat ranges of many fish species to shift, which could disrupt ecosystems. Overall, climate change would adversely affect over crops, animals and marine life.
- Higher CO_2 levels can affect crop yields. The elevated CO_2 levels can increase plant growth particularly in plants with C_3 photosynthetic mechanism. Increased CO_2 level also reduces the quality of produce by reducing protein, nitrogen and minerals content in most plant species, including wheat, soybeans and rice; this will threaten human health.

- Climate change also increased pest pressures and reductions in the efficacy of pesticides; thus more pesticides will be required in future which also threatened human health (Ziska *et al.*, 2016).
- Reduced grain and forage quality can reduce the ability of pasture and rangeland to support grazing livestock (Hatfield *et al.*, 2014).
- Climate change also affects weeds, pests and fungi infestation due to an increase in host susceptibility. The ranges and distribution of weeds and pests are likely to increase with climate change. This could cause new problems for farmers' crops previously unexposed to these species. Both historical and future estimated impacts of climate change on cereal crop yields in different locations indicate that the yield loss can be up to −35% for rice, −20% for wheat, −50% for sorghum, −60% for maize and −13% for barley depending on the location, future climate scenarios and projected year (Porter *et al.*, 2014).
- Heat stress also affects animals both directly and indirectly. Over time, heat stress can increase vulnerability to disease, and reduce fertility and milk production.
- Climate change also affects rainfall pattern and distribution; due to this, many areas are facing a problem of drought stress which may threaten pasture and feed supplies. Drought reduces both quantity and quality of forage available to grazing livestock. Some areas could experience longer, more intense droughts, resulting from higher summer temperatures and reduced precipitation.
- Changes in crop production due to drought could also become a problem for animals that rely on grain. Climate change may increase the prevalence of parasites and diseases that affect livestock. More variability in weather pattern could provide the chance of insect, pest infestation in crop plant (CCSP, 2008). Pasture productivity is increased by elevated CO_2 but its quality may be reduced due to high lignin concentration, low protein and digestibility. However, the quality of some of the forages found in pasturelands decreases with higher CO_2. As a result, the overall performance of cattle would be affected and require eating more to get the same nutritional benefits.
- Fisheries and aquaculture, which provide at least 50% of animal protein to millions of people in low-income countries, are already under multiple stresses, including overfishing, habitat loss and water pollution (FAO, 2012).

- Changes in temperature and seasons can also affect the timing of reproduction and migration fishes and other aquatic life. Lifecycle of aquatic animals is controlled by the temperature of aquatic system and the change in the seasons. The increased intensity and frequency of storms, hurricanes and cyclones will adversely affect coastal fisheries, aquaculture and mangroves.
- Forests provide employment and livelihood for more than 100 million people of the world's rural poor. Forests are home to more than 80% of the world's earth biodiversity and provide ecosystem services like food, fuel, timber, medicines and shelter. Aberrant weather limits the capacity of forests to produce goods and services and also affect the people who depend on them directly or indirectly. On the other hand, temperate forest region communities will be benefitted from elevated CO_2 and temperature. While due to changes in occurrence, distribution and frequency of precipitation, this will more likely affect natural biodiversity and also adversely affect crop growth, productivity and ultimately yield (FAO, 2013).
- India and China, the most densely populated countries of the world, to keep pace with the growing population and to sustain world food security will require maintaining at least 4–5% annual growth rate in agriculture. India supports about 17% of human and 11% of livestock population of the world just on 4.2% of water resources and 2.8% land (Singh, 2015). As per recent estimates, India will need to produce about 281 million tonne (mt) food grains, 53.7 mt oilseeds, 22 mt pulses, 127 mt vegetables and 86 mt fruits by 2020–2021 (Singh, 2015).. In India, the average food consumption at present is 550 g per capita per day, whereas in China and USA are 980 and 2850 g, respectively Mall *et al.*, 2005& 2006). To meet the demand for food from this increased population, the country's farmers need to produce 50% more grain by 2020 (Kumar and Gautam, 2014). To enhance food production is under changing climatic conditions like aberrant weather, rising CO_2 level, rising temperature and rising sea level, and we require the reorientation of agriculture from current practices to more sustainable and environmental friendly practices. In this context, scientists are more focused on climate smart or resilience agriculture (CSA). Climate resilience agriculture helps to improve food security under changing climatic situation while also reducing food wastes globally (CCAFS, 2013) and minimizing danger on natural resources.

Some CSA Practices as Reflected in the Three Pillars

CSA Practices	Description	Relevance	Adaptation	Mitigation	Production
Conservation Agriculture	Reduced tillage; crop residue management and intercropping; crop rotation with cereals and legumes	Carbon sequestration; reducing existing emissions	Increased water retention; reducing crop losses	Promoting carbon storage in soil; water retention increased, thus reducing energy for irrigation	Increased productivity due to higher nutrients in soil, healthier soil, reduce soil erosion
Integrated fertility Management	Compost and manure management	Reduce N_2O and CH_4 emissions, improving soil structure	Food security	Reducing pollution	Creating carbon sink; improved soil fertility
Small-scale irrigation	New crop varieties	Improved production; sustainable food security	Reducing total crop failures		Sustainable food production; Creating carbon sink
Crop diversification	New crop varieties	Pest resistance, high yields, drought tolerance	Enhances resilience in agriculture		Sustainable food production; healthier diets
Drought tolerant Varieties	Moisture retention	Water and energy smart	Reducing total crop failure	Short lifecycle reduces environmental impact	Increased incomes and higher yields, climate resilience
Other practices	Early warning system; in situ conservation /harvesting; crop insurance	Reduction of risk Sustainable food security	Sustainable production		Drought resilience agriculture; reduced climate risk; reduced emissions and increased incomes

Climate-Smart Agriculture Indicators

Theme/Pillar	Sub-Theme	Indicator
Productivity	Crop system	Increase in yields, reduction in risk of water/wind erosion, enhances in-soil fertility
	Water use	Increase in irrigated land area, reduction in total water withdrawal, increase in water productivity
	Energy	Reduction in agricultural energy use, increase in use of renewable energy

Resilience	Robustness	Increases stability of production, promotes income diversification, incorporate site-specific knowledge
	Cropping system	Increases resilience to drought
	Livestock system	Increases resilience to drought
Mitigation	Emissions intensity	Reduces GHG emissions
	Sequesters carbon	Increases carbon sequestration

Climate Smart Strategies

Many agricultural technologies and practices such as minimum tillage, different methods of crop establishment, nutrient and irrigation management and residue management can improve crop yields: nutrient and water use efficiency and reduced greenhouse gas (GHG) emissions from agricultural activities (Sapkota *et al.*, 2015). Similarly, the use of improved seeds, rainwater harvesting (RH), Information and Communication Technologies (ICTs) -based agro-advisories and crop/livestock insurances can also help farmers to reduce the impact of climate change and variability (Mittal, 2012). In general, the CSA options integrate innovative and traditional technologies, practices and services that are relevant for particular location and reduce the effect of climate change and provide the opportunities to stand such changing scenario. Adaptation and mitigation are complementary strategies for reducing and managing the risks of climate change. Substantial reduction in GHG emission over the next few decades can reduce the occurrence of climatic variability in the twenty-first century and beyond, increase prospects for effective adaptation, reduce the costs and challenges of mitigation in the longer term and contribute to climate-resilient pathways for sustainable agriculture.

What is Conservation Agriculture (CA)?

Conservation agriculture is a farming approach that can sustainably increase yields from cereals, legumes, fodder, and cash crops. The various practices counted to conservation agriculture are characterized by reducing production costs while following the 4 key principles:

1. Minimal soil disturbance;
2. Permanent ground cover-maintenance of carbon-rich organic matter covering and feeding the soil (e.g. straw and/or other crop residues including cover crops);
3. Crop rotation or sequences and associations of crops and nitrogen-fixing legumes;
4. Balanced application of chemical inputs (FAO 2015, FAO 2016).

Mitigation Strategies

Mitigation practices help to reduce the emission of greenhouse gases from agriculture system because elevated greenhouse gases ultimately lead to climate change.

Mitigation through Farming Practices

A change in the system of cultivation, cultural practices, nutrient management and water management helps to reduce the emission of greenhouse gases.CA practices include soil and water conservation mechanisms such as:

i. Zero or minimum or reduced land tillage techniques like: ripping, sub-soiling, planting basins or strong rooted cover crops or trees: Protective tillage mechanisms not only prepare the soil perfectly for planting while minimizing carbon dioxide losses but even correct compaction and hardpans. They also support the increase of organic soil matter.

ii. Conservation Agriculture with Trees (CAWT): Control pests and weeds thereby ensuring good harvests and reducing post-harvest losses. Serve as vegetative barriers to fight soil loss and facilitate water percolation. Provide fodder, fuel, construction materials, etc. and fixate atmospheric nitrogen enriching the soil with nitrates.

iii. Crop rotation (changing of crop sequences) and association (intercropping with leguminous crops): Crop rotation and association decrease the occurrence of pests, diseases and weeds and protect the soil from leaching. By hosting nitrogen-fixing bacteria, leguminous crops also contribute to good plant growth without (high) fertilizer use.

iv. Permanent soil cover (cover cropping) with mulch or live cover crops (either non-legumes – like sweet potato, etc. or legumes, that fix nitrogen like Cowpeas, Pigeon peas and Desmodium: Permanent soil cover shields the soil surface from heat, wind and rain, keeps it cooler and reduces moisture losses by evaporation. Mulch layers also improve water and nutrients in the soil and contribute to net increase of soil organic matter.

v. Pest and disease management using knapsack sprayer, shallow weeder, contact applicator (zam wipe) and cover crop: CA pest and disease management techniques avoid disturbance of deeper soil layers and field ecosystems thereby protecting beneficial organisms such as bees and avoiding the evaporation of soil-water (FAO 2015).

vi. Crop diversification and crop intensification: Improves soil and water quality and also reduces emission and water requirement and provides food security.

vii. Rangeland and pasture management through rotational grazing and improved forage: Reduces water requirement, helps in withstand drought. Increases long-term grassland productivity.

Methane Emission

Methane is about 25 times more effective as a heat-trapping gas than CO_2. The main sources of methane are wetlands, organic decay, termites, natural gas and oil extraction, biomass burning, rice cultivation, enteric fermentation and refuse landfills. CH_4 emissions can be controlled by the following practices:

- Controlling production, oxidation and transport of CH_4 from paddy field by alteration in water management, particularly promoting mid-season aeration, by short-term drainage, direct-seeded rice (DSR) cultivation. The IPCC, (2016) has estimated that rice cultivation is a major contribution to global warming. Rice cultivation contributes 23% of total greenhouse gas emission from agriculture sector. Under continuous flooding, dry-DSR emitted 24–79% of CH_4 as compared with 8–22% in wet-DSR, whereas, under intermittent irrigation, the reduction ranged from 43 to 75% in dry-DSR compared with conventional tilled transplanted rice (CT-TPR). However, reduction in CH_4 emissions increased further when DSR was combined with mid-season drainage or intermittent irrigation, compared with flooded CT-TPR. The reason for low CH_4 emissions from dry-DSR is aerobic conditions, especially during the early growth stage (Kumar and Ladha, 2011). Methane emission starts at a redox potential of soil below −150 mV and is stimulated at less than −200 mV (Jugsujinda *et al*., 1996).

- Improved organic matter management by promoting aerobic degradation through incorporating or composting it into soil during off-season-drained period.

- Use of rice cultivars with few unproductive tillers, high root oxidative activity and high harvest index.

- Improved management of livestock diet through the use of improved feed additives, substitution of low digestibility feeds with high digestibility ones.

Nitrous Oxide

As a greenhouse, nitrous oxide is 298 times more effective than CO_2. Forests, grasslands, oceans, soils, nitrogenous fertilizers and burning of biomass and fossil fuels are the major sources of nitrous oxide. Site-specific nutrient management (SSNM), integrated nutrient management (INM), use of slow-release nitrogen fertilizers and nitrification inhibitors, and placement of fertilizer in reduced zone helps to reduce emission of NO_2.

Carbon Dioxide

The main sources of carbon dioxide emission are decay of organic matter, forest fires and eruption of volcanoes, burning of fossil fuels, deforestation and land-use changes. CO2 emission mitigates by carbon sequestration, soil management practices and afforestation. Carbon sequestration is one of the best strategies to mitigate CO_2 emission, which is the removal of atmospheric CO_2 into long-lived stable form.

- Carbon sequestration in soil through manipulation of soil moisture and temperature and restoration of soil carbon on degraded lands. Trees, plants and crops absorb CO_2 naturally through photosynthesis and stored as carbon in biomass in tree trunks, branches, foliage and roots and soils (Philippot and Hallin, 2011). Forests and stable grasslands are good sink of carbon because they can store large amounts of carbon in their vegetation and root systems in long-life stable form as compared to field crops. Soils are the largest terrestrial sink for carbon on the planet.

- Soil management practices such as reduced tillage and manuring residue incorporation. Improving soil biodiversity, micro-aggregation and mulching.

Mitigation through Transgenic Crops

Agriculture contributes significantly to greenhouse gas (GHG) emissions. As indicated by researchers, there is a need to give priorities to develop new varieties of crops cultivars that can reduce GHG emissions in farming systems (Barfoot and Brookes, 2014). In this regard, GM crops have been contributing to lower GHG emissions through reducing fuel use, due both to less pesticide applications and increasing the area grown under conservation agriculture (CA), which involves practices such as 'no-till' or 'reduced-till'. It is reported that, in 2012, GM crops were grown roughly on 12% of the world's arable land, which reduced over 26.7 billion kg of carbon dioxide (CO_2) or the equivalent of removing nearly 12 million cars from circulation with a total reduction (Lybbert and Sumner, 2010). Genetically modified crops such as Roundup Ready TM modify the plants through genetic engineering so that they can take more carbon from atmosphere and convert it to oxygen (Shrawat, 2008).

Crops are bred for N-use efficiency (NUE) because this trait is a key factor for reducing N fertilizer pollution, improving yields in N-limited environments and reducing fertilizer costs.

There are various genetic engineering activities which improve NUE in crops (Subbarao *et al.*, 2007). The alanine aminotransferase gene from barley, which catalyses a reversible transamination reaction in the N-assimilation pathway and enhances NUE, appears to be a target for plant breeding option. Keeping N in ammonium form will affect how N remains available for crop uptake and will improve N-recovery, thus reducing groundwater and the atmosphere pollution. There are genes in tropical grasses such as Brachiaria humidicola and in the wheat wild relative Leymus racemosus that inhibit soil nitrification by releasing inhibitory compounds from roots and suppressing nitrosomonas bacteria (Lutz, 2013). Engineering nitrogen fixation for self-fertilizing crops can also help to reduce the use of synthetic fertilizer. The ultimate goal is to transfer the molecular machinery for nitrogen fixation into non-N_2-fixing plants so that they will fix atmospheric N_2 (OECD, 2014). The process is very complex and challenging because N_2-fixing nitrogenase enzyme is adversely affected by oxygen produced during photosynthesis. But agriculture would be revolutionized if plants can be engineered to fix their own nitrogen; this would free agriculture from synthetic nitrogenous fertilizers and significantly decouple it from the fossil fuel industry (Alford *et al.*, 2006).

About one-fifth of global CH_4emissions are from enteric fermentation in ruminant animals. Apart from various rumen manipulations and emission control strategies, genetic engineering is a promising tool to reduce these emissions. The most practical and rapid mitigation procedure may be to reduce the per cow CH_4 emission through animal breeding and genetic selection for feed efficiency as it is permanent and cumulative (Dwivedi *et al.*, 2013). Other options like manipulation in diet composition, supplementation of feed additives and selection of forage plants of high quality for breeding provide solution for reduced CH_4 emission from enteric fermentation.

Adaptation Strategies

Adaptation is a key factor that will shape the future severity of climate change impacts on food production. To deal with the impact of climate change, the potential adaptation strategies are as follows:

Adaptation through Transgenic Approaches

Transgenic approaches are one of the many tools available for modern plant improvement programmes. This approach involves manipulation of genetic material and the fusion of cells beyond normal breeding barriers. The most evident example of genetic engineering is 'transgenic' technology which involves the insertion or deletion of genes and creating genetically modified organisms (GMOs). In genetic engineering or genetic transformation, the genetic material is modified by artificial means. It involves isolation and cutting of a gene at a precise location by using specific enzymes. Identification and transfer of stress-tolerance gene into plant provide improved productivity and adaptation to abiotic stresses (Fig 12.1). Hence, the transformation of major crop species with genes from any biological source (plant, animal and microbial) is the most powerful tool for molecular plant breeding. Transgenic plant can be used as sources of new cultivars and they are also extremely useful as proof-of-concept tools to dissect and characterize the activity and interplay of gene networks for abiotic stress resistance. Transcription factor (TF), and genes involved in adaptation to extreme temperature, drought or salinity in crops are (Hasanuzzaman *et al.*, 2013) as follows:

1. Involved in sense, percept and transduction of signal.
2. Mechanisms of stress response for adaptation and abscisic acid (ABA) biosynthesis for enhancement for drought adaptation.
3. Genes involved in transportation, detoxification and signal transduction as well as TF tolerance to salinity.
4. Some associated to reactive oxygen species (ROS), modification of membrane and production of chaperon, late embryogenesis abundance (LEA) proteins, osmoprotectants/ compitable solutes and TF are continued in crop genetic engineering for temperature extremes.

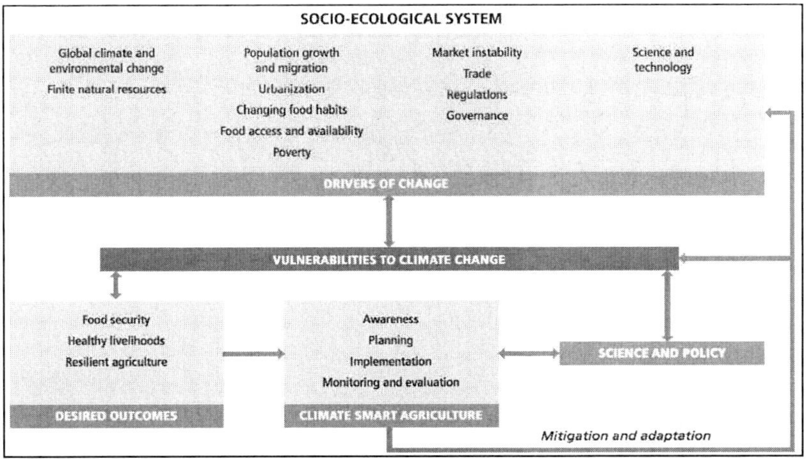

Fig 12.1: Diagram illustrating how climate-smart agriculture can be utilized as an agent for developing resilience, mitigation and adaptation within the socio-ecological system.

Heat

Plant responses to high temperature vary with the degree and duration of high temperature and the plant type. Plants have various mechanisms to cope with high temperatures, for example, by maintaining membrane stability, or by ion transporters, proteins, osmoprotectants, antioxidants and other factors involved in signalling cascades and transcriptional control (Tuteja, 2005). The activity of the enzymes CAT, glutathione S-transferase (GST) and APX was more enhanced in the cultivar and showed better tolerance to heat stress and projection against ROS production. Upregulation of different genes helps the plant to withstand the stress conditions and also adaptation to stress (Kaur and Gupta, 2005). Stress plants receive the internal and external signals by different interrelated or independent pathways which are used to control various responses for its tolerance development (Gusta, 2012). Some stress-associated genes such as ROB5-a stress-inducible gene isolated from bromegrass-enhanced the performance of transgenic canola and potato at high temperatures (Katiyar *et al*., 2003). Likewise, researchers introduced hsp101-a heat shock protein gene from Arabidopsis-in basmati rice. This transgenic rice had a better growth in the recovery phase after suffering heat stress (Kuromori *et al*., 2014).

Drought

Several drought-tolerant transgenic plants, including rice, tomato, soybean, maize, barley and Arabidopsis have been developed. Such genetically engineered plants have generally been developed using gene-encoding proteins that control drought-regulatory networks. Stress signaling networks in drought responses are made of intercellular communication systems, intracellular signalling systems and transcriptional-regulatory complexes (Todaka *et al.*, 2015). Some important proteins like protein kinases, receptor-like kinases, transcription factors, some enzymes related to osmoprotectant or plant hormone synthesis, and other functional and regulatory proteins (Golldack *et al.*, 2014) play an important role in plant defence during drought stress. NAC, AP2/ERF, bZIP and MYC orchestrate regulatory networks are major transcription factor families of plants, which underlie drought stress tolerance (Bhardwaj and Yadav, 2012). Enhanced adaptation to drought could be provided directly by metabolites such as trehalose, mannitol, and glycinebetaine or indirectly through regulation of gene expression through TF and kinases in signal transduction (Karaba *et al.*, 2007). The Arabidopsis' HARDY (HRD) gene improved WUE by enhancing photosynthetic assimilation and reducing transpiration (Castiglionit *et al.*, 2008). Further research by a multinational seed company (Tester and Davenport, 2003) revealed that the encoding gene cold shock protein B (CspB) from Bacillus subtilis, a soil bacterium, allows the transgenic maize plant to react more quickly to drought, slowing its growth and conserving water, thereby making water available for key plant functions after the onset of drought stress. The DroughtGard™ hybrid maize was bred and released for farming in the USA in 2013. Under stress, a DroughtGard™ hybrid used 261 mm of water from the soil while the control used 338 mm of water from the soil. Under drought stress, a maize plant begins producing sugars such as trehalose, which can be broken by the glycoside hydrolase enzyme trehalase. When this sugar is not destroyed, the plant shows enhanced adaptation to drought stress.

Salinity

The damaging effects of salt accumulation in agricultural soils have severely affected agricultural productivity in large swathes of arable land throughout the world (Chinnusaamy *et al.*, 2005). Salt-affected land accounts for more than 6% of the world's total land area. Climate change also causes salinity due to effect on soil water and increase in temperature. Therefore, there is an urgent need to develop salt-tolerant varieties of crop through conventional breeding and transgenic approaches. Genes enhance tolerance in plant against salinity by employing several mechanisms such as by limiting uptake rate of salts from soil and further transport of salts inside plant system, regulate leaf development and the onset of plant senescence and adjust the ionic and osmotic balance of cells in roots and shoots. Salt tolerance mechanism varies with the plant species; however, the capacity to maintain low cytosolic Na^+ is thought to be one of the key determinants of plant salt tolerance (Ortiz *et al.*, 2007). Transgenic plants accumulated more K^+, Ca^{2+} and Mg^{2+} and less Na^+ in their shoots compared with non-transformed controls. The most promising genes for the genetic engineering of salinity tolerance in crops, as noted by scientist (Ortiz *et al.*, 2007), are related to ion transporters and their regulators, as well as the C-repeat-binding factor (He *et al.*, 2005). The successful use of transporter genes has been observed in several plants. He *et al.*, developed transgenic cotton plants with AtNHX1 expression and revealed that plants produced more fibre content and generated more biomass under salt stress in a greenhouse condition (Moghaieb *et al.*, 2011).

Developing Climate-Ready Crops

Development of new crop varieties with higher yield potential and resistance to multiple stresses (biotic and abiotic) will be the key to maintain yield stability. Several drought tolerances varieties have been released in South and Southeast Asia. For example, in 2010 the variety Sahbhagi Dhan, released and notified in India, showed a consistently good performance under transplanted low-land conditions and rain-fed direct-seeded upland (Singh *et al.*, 2014). Nitrogen use efficiency may be reduced under the climate change scenarios because of heavy precipitation and high temperature which increase leaching and volatilization losses, respectively. But at the same time, elevated CO_2 concentration causes plant growth and development which leads to increased demand for nitrogen. So, there is a need to develop the improved root system of plant for nutrient and water absorption. Genetic engineering is the best option for pooling all the desirable traits in a plant to get the ideal plant type for adverse climatic situation.

Crop Diversification

Major shift in terms of diversification of agriculture into crops, commodities, enterprises and cropping/ farming systems is called upon to revert the process of degradation of natural resources, rejuvenations of waste lands and also to make agriculture a profitable business. Diversified agricultural systems may be a productive way to build resilience into agricultural systems. Crop diversification helps farmer against aberrant weather conditions like early season drought, late season drought and dry spell during crop growth season. Intercropping of soybean + pigeon pea (4:2), pearl millet + pigeon pea (3:3), pigeon pea + green gram (1:2) and cotton + green gram (1:1) are more economic than mono-cropping (Prasad *et al*., 2014). Due to growing income and urbanization, demand for high-value food products, such as fruits, vegetables, dairy, meat, eggs and fish, is increasing.

This is reducing the demand for traditional rice and wheat. Diversification of rice-wheat system (RWS) towards high-value commodities will increase income and result in reduced water, fertilizer and other resource uses. Pulses have a positive impact on soil quality because they help fix nitrogen in the soil. Nitrogen-fixing pulse crops have a lower carbon and water footprint as compared to other crops. For instance, the water footprints to produce a kilogram of beef, pork, chicken and soybeans are 43, 18, 11 and 5 times higher than the water footprint of pulses. In fact, one study showed that 1 kg of legume only emits 0.5 kg in CO_2 equivalent, whereas 1 kg of beef produces 9.5 kg in CO_2 equivalent. So the inclusion of pulses in crop rotation and intercropping system shows better productivity, sustainability and environmental safety benefits (IYP, 2016).

Alteration in Land-use Pattern

Change in location of crop and livestock, adjustment in cropping pattern, planting time and methods, fertilizer and pesticide use pattern, and other management practices help to reduce the risk of climate. Alternate land-use management practices also reduce disease and pest outbreak and provide remunerative production under aberrant weather situation.

Changing Cropping Season

Yield instability is reduced by changing planting time which reduce the impact of temperature increase-induced spikelet sterility and also avoid flowering period to coincide with the hottest period. Crop calendar provides the information about crop location and cropping pattern based on weather pattern which helps the farmer for growing crop according to the occurrence of weather events. Cropping systems may have to be altered to the growth of suitable cultivars, increasing crop intensities (i.e. the number of successive crop produced per unit area per year) or planting different types of crops.

How Does CA Contribute to CSA?

Conservation Agriculture practices' contribution to the three pillars of climate change differ significantly:

1. Increasing agricultural productivity and income: Conservation Agriculture ensures optimum plant growth without increased fertilizer use by carefully protecting soil quality and nutrients. This way soils stay productive and income earned does not need to be spent on extra fertilizers.

2. Enhancing resilience or adaptation of livelihoods and ecosystems towards climate extremes CA needs less water and helps to stabilize yields in years of extreme weather, pests, and diseases. It does so by protecting the soil, its surrounding ecosystems and wisely combining crops, legumes and trees.
3. Reducing and removing GHG emissions from the atmosphere CA practices such as minimum tillage mechanisms have great potential to mitigate GHG emissions of nitrogen and CO_2 and even help sequestrating them in the soil (World Bank 2015, FAO 2015, FAO 2016).

References

1. Alford AR, Hegarty RS, Parnell PF, Cacho OJ, *et al.*, The impact of breeding to reduce residual feed intake on enteric methane emissions from the Australian beef industry. Australian J Exp Agri. 2006; 46:813-820
2. Alexandratos N, Bruninsma J. World agriculture towards 2030/2050: The 2012 revision. ESA Working Paper No. 12-03. Rome, Italy: Food and Agriculture Organization of the United Nations (FAO); 2012
3. Barfoot P and Brookes G. Key global environmental impacts of genetically modified (GM) crop use 1996-2012. GM Crops and Food: Biotechnology in Agriculture and the Food Chain. 2014; 5(2):149-160
4. Bhardwaj J, and Yadav SK. Genetic mechanisms of drought stress tolerance, implications of transgenic crops for agriculture. In: Lichtfouse E, editor. Agroecology and Strategies for Climate Change: Springer, 2012, 213-235
5. Brown ME, Antle JM, Backlund P, Carr ER, *et al.*, Climate Change, Global Food Security, and the U.S. Food System. USDA, 2015, 146
6. CCAFS. Big Facts on Climate Change, Agriculture and Food Security. Copenhagen, Denmark: CGIAR Research Program on Climate Change, Agriculture and Food Security (CCAFS); 2013
7. CCSP. The Effects of Climate Change on Agriculture, Land Resources, Water Resources, and Biodiversity in the United States. A Report by the U.S. Climate Change Science Program and the Subcommittee on Global Change Research. In: Backlund P, Janetos A, Schimel D, Hatfield J, Boote K, Fay P, Hahn L, Izaurralde C, Kimball BA, Mader T, Morgan J, Ort D, Polley W, Thomson A, Wolfe D, Ryan M, Archer S, Birdsey R,

Dahm C, Heath L, Hicke J, Hollinger D, Huxman T, Okin G, Oren R, Randerson J, Schlesinger W, D. Lettenmaier, D. Major, L. Poff, S. Running, L. Hansen, Inouye D, Kelly BP, Meyerson L, Peterson B, Shaw R. Washington, DC, USA: U.S. Environmental Protection Agency; 2008

8. Castiglionit P, Warner D, Bensen RJ, Anstrom DC, *et al.*, Bacterial RNA chaperones confer abiotic stress tolerance in plants and improved grain yield in maize under water–limited conditions. Plant Physio. 2008; 147:446-455

9. Chinnusaamy V, Jagendorf A, Zhu JK. Understanding and improving salt tolerance in plants. Crop Sci. 2005; 45:437-448

10. Dwivedi SL, Sahrawat K, Upadhyaya H, Ortiz R. Food, nutrition and agro-biodiversity under global climate change. Adv Agro. 2013; 120:1-118

11. FAO. Rome: The State of World Fisheries and Aquaculture. World review of fisheries and aquaculture; 2012

12. FAO. Climate Change Guidelines for Forest Managers. FAO Forestry Paper 172. Rome. 2013

13. FAO. Climate change, Agriculture and Food Security. Rome, Italy; Food and Agriculture Organization of the United Nations (FAO); 2016

14. Golldack D, Li C, Mohan H, and Probst N. Tolerance to drought and salt stress in plants: Unraveling the signaling networks. Frontiers Plant Sci. 2014; 5:151

15. Gusta L. Abiotic stresses and agricultural sustainability. J.Crop Improvement. 2012; 26:415-427

16. Hasanuzzaman M, Nahar K, Alam MdM, Roychowdhury R, Fujita M. Physiological, biochemical, and molecular mechanisms of heat stress tolerance in plants. International J Molecular Sci. 2013; 14:9643-9684

17. Hatfield JG, Takle R, Grotjahn P, Holden RC, *et al.*, In: Melillo JM, Terese Richmond TC, Yohe GW, editors. Ch. 6: Agriculture. Climate Change Impacts in the United States: The Third National Climate Assessment. U.S. Global Change Research Program, 2014, 150-174.

18. He C, Yan J, Shen G, Fu L, *et al.*, Expression of an Arabidopsis vacuolar sodium/proton antiporter gene in cotton improves photosynthetic performance under salt conditions and increases fiber yield in the field. Plant Cell Physio. 2005; 46:1848-1854.

19. IPCC: Intergovernmental Panel on Climate Change. Six assessment report, 2016.

20. IYP. International Year of Pulses, Theme: Productivity & Environmental Sustainability, 2016.

21. Jugsujinda A, Delaune RD, Lindau CW. Factors controlling carbon dioxide and methane production in acid sulfate soils. Water Air Soil Pollution. 1996; 87:345-355.

22. Karaba A, Dixit S, Greco R, Aharoni A, *et al.*, Improvement of water use efficiency in rice by expression of HARDY, an Arabidopsis drought and salt tolerance gene. Proceedings of the National Academy of Sciences of the United States of America. 2007; 104:15270-15275.

23. Katiyar-Agarwal S, Agarwal M, Grover A. Heat tolerant basmati rice engineered by over-expression of hsp101. Plant Molecular Bio. 2003; 51:677-686.

24. Kaur N, Gupta AK. Signal transduction pathways under abiotic stresses in plants. Curr. Sci. 2005; 88:1771-1780.

25. Kumar R, Gautam HR. Climate change and its impact on agricultural productivity in India. J Climatology Weather Forecasting. 2014; 2:109.

26. Kumar V, Ladha JK. Direct seeding of rice: Recent developments and future research needs. Adv Agron. 2011; 111:29-413.

27. Kuromori T, Mizoi J, Umezawa T, Yamaguch-Shinozaki K, Shinozaki K. Drought stress signaling network. Molecular Bio. 2014; 2:383-409.

28. Lutz D. Creating plants that make their own fertilizer. Washington University in St. Louis News Release, 2013.

29. Lybbert T, and Sumner D. Agricultural technologies for climate change mitigation and adaptation in developing countries: Policy options for innovation and technology diffusion. ICTSD-IPC Platform on Climate Change, ATS Policy Brief, 2010.

30. Mall RK, Gupta A, Singh R, Singh RS, Rathore LS. Water resources and climate change: An Indian perspective. Curr Sci. 2005; 90:1610-1626.

31. Mall RK, Singh R, Gupta A, Singh RS, Srinivasan G. Impact of climate change on Indian agriculture: A review. Climate Change. 2006; 78:445-478.

32. Mittal S. Modern ICT for Agricultural Development and Risk

Management in Smallholder Agriculture in India. CIMMYT. Socio-Economic Working Paper 3. Mexico, D.F.: CIMMYT; 2012

33. Moghaieb RE, Nakamura A, Saneoka H, and Fujita K. Evaluation of salt tolerance in ectoinetransgenic tomato plants (Lycopersicon esculentum) in terms of photosynthesis, osmotic adjustment, and carbon partitioning. GM He Crops. 2011; 2:58-65

34. OECD. Synthetic biology: A new and promising technology. In: OECD, Emerging Policy Issues in Synthetic Biology. Paris: OECD Publishing; 2014.

35. Ortiz R, Iwanaga M, Reynolds MP, Wu X, Crouch JH. Overview on crop genetic engineering for drought-prone environments. J. Semi-Arid Tropical Agri. Res. 2007; 4:1-30.

36. Philippot L, Hallin S. Towards food, feed and energy crops mitigating climate change. Trends Plant Sci. 2011; 16:476-480

37. Porter JR, Xie L, Challinor AJ, Cochrane K, *et al.*, Food security and food production systems. In: Field *et al.*, editors. Climate Change 2014: Impacts, Adaptation, and Vulnerability. Part A: Global and Sectoral Aspects.Contribution of Working Group II to the Fifth Assessment Report of the Intergovernmental Panel on Climate Change. Cambridge, UK and New York, USA: Cambridge University Press, 2014, 485-533

38. Prasad YG, Maheswari M, Dixit S, Srinivasarao CH, *et al.*, Smart Practices and Technologies for Climate Resilient Agriculture. Hyderabad: Central Research Institute for Dryland Agriculture (ICAR), 2014, 76

39. Sapkota TB, Jat ML, Aryal JP, Jat RK, Khatri-Chhetri A. Climate change adaptation, greenhouse gas mitigation and economic profitability of conservation agriculture: Some examples from cereal systems of Indo-Gangetic Plains. J Integrative Agri. 2015; 14(8):1524-1533

40. Shrawat AK. Good AG. Genetic engineering approaches to improving nitrogen use efficiency. ISB Report, 2008.

41. Singh A, Yadav A, Matthew K, Varrieur V, Singh S. Success of IRRI stresses tolerant rice varieties in Bihar, India-Experience of IRRAS farmers. In: Poster Presented at Fourth International Rice Conference; 27 October to 01 November 2014; Bangkok, Thailand, 2014.

42. Singh G. Agriculture diversification for food, nutrition, livelihood and environmental security: Challenges and opportunities. Indian J Agron. 2015; 60 (2):172-184.

43. Subbarao GV, Ban T, Kishii M, Ito O, *et al.*, Can biological nitrification inhibition (BNI) genes from perennial Leymus racemosus (Triticeae) combat nitrification in wheat farming? Plant Soil. 2007; 299:55-64.

44. Tester M, Davenport R. Na^+ tolerance and Na^+ transport in higher plants. Annals Bot. 2003; 91:503-527.

45. Todaka D, Shinozaki K, Yamaguchi-Shinozaki K. Recent advances in the dissection of drought-stress regulatory networks and strategies for development of drought-tolerant transgenic rice plants. Frontiers Plant Sci. 2015; 6:1-20.

46. Tuteja N. Integrated calcium signaling in plants. In: Baluska F, Mancuso S, editors. Signaling in Plants I. Heidelberg, Germany: Springer, 2005, 29-49.

47. United Nations. Department of Economic and Social Affairs Population Division. World Population Prospects: The 2015 Revision. Working Paper No. ESA/P/WP.241. New York, NY: The Department of Economic and Social Affairs of the UN Secretariat; 2015.

48. Ziska L, Crimmins A, Auclair S, DeGrasse JF, *et al.*, Ch. 7: Food Safety, Nutrition, and Distribution. The Impacts of Climate Change on Human Health in the United States: A Scientific Assessment. Washington, DC: U.S. Global Change Research Program, 2016, 189-21.

Chapter - 13
Climate Change on Carbon Sequestration in Indian Soils and Food Security

Authors

Dr. R.K. Naresh
Sardar Vallabhbhai Patel University of Agriculture & Technology, Meerut, Uttar Pradesh, India

K.S. Krishna Prasad
Sardar Vallabhbhai Patel University of Agriculture & Technology, Meerut, Uttar Pradesh, India

Chapter - 13

Climate Change on Carbon Sequestration in Indian Soils and Food Security

Dr. R.K. Naresh and K.S. Krishna Prasad

Changes in cropland management have a high potential to sequester soil organic carbon and can be a win-win strategy for increased soil fertility, soil quality and food production and for mitigating climate change. However, local increases in soil carbon are not necessarily true carbon sequestration. Therefore, it is important to perform life-cycle analysis to include all emissions involved with changed cropland management to sequester soil organic carbon. Only maximizing carbon sequestration in croplands may not always be the best policy option; substitution of fossil fuels by using crop residues for bio energy may be as effective for reducing atmospheric CO_2. Changes in land use, climate and farm management can have significant effects on carbon sequestration in cropland. Abandonment of agricultural land can result in large carbon sequestration. Adaption of less intensive tillage practices works to sequester organic carbon in croplands mostly in warmer climates – but less so in colder climates. Fertilization generally increases carbon stocks in croplands since carbon inputs to soil increase, but this can be counteracted by increased N_2O emissions. Application of manure and crop residue return increases carbon stocks in cropland. Studies show that an increase in aridity, due to climate change, can cause significant decrease in soil organic carbon stocks. However, this could be mitigated by land use change.

Soil carbon plays a key role in all the four classes of soil ecosystem services. The flows arising from environmental processes depend on ecosystem structure, where soil carbon is a key component, along with the environmental conditions and human interventions that can strongly influence the services, goods and benefits produced (Fig 13.1a).

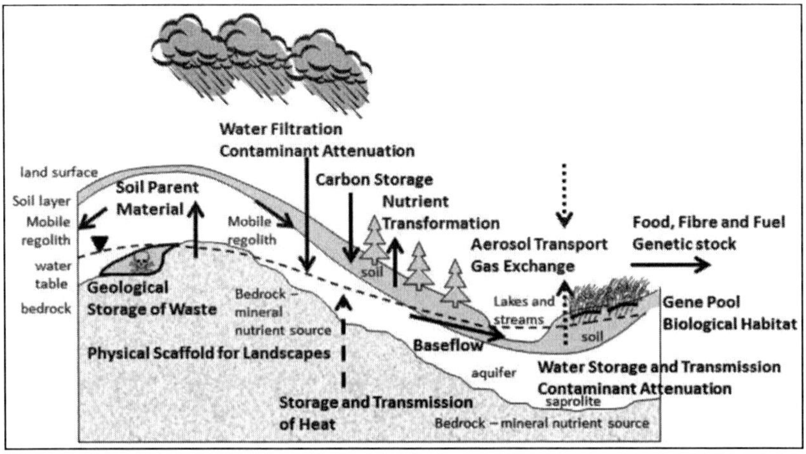

Fig 13.1a: Soil functions and ecosystem services are at the heart of Earth's critical zone. Reproduced with permission from Banwart *et al.*, 2012

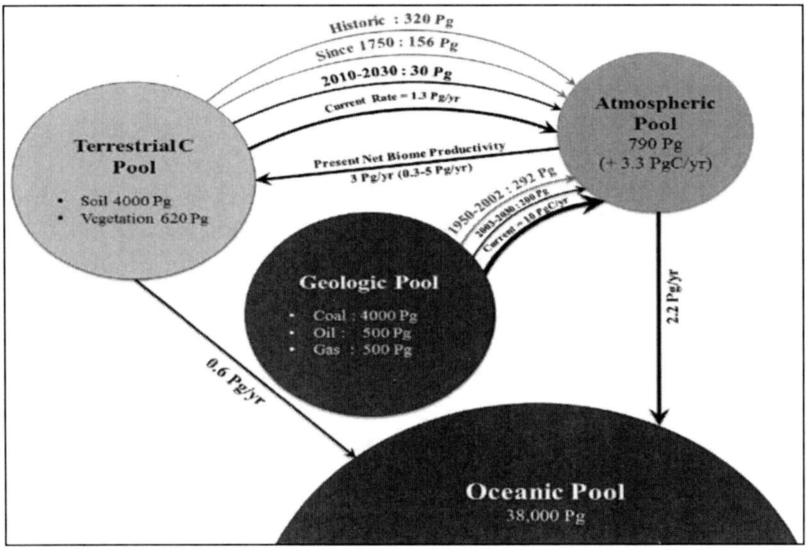

Source: Jansson *et al.*, 2010; WMO, 2011

Fig 13.1b: Principle components of the global carbon cycle

There are many goals and actions that must be addressed to meet the growing human demands for food, water, energy, climate change mitigation and biodiversity in the coming decades. Soil organic carbon (SOC) is central to these being a potentially important determinant of the maintenance, buffering and enhancement of the supply of many ecosystem goods and services under changing socioeconomic and environmental conditions (Fig 13.1b). In the next few decades, an increase in SOC has the potential to improve the five essential ecosystem services (Fig 13.2a & Fig 13.2b). However, this potential is dependent on time and is constrained by varying factors. It is known that under given climatic, substrate, relief and hydrological conditions there are biophysical limits to how much carbon a soil can store. However, there is little information on the inherent capacity of many soils to sequester carbon since native reference soils no longer exist.

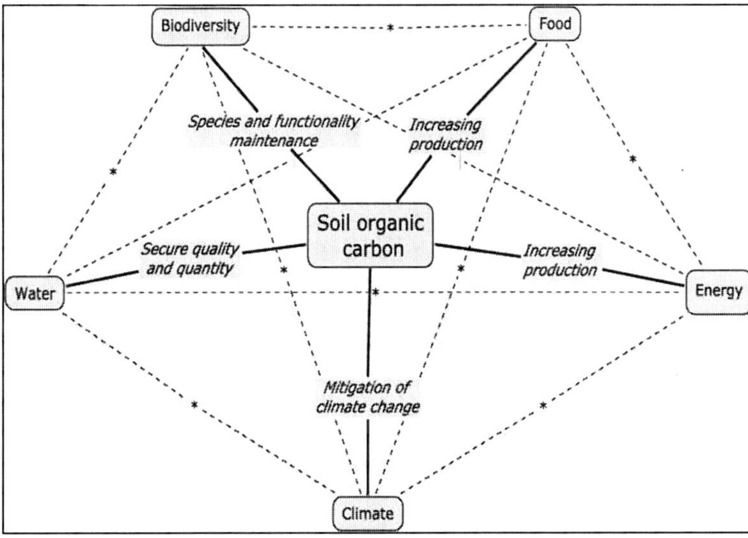

Fig 13.2a: Interactions between soil organic carbon and the five essential services. Reproduced with permission from Nziguheba *et al.*, 2014

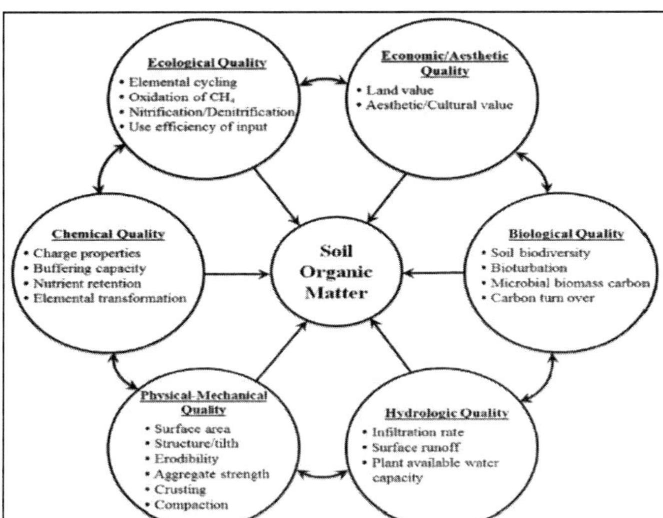

Fig 13.2b: Effects of soil organic matter content on soil quality

The global soil carbon (C) pool of 2500 gigatons (Gt) includes about 1550 Gt of soil organic carbon (SOC) and 950 Gt of soil inorganic carbon (SIC). The soil C pool is 3.3 times the size of the atmospheric pool (760 Gt) and 4.5 times the size of the biotic pool (560 Gt, (Fig 13.3a). The SOC pool to 1-m depth ranges from 30 tons/ha in arid climates to 800 tons/ha in organic soils in cold regions, and a predominant range of 50 to 150 tons/ha. The SOC pool represents a dynamic equilibrium of gains and losses (Fig 13.3a).

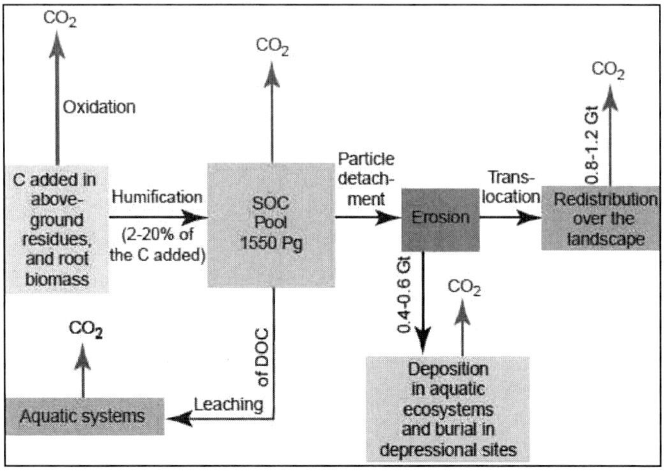

Fig 13.3a: Processes affecting soil organic carbon (SOC) dynamics. Arrows pointed upward indicate emissions of CO_2 into the atmosphere. There may also be emission of CH_4 under anaerobic conditions, although most well-drained soils are a sink of CH_4. DOC dissolved organic carbon.

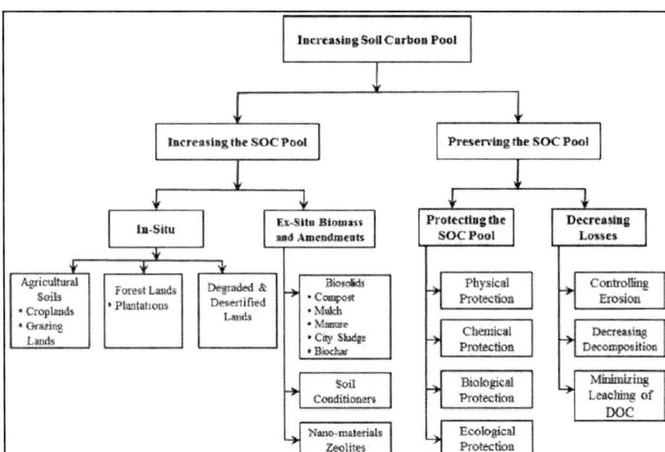

13.3b: Principal strategies on converting soils of agroecosystems from C source to sink.

Conversion of natural to agricultural ecosystems causes depletion of the SOC pool by as much as 60% in soils of temperate regions and 75% or more in cultivated soils of the tropics. The depletion is exacerbated when the output of C exceeds the input and when soil degradation is severe. Some soils have lost as much as 20 to 80 tons C/ha, mostly emitted into the atmosphere. Severe depletion of the SOC pool degrades soil quality, reduces biomass productivity, and adversely impacts water quality, and the depletion may be exacerbated by projected global warming (Fig 13.3b).

During the preindustrial era, the total C emission from terrestrial ecosystems was supposedly about twice (320 Gt or 0.04 Gt C/year for 7800 years) that of the industrial era (160 Gt or 0.8 Gt C/year for 200 years)

(Ruddiman, 2003). Between 1850 and 1998, the emission from fossil-fuel combustion (270 ± 30 Gt) was about twice that from the terrestrial ecosystems (136 ± 55 Gt) (IPCC, 2000). The latter includes 78 ±12 Gt from soil, of which about one-third is attributed to soil degradation and accelerated erosion and twothirds to mineralization (Table 1). The estimates of historic SOC loss range widely, from 44 to 537 Gt, with a common range of 55 to 78 Gt (Lal, 2001).

Soil Carbon Sequestration

Carbon sequestration implies transferring atmospheric CO_2 into long-lived pools and storing it securely so it is not immediately reemitted. Thus, soil C sequestration means increasing SOC and SICS stocks through judicious land use and recommended management practices (RMPs). The potential soil C sink capacity of managed ecosystems approximately equals the cumulative historic C loss estimated at 55 to 78 Gt. The attainable soil C sink capacity is only 50 to 66% of the potential capacity. The strategy of soil C sequestration is cost-effective and environmentally friendly (Table 13.1).

Table 13.1: Estimates of pre- and postindustrial losses of carbon from soil and emission from fossil-fuel combustion. Data were compiled from diverse sources (IPCC, 2000; Lal, 2001). Ruddiman, 2003 estimated the emission from land-use conversion during the postindustrial era at 0.8 Gt C/year for 200 years at 160 Gt C.

Source	Historic Carbon Emission (Gt)
Preindustrial Era	
Fossil-fuel combustion	0
Land-use conversion at 0.04 Gt C/year for 7800 years	320
Postindustrial Era	
Fossil-fuel combustion (since 1850)	270 ± 30

Land-use conversion	136 ± 5
Soil cultivation	78 ±12
Erosion	26 ±9
Mineralization	52 ± 8

The rate of increase in the SOC stock, through land-use change and adoption of RMPs, follows a sigmoid curve, attains the maximum 5 to 20 years after adoption of RMPs, and continues until SOC attains equilibrium. Observed rates of SOC sequestration in agricultural and restored ecosystems depend on soil texture, profile characteristics, and climate, and range from 0 to 150 kg C/ha per year in dry and warm regions Armstrong *et al*., (2003), and 100 to 1000 kg C/ha per year in humid and cool climates (Grace, *et al*., 1995; West and Post, 2002) (fig 4a). With continuous use of RMPs, these rates can be sustained for 20 to 50 years or until the soil sink capacity is filled (West and Post, 2002; Sauerbeck, 2001).

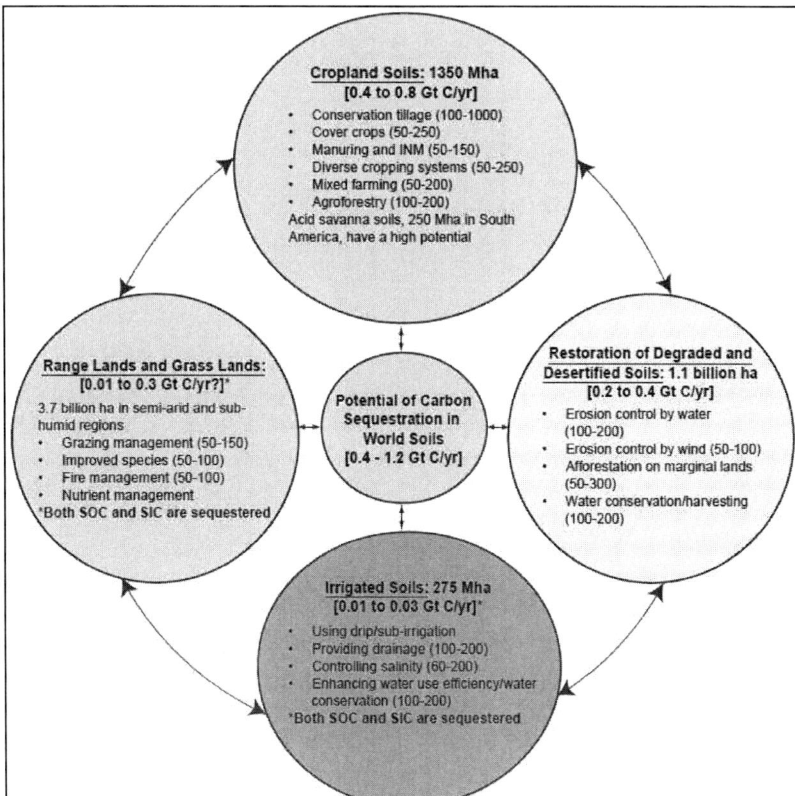

Fig 13.4a: Ecosystems with a high and attainable soil C sequestration potential are cropland, grazing/range land, degraded/desertified lands, and irrigated soils

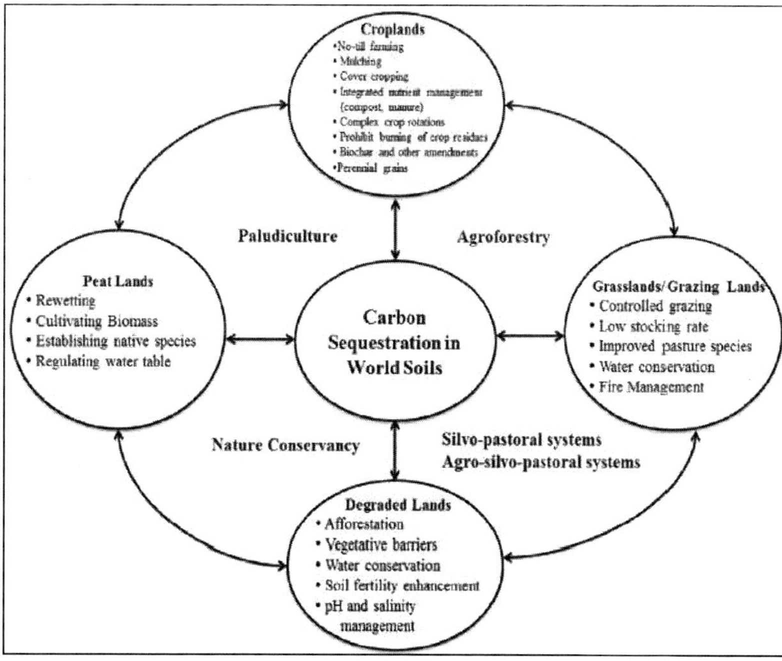

Fig 13.4b: Priority biomes and some recommended technological options for carbon sequestration in world soils

The SOC sequestration is caused by those management systems that add high amounts of biomass to the soil, cause minimal soil disturbance, conserve soil and water, improve soil structure, enhance activity and species diversity of soil fauna, and strengthen mechanisms of elemental cycling (Fig 13.4a).

Soil Carbon Sequestration for Mitigating Climate Change

Estimates of the total potential of C sequestration in world soils vary widely from a low of 0.4 to 0.6 Gt CYr^{-1} Sauerbeck, (2001) to a high of 0.6 to 1.2 Gt C yr^{-1} Lal, (2002) (Fig 13.4b). Thus, the potential is finite in capacity and time. Nonetheless, soil C sequestration buys us time until the alternatives to fossil fuel take effect. Some issues related to this strategy are as follows:

1) Agricultural chemicals. Most RMPs involve C-based input. It takes 0.86 kg C/kg N, 0.17 kg C/kg P_2O_5, 0.12 kg C/kg K_2O, 0.36 kg C/kg lime, 4.7 kg C/kg of herbicides, 5.2 kg C/kg of fungicides, 4.9 kg C/kg of insecticides West, G. Marland (2002), and 150 kg C/ha for pumping groundwater for irrigation Follett, (2001). Tillage operations emit 15 kg C/ha for moldboard plowing, 8 kg C/ha for

chisel plowing and heavy tandem disking, 6 kg C/ha for light tandem disking, 11 kg C/ha for subsoiling, 4 kg C/ha for cultivation, and 2 kg/ha for rotary hoeing Lal, (2003). Therefore, conversion from conventional till to no-till farming reduces emission by 30 to 35 kg C/ha per season Lal, (2003). Similarly, a judicious use of C-based inputs is essential to enhancing use efficiency and minimizing losses. However, inputs are needed not for soil C sequestration per se, but for increasing food production and ensuring sustainable use of soil and water resources.

2) Nutrients required. Carbon is only one of the elemental constituents of humus. It is estimated that sequestration of 1 Gt of C in world soils would require 80 million tons (Mt) of N, 20 Mt of P, and 15 Mt of K. In comparison, the global fertilizer use in 2000 was 136 Mt IFDC, (2000). However, there are several sources of nutrients for C sequestration, including biological nitrogen fixation, recycling from subsoil, aerial deposition, use of biosolids, and crop residues. One ton of cereal residue contains 12 to 20 kg N, 1 to 4 kg P, 7 to 30 kg K, 4 to 8 kg Ca, and 2 to 4 kg Mg. Annually, 3 Gt of residues of grain crops are produced globally, which if recycled rather than removed for fuel and other uses, would improve soil quality and sequester C. Crop residue is also a potential source of energy by direct combustion, or for production of ethanol or H_2. It can be used either for biofuel production or to sequester C and improve soil quality. The economics of these two competing uses need to be assessed.

3) Soil erosion and deposition. The SOC is preferentially removed by wind- and water-borne sediments through erosional processes. Some of the SOC-enriched sediments are redistributed over the landscape, others are deposited in depressional sites, and some are carried into the aquatic ecosystems (Fig 13.3a). Although a part of the C translocated by erosion may be buried and redistributed Smith *et al.*, (2001), the rest is emitted into the atmosphere either as CO_2 by mineralization or as CH_4 by methanogenesis. Erosion-induced deposition and burial may be 0.4 to 0.6 Gt C/year compared with perhaps 0.8 to 1.2 Gt C yr^{-1} emitted into the atmosphere (Fig 13.3a) Lal, (2003). Quantification of emission versus burial of C is a high priority. Yet, an effective soil erosion control is essential to sustainable use of agricultural soils and improving environment quality.

4) Extractive farming practices. The annual depletion rate of nutrients for sub- Saharan Africa (SSA) caused by low-input/ subsistence farming is estimated to be 40 kg of NPK/ha of cultivated land since the mid- 1960s Sanchez, (2002). Mining SOC from soil for nutrients through organic-matter decomposition has an effect on the atmosphere similar to that of fossil-fuel combustion. Therefore, RMPs must enhance rather than deplete SOC pool and soil fertility, increase rather than maintain or decrease crop yield per unit use of fertilizer and other inputs, and improve rather than degrade soil quality.

5) Societal value and hidden benefits. Commodification of soil C is important for trading C credits. Carbon trading markets have existed since 2002, especially in European Union countries Johnson and Heinen, (2004). The low current price ($1/ton CO_2) of SOC may increase with emission cap and regulation. For the concept of SOC credits trading to become routine as a part of the solution to mitigate climate change, the ability to measure short-term (3- to 5-year) changes in SOC pool exists Lal, (2001), but the price of soil C must be based on both on-site and off-site societal benefits. Undervaluing soil C can lead to a "tragedy of the commons."

6) Hydrologic and carbon cycles. Because renewable freshwater is scarce, a projected increase in cereal production by 56% between 1997 and 2050 Rosegrant and Cline, (2003) must occur on the same or smaller land area and with the same or less water. Thus, linking the hydrologic and C cycles through conservation of water resources is crucial to improving agronomic yields and to soil C sequestration in dryland. The low SOC stock in rainfed agriculture can be enhanced through water conservation, water harvesting and water-efficient farming systems.

   Enhancing SOC stock in dryland ecosystems through notill farming is important to drought management: a truly win-win option Lal, (2004).

7) Soil C sequestration and global warming. Global warming is a "century-scale" problem and a "global commons" issue. Soil C sequestration is a related but separate issue with its own merits of increasing productivity, improving water quality, and restoring degraded soils and ecosystems, irrespective of the global-warming debate. Offsetting fossil-fuel emissions by achievable SOC potential provides multiple biophysical and societal benefits.

Furthermore, soil C sequestration is a bridge across three global issues-climate change, desertification, and biodiversity-and a natural link among three UN conventions.

8) Other greenhouse gases. Enhancing SOC stock increases the soil's capacity to oxidize CH_4, especially under no-till farming Six *et al.*, (2002), but may also exacerbate emission of N_2O Smith *et al.*, (2001). Fluxes of CH_4 and N_2O may change the CO_2-mitigation potential of soil management practices and must be considered along with SOC sequestration.

9) Soils of the tropics. Because of its severe depletion and degradation, the C sink capacity of soils of the tropics may be high, but the rate of sequestration can be low. The need for enhancing soil quality is also more urgent in soils of the tropics than in soils of high latitudes because of low crop yields. Yet, the challenge is greater because of weak institutions, limited infrastructure, and predominantly resource-poor agriculture systems. Soil-restorative farm policies must be implemented to mitigate soil-degradative trends.

10) Permanence. Soil C sequestration is a natural, cost-effective, and environmentfriendly process. Once sequestered, C remains in the soil as long as restorative land use, no-till farming, and other RMPs are followed. Soil sink capacity and permanence are related to clay content and mineralogy, structural stability, landscape position, moisture and temperature regimes, and ability to form and retain stable microaggregates.

Soil Carbon Sequestration and Global Food Security

Global hotspots of soil degradation with a high priority for soil restoration and C sequestration include SSA, central and south Asia, China, the Andean region, the Caribbean, and the acid savannas of South America. Complete residue removal for fodder and fuel is a norm in south Asia and Africa. Thus, depletion of SOC stock from the root zone has adversely affected the soil productivity and environmental quality of these regions. Simply put, poor farmers have passed on their suffering to the land through extractive practices. They cultivate marginal soils with marginal inputs, produce marginal yields, and perpetuate marginal living and poverty.

As a source of nutrients for growing crops, the SOC pool is a mean of production in subsistence farming systems of SSA, which accounts for only 2.5% of the fertilizer consumption and 2% of the world's irrigated land area, both essential to SOC sequestration. Benefits of RMPs cannot be realized in severely degraded soils depleted of their SOC stock-soil's life support system. An optimum level of SOC stock is needed to hold water and nutrients, decrease risks of erosion and degradation, improve soil structure and tilth, and provide energy to soil microorganisms.

The SOC is a biomembrane that filters pollutants, reduces sediment load in rivers, decreases hypoxia in coastal ecosystems, degrades contaminants, and is a major sink for atmospheric CO_2 and CH_4. Fertilizer application is an important strategy of increasing crop yield in SSA Pieri, (1986), but its effectiveness is enhanced when used in conjunction with crop residue mulch Yamoah *et al.*, (2002) or trees Sanchez, (2002). Increase in SOC stock increases crop yield even in high-input commercial agriculture Bauer and Black, (1994), but especially in soils where it has been depleted Johnston, (1986). An increase of 1 ton of SOC increased wheat grain yield by 27 kg/ha in North Dakota, United States Bauer and Black, (1994), and by 40 kg/ha in semi-arid pampas of Argentina Diaz-Zorita *et al.*, (2002), 6 kg/ha of wheat and 3 kg/ha of maize in alluvial soils of northern India Kanchikerimath and Singh, (2001), 17 kg/ha of maize in Thailand Petchawee and Chaitep,(1995), and 10 kg/ha of maize and 1 kg/ha of cowpea in western Nigeria Lal, (1981). High SOC stock is also needed to maintain consistent yields through improvements in water and nutrient holding capacity, soil structure, and biotic activity. The critical limit of SOC concentration for most soils of the tropics is 1.1% Aune and Lal (1987). Increasing SOC concentration from a low of 0.1 to 0.2% to a critical level of 1.1% is a major challenge for tropical ecosystems. Yet, a drastic reduction in the SOC pool in SSA and elsewhere must be reversed in order to advance food security. The vicious cycle of declining productivity– depleting SOC stock–lower yields will have to be broken (Fig 13.5b) by improving soil quality through SOC sequestration in order to free much of humanity from perpetual poverty, malnutrition, hunger, and substandard living.

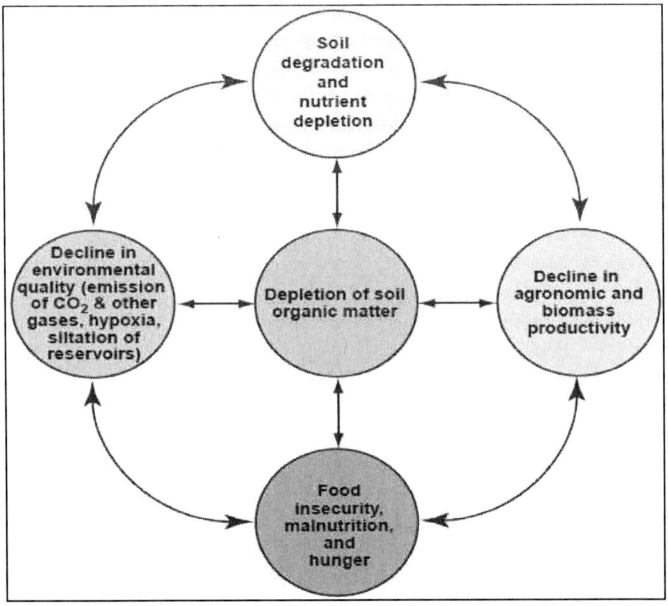

Fig 13.5a: The vicious cycle of deletion in soil organic matter-decline in crop yield-food insecurity-soil and environmental degradation can be broken by improving soil fertility through enhancement of the soil organic matter pool, which requires use of sustainable agricultural technologies for water and nutrients management, including no-till farming, composts and mulching, leguminous cover crops, water harvesting, agroforestry, and integrated farming systems, along with judicious use of chemicals. Strategy can break the tyranny of hunger.

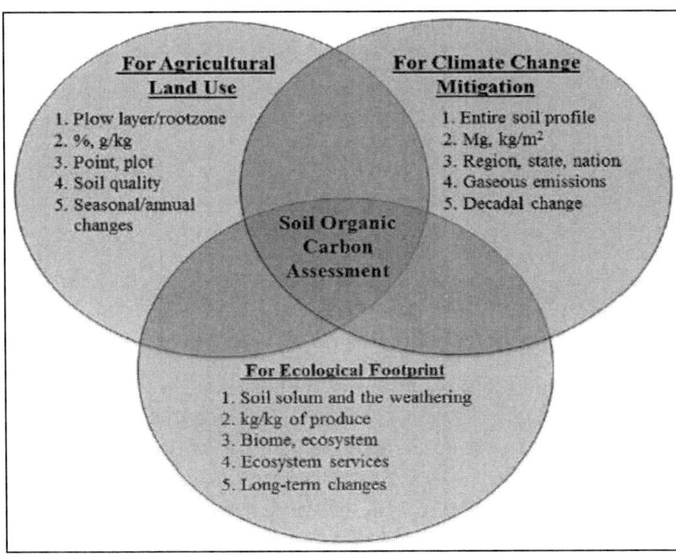

13.5b: Measurement of soil organic carbon for specific objectives

Climate Change and Global Food Security

Settled or the intentional agriculture 10-12 millennia ago was the defining moment in human history. Domestication of plants and animals began independently at numerous sites. Two among other factors responsible for origin of agriculture were

i. Increase in global temperature, and
ii. Increase in atmospheric concentration of CO_2 from 180 ppmv to 280 ppmv. Along with the increase in mean global temperature by as much as 5°C, the increase in atmospheric CO_2 concentration enhanced biomass production of C_3 plants (e.g., wheat, barley) dramatically and that of C_4 plants (e.g., corn) moderately. Another impact was the increase in biological nitrogen fixation (BNF) by legumes, which enhanced soil fertility. With progressive developments in agriculture, both the human population and the ones under agricultural land use (cropland and grazing land) increased. For example, the agricultural land area in 1 AD corresponding with a human population of 188 million was 240 million ha (Mha). The agricultural area increased to 930 Mha for population of ~1 billion in 1800, and to 4738 Mha for population of 7 billion in 2010.

Challenges to achieving food security for projected population of 9.2 billion by 2050 include land degradation already estimated at 3500 Mha (23.5% of Earth's land area) (Bai *et al.*, 2008) and affecting an additional 5-10 Mha yr^{-1}, urbanization needing 3 Mha yr^{-1} of agricultural land, land area needed for establishing biofuel plantations, and mitigating the climate change. While food grain production must be increased 70% by 2050, it is also widely believed that 1°C increase in mean temperature can reduce global grain yield by 10-17%. Thus, over and above the sustainable intensification of land already allocated to agroecosystems, additional land area would be needed to meet the food demands of increasingly affluent population with growing preference towards animal-based diets.

Changing and Uncertain Climate Affects food Security in Numerous Ways: Direct and indirect; and positive and negative. However, in some site-specific cases Abrupt climate change (ACC) may also enhance agronomic/food production. Over and above the positive effects of prolonging the growing season in higher latitudes, some geo-engineering techniques that reflect sunlight may enhance crop performance through the CO_2 fertilization effects (Pongratz *et al.*, 2012). For example, despite the adverse impact of climate change, rice productivity in Sub-Saharan Africa (SSA) increased by 9.5% yr^{-1} between 2007 and 2011 (Seck *et al.*, 2012), compared to merely 1.6% yr^{-1} in Asia. Thus, filling the yield gap, as has been done with rice in SSA, is a high priority to increasing global food production. Air quality, concentration of trace gases and dust/particulate materials, affect productivity through their effects on plant growth and also by changing precipitation, temperature and the growing season. Air pollutants (e.g., heavy metals and trace gases) affect plant growth directly and through their impact on abiotic and biotic stresses (Bender and Weigel 2011).

The present and projected climate change is affecting food security globally, but especially in regions with scarce resources and large population (Table 13.2). Important among these are biomes/ecoregions with predominantly small landholders and resource-poor farmers (Altieri *et al.*, 2012). The overall effect of climate change in West Africa shows a yield decline of -11% to -15%, with -18% in northern West Africa, -13% in southern West Africa, and -15% in regions with intense warming (Roudier *et al.*, 2011). Globally, major food crops vulnerable to climate change include wheat (Asseng *et al.*, 2011, Song and Zhao 2012), rice (Seck *et al.*, 2012), maize (Bellon *et al.*, 2011), lentils and others (Table 13.2). It is estimated that an additional 116 million Mg of rice will be needed by 2035 to feed the growing population (Seck *et al.*, 2012). In SSA, on the contrary, 30 million Mg more rice will be needed by 2035—an increase of 130% and about one-third of this extra need is for Nigeria. Globally, rice production will have to be increased by 1.2% to 1.5% or an average yield increase of 0.6 Mg ha^{-1}. In SSA, yield of food crops are projected to decline from 8% to 22% depending on crop type (Schlenker and Lobell 2010).

Table 13.2: Examples of regions and crops prone to food insecurity aggravated by the changing climate

S. No.	Country/Region	Crop	Climatic Factor	Time Period	Reference
1.	China	Winter wheat	Drought in winter wheat	Near future (10-30 yrs)	Song and Zhao (2012)
2.	Russia/ Ukraine/ Kazakhstan	Wheat	Climate variability	Next decades	Lioubimtseva *et al.*, (2012)
3.	Africa	Rice	Coping with climate change	By 2035	Seck *et al.*, (2012)
4.	South Asia	Dryland crops	Managing water and SNRM	Next decades	Venkaterswarlu and Shanker (2012)
5.	Mexico	Maize	Drying and	Near future	Shiefaw *et al.*,

			warming trends (Highlands)		(2011)
6.	South Asia	Lentil	Drought stress	2010-2050	Eskine *et al.*, (2011)
7.	West Africa	Rainfed crops	Drought, warming	Near future effects: Sudano-Sahelian= -18% Guinean Zone = -13% Overall = -11%	Roudier *et al.*, (2011)
8.	Sub-Saharan Africa	Food crops	Drought and temperature	Near future effects: Maize = -22% Sorghum = -17% Millet = -17% Groundnut = -18% Cassave = -8%	Schlenker and Lobell (2010)

The dryland (rainfed) agroecosystems, especially in South Asia (SA) and SSA, are prone to vagaries of changing climate (Venkateswarlu and Sankar 2012). Maize, which supplies 30% of the food calories to >4.5 billion people in 94 developing countries (Bellon *et al.*, 2011), may be adversely affected by the climate change. Among legumes, lentils are also an important source of protein in South Asia. The annual rate of increase in grain yield of lentil has been hardly 8.6 kg $ha^{-1}yr^{-1}$ between 1961 and 2008 (Erskine *et al.*, 2011), and adaptation to ACC would boost production by filling the yield gap.

Yet, there exists a yield gap between attainable and actual national yield and disseminated.The yield gap can be abridged by a adopting of BMP's of improving soil quality by enhancing the SOC pool. The strategy of sustainable soil management is to replace what is removed, respond wisely to what is changed, and predict changes in soil quality by natural and anthropogenic perturbations so that adaptive systems can be implemented. Four Ps of bridging the yield gap are :1) "Policy" specifically focused on soil quality improvement through enhancing the SOC pool, 2)"People" with passion for soil improvement and scientific/traditional knowledge for good governance,3)"Procedure" consistent and pragmatic in implementing the policy, and 4)"Pricing" based on societal value of the scarce resource. Similarly, three Cs of bridging the yield gap are: 1)" Commitment "of the people involved, 2)"Continuity "of the program over long time, and 3)"Coherence" in program implementation. These Ps and Cs are especially relevant to bridging the yield gap in South Asia and SSA.

The world population is projected to grow from ~7 billion in 2011 to 9.3 billion by 2050, with almost all the increase to occur in developing countries. Merely a handful of developing nations (e.g., India, China, Nigeria, Brazil,

Bangladesh, Indonesia and Pakistan) account for half of the world population but have finite natural resources (low per capita arable land and renewable freshwater). Furthermore, there is a strong competition for land among principal uses such as agriculture, biofuel plantations, urbanization, nature conservancy, and cultural and aesthetical uses. In view of the immediate future (3 to 4 decades or up to 2050), important issues are: (i) how to adapt agriculture to an uncertain climate, (ii) how to synchronize responses to local vulnerabilities with global disparities, and (iii) how to prioritize the adaptions strategies (Iglesias *et al.*, 2011). It is in the context that the importance of improving soil quality and enhancing SOC concentration/ pool cannot be over emphasized.

Soils of India and Their Soil Carbon Pool

In accord with a wide range of ecoregions (Sehgal *et al.*, 1990), India is also endowed with diverse soils of varying characteristics (Table 13.3). Out of the total land area of 297.3 Mha, the principal soil types include 81.1 Mha of Alfisols (27.3%), 60.4 Mha of Vertisols (20.3%), 51.7 Mha of Inceptisols (17.4%), 36.6 Mha of Ultisols (12.3%), 24.8 Mha of Entisols (8.3%), 18.3 Mha of Aridisols (6.2%), 1.8 Mha of Mollisols (0.6%), and 0.8 Mha of Gelisols (0.27%). These diverse soils are also characterized by a wide range of SOC concentration, which is generally related to clay content (Ali *et al.*, 1966) and climate (Jenny and Raychaudhary, 1960). In general, SOC concentration increases with increase in clay content and rainfall, and decreases with increase in mean annual temperature. Some of these soils have been cultivated for centuries, and often with low off-farm input, based on systems that involve removal of crop residue and dung for fuel and other purposes.

Consequently, SOC concentration of most soils is low. The data in Table 13.3 show that SOC concentration of most soils is less than 10 g/kg, and is generally less than 5 g/kg. Because of the low clay contents, the SOC concentration is especially low in alluvial soils of the Indo-Gangetic Plains, coarse-textured soils of southern India, and arid zone soils of northwestern India (Dhir *et al.*, 1991).

Table 13.3: Soil organic carbon (SOC) concentration of some soils in India (adapted from Nambiar, 1994)

Location	Soil Type	Texture	SOC Content (g/kg)
Bangalore, KT	Haplustalf	Sandy loam	5.5
Barrackpore, WB	Eutrochrept	Sandy loam	7.1

Bhubaneswar, O	Haplaquept	Sandy	2.7
Coimbatore	Vertic Ustochrept	Clay loam	3.0
Delhi	Ustochrept	Sandy loam	4.4
Hydrabad, AP	Tropaquept	Sandy clay loam	5.1
Jabalpur, MP	Chromustert	Clayey	5.7
Ludhiana, Pb	Ustochrept	Loamy sand	2.1
Palampur, HP	Hapludalf	Silty clay loam	7.9
Pantnagar, UP	Hapludoll	Silty clay loam	14.8
Rauchi, B	Haplustalf	Silty clay	4.5

Table 13.4: Soil organic carbon (SOC) concentration of soils of India in relation to the rainfall regime and temperature (Sekhon and Meelu, 1994)

Rainfall (mm/y)	Mean Annual Temperature (°C)	SOC Content (g/kg)	
		Surface	Sub-Soil
<500	25.9–26.7	1.2–8.0	1.2–4.0
500–1000	23.6–27.9	1.8–12.5	0.7–11.7
>1000	24.4–27.2	2.6–9.0	2.3–8.4

The prevalent low levels of SOC concentrations are attributed to soil-mining practices of excessive tillage, imbalance in fertilizer use, little or no crop residue returned to the soil, and severe soil degradation. Consequently, even the wellestablished relationship between climate (temperature and precipitation) (Jenny and Raychaudhary, 1960) and SOC concentration does not exist. The data in Table 13.4 show consistently low SOC concentration in soils with rainfall regime of 500 to 1000 mm/y. Under native system and undisturbed soils, however, the SOC concentration of most soils is high (Jenny and Raychaudhuri, 1960). Total SOC pool in soils of India is estimated at 21 Pg to 30 cm depth and 63 Pg to 150 cm depth (Table 13.5). The SOC pool in soils of India is 2.2% of the world pool for 1 m depth and 2.6% to 2 m depth. The data in Table 6 shows a decline in SOC concentration of cultivated soils by 30 to 60% compared with the antecedent level in undisturbed ecosystems even by 1960. The total soil C pool also comprises the soil inorganic carbon (SIC), which is generally high in calcareous soils of arid and semi-arid regions. Calcareous soils are widely distributed covering 54% of the geographical area of India, but especially occur in Rajsthan, Gujrat, Punjab, Haryana, Uttar Pradesh, Maharashtra, Karnatka, Tamil Nadu, Andhra Pradesh and parts of Madhya Pradesh and

Bihar (Pal *et al.*, 2000). Total SIC pool in soils of India is estimated at 196 Pg to 1-m depth (Pal *et al.*, 2000) (Table 13.6). The SIC pool in world soils is estimated at 722 Pg to 1-m depth (Batjes, 1996).Therefore, the SIC pool in soils of India comprises about 27% of the world total. Pedogenic or secondary carbonates play a significant role in C sequestration through formation of $CaCO_3$ or $MgCO_3$ and leaching of Ca $(HCO_3)_2$ especially in irrigated systems. The rate of formation of secondary carbonates may range from 30 to130 kg/ha/y (Pal *et al.*, 2000).

Table 13.5: Organic carbon pool in soils of India and the world (adapted from Velayuthan *et al.*, 2000)

Soil Order	India		World	
	0–30 cm (Pg)	0–150 cm (Pg)	0–25 cm (Pg)	0–100 cm (Pg)
Alfisols	4.22	13.54	73	136
Andisols	–	–	38	69
Aridisols	7.67	20.30	57	110
Entisols	1.36	4.17	37	106
Histosols	–	–	26	390
Inceptisols	4.67	15.07	162	267
Mollisols	0.12	0.50	41	72
Oxisols	0.19	0.49	88	150
Spodosols	–	–	39	98
Ultisols	0.14	0.34	74	101
Vertisols	2.62	8.78	17	38
Total	20.99	63.19	652	1555

Table 13.6: Depletion of soil organic carbon concentration of cultivated compared with that in undisturbed soils (adapted from Swarup *et al.*, 2000)

Region	SOC Content		Percent Reduction
	Cultivated (g/kg)	Native (g/kg)	
1. Northwest India			
• Indo-Gangetic Plains	4.2± 0.9	104.±3.6	59.6
• Northwest Himalaya	24.3±8.7	34.5±11.6	29.6
2. Northeast India	23.2±10.4	38.3±23.3	39.4
3. Southeast India	29.6±30.1	43.7±23.4	32.3
4. West coast	13.2±8.1	18.6±2.1	29.1
5. Deccan Plateau	7.7±4.1	17.9±7.6	57.0

Soil Quality and Soil Organic Carbon Concentration

Soil organic matter is an important component necessary to the formation of both micro- and macro-aggregates. The SOC thus encapsulated is protected from the microbial processes, and is sequestered until aggregates are broken or disrupted by plowing and other disruptive processes such as raindrop impact and shearing force of running water. The degree of aggregation and the stability of aggregates are directly proportional to SOC concentration. Because of high aggregation, soils with high SOC concentration have high available water holding capacity, low susceptibility to soil erosion, and have low losses of plant nutrients into the ground water. Use efficiency of fertilizer, irrigation and other input is high in soils with high SOC concentration. All other factors remaining the same, soils with high SOC concentration have more agronomic/biomass productivity thanthosewithlow SOC concentration (Sandhu *et al.*, 1996). High crop yields in coarse-textured soils of Punjab are due to high input of water and fertilizers. Therefore, enhancement and management of SOC concentration are important to sustainable management of soil and water resources.

The totalpotential ranges from12.7 to16.5 Tg C/y. Included inthispotential isalsothat of the restoration of degraded soils and ecosystems estimated at 7.2 to 9.8 Tg C/y.Therefore, potential ofagricultural intensification forSOCsequestration is 5.5 to 6.7 Tg C/y. In addition, there is also a potential of SIC sequestration estimated at 21.8 to 25.6 Tg C/y (Table 13.7). With reduction in erosion-induced emission of 4.3 to 7.2 Tg C/y, total potential of soil C sequestration in India is 39.3 to 49.3 Tg C/y (44.3±7.1) (Table 13.8). There are numerous agricultural sources of GHG emissions (Duxbury, 1994) with hidden C costs of tillage, fertilizer, pesticide use and irrigation. In general, net C sequestration must take into account these costs. It is assumed, however, that these inputs are needed for enhancing agricultural production to meet food demands of increasing population. Therefore, soil C sequestration is a by-product of adopting RMPs on agricultural land and restoring degraded soils.

Table 13.7: Potential of sequestration of secondary carbonates

Soil	Area (Mha)	Rate of SIC Sequestration (kg/ha/y) [a]	Total Potential (Tg C/y)
Vertisols	60.4	35–40	2.11–2.42
Alluvial	94.8	125–133	11.85–12.61
Ferriginous	116.5	28–32	3.26–3.73
Irrigated	57.0	80–120	4.56–6.84
Total	328.7	-	21.78–25.60

[a] Pal *et al.*, (2000), and Nordt *et al.*, (2000).

Enhancing soil quality is important to increasing use efficiency of inputs (e.g., fertilizers, irrigation), increasing biomass/agronomic yields, and improving the environment. Improving quality and quantity of SOC concentration are important to enhancing soil quality. In fact, there is a strong linkage between low SOC concentration in soils of India and the widespread problem of soil degradation. Therefore, reversing soil degradation trends necessitates increasing SOC concentration through adoption of no-till farming, use of crop residue mulch and compost on soil, and legume-based rotations. A major constraint in adopting conservation tillage and mulch farming in India is the non-availability of crop residue for returning to the soil. Most of the crop residues are removed from the fields for use as fodder and fuel. Dung is also used as fuel for cooking. Thus, adoption of mulch farming techniques is possible only if economic sources of fuel and alternative sources of fodder are identified. Emissions from fossil fuel combustion in India are increasing. The soil C sequestration potential of 39.3 to 49.3 Tg C/y (mean of 43.3 Tg C/y) can be significant towards reducing the net emission from fossil fuel combustion. Further, there is an additional potential of C sequestration in biomass especially by forest and other biota. This potential is considerable in terms of the negotiation under the provision of Clean Development Mechanisms under IPCC, and for trading C in the national and international markets. Biosequestration of C, both by soil and biota, is a truly win-win situation. While improving agronomic/biomass productivity, these options also improve water quality and mitigate climate change by decreasing the rate of enrichment of atmospheric CO_2. Realization ofthisvastpotential, whichisininterest ofIndia, requires adoption ofrecommended management practices including theuse ofmulch farming and conservation tillage, integrated nutrient management and manuring, agroforestry systems, restoration of eroded and salinized soils, and conversion of agriculturally marginal lands into restorative land uses.

Table 13.8: Total potential of carbon sequestration in soils of India

Process	Potential (Tg C/y)
A. Soil organic carbon (SOC)	
▪Restoration of degraded soils	7.2–9.8
▪Agricultural intensification	5.5–6.7
B. Secondary carbonates	21.8–25.6
C. Erosion control	4.8–7.2
Total	39.3–49.3

References

1. Altieri MA, Funes-Monzote FR, Petersen P. Agroecologically efficient systems for smallholder farmers: contributions to food sovereignty. Agron Sustain Develop, 2016; 32:1-3.
2. Asseng S, Fooster I, Turner NC. The impact of temperature variability on wheat yields. Global Change Bio. 2011; 17:997-1012.
3. Aune JA, Lal R. Agricultural productivity in the tropics and critical limits of properties of oxisols, ultisols, and alfisols. Trop. Agric. 1997; 74:96-103.
4. Bai ZG, Dent DL, Olsson L, Schaepman ME. Proxy global assessment of land degradation. Soil Use & Manag. 2008; 24(3):233-234.
5. Banwart SA, Menon M, Bernasconi SM *et al.*, Soil processes and functions across an international network of critical zone observatories: introduction to experimental methods and initial results. Comptes Rendus Geosci. 2012; 344:758–772.
6. Bauer A, Black AL. Quantification of the effect of soil organic matter content on soil productivity. Soil Sci. Soc. Am. J, 1994; 58:185-193.
7. Bellon MR, Hodson D, Hellin J. Assessing the vulnerability of traditional maize seed systems in Mexico to climate change. Proc Nat Acad Sci USA. 2011; 108(33):13432-13437.
8. Bender J, Weigel HJ. Changes in atmospheric chemistry and crop health: a review. Agron Sustain Develop. 2011; 31:81-89.
9. Campbell CA, Zentner RP, Selles F, Biederbeck VO *et al.*, Quantifying short-term effects of crop rotations on soil organic carbon in southwestern Saskatchewan. Can. J Soil Sci. 2000; 80(1):193-202.
10. Diaz-Zorita M, Gustavo A, John D, Grove H. A review of no-till systems and soil management for sustainable crop production in the subhumid and semiarid Pampas of Argentina. Soil Tillage Res. 2002; 65(1):1-18.
11. Erskine W, Sarker AK. Crops that feed the world 3. Investing in lentil improvement toward a food secure world. Food Sec. 2011; 3:127-139.
12. Follett RF. Soil management concepts and carbon sequestration in cropland soils. Soil Till. Res. 2001; 61(1-2):77-92.
13. Grace PR, Oades JM, Keith H, Hancock TW. Trends in wheat yields and soil organic carbon in the permanent rotation trial at the Waite

Agricultural Research institute, South Australia Aust. J Exp. Agric. 1995; 35:857-864.

14. Iglesias A, Quiroga S, Diz A. Looking into the future of agriculture in a changing climate. Eurp Rev Agric Econ. 2011; 38(3SI):427-447.
15. International Fertilizer Development Center (IFDC), World Fertilizer Consumption (IFDC, Muscle Shoals, AL), 2000.
16. Intergovernmental Panel on Climate Change, Land Use, Land Use Change and Forestry (Cambridge Univ. Press, Cambridge), 2000, 181-281.
17. Johnston AE. Soil organic matter, effects on soils and crops. Soil Use Manage. 1986; 2(3):97-105.
18. Johnson E, Heinen R. Carbon trading: time for industry involvement. Environ. Int. 2004; 30:279-288.
19. Kanchikerimath M, Singh D. Soil organic matter and biological properties after 26 years of maize-wheat-cowpea cropping as affected by manure and fertilization in a Cambisol in semiarid region of India. Agric. Ecosyst. Environ. 2001; 86(2):155-162.
20. Lal R. Soil erosion and the global carbon budget. Environ. Int. 2003; 29:437-450.
21. Lal R. Soil Carbon Sequestration Impacts on Global Climate Change and Food Security Sci. 2004; 304:1623-1627.
22. Levy R. In Chemistry of Irrigated Soils, Levy R, Ed. (Van Nostrand-Reinhold, New York), 1984, 182–229.
23. Lioubimtseva E, Henebry GM. Grain production in trends in Russia, Ukraine, and Kazakhstan: new opportunities in an increasingly unstable world? Front Earth Sci. 2012; 6:157-166.
24. Lobell DB, Cassman KG, Field CB. Crop yield gaps: their importance, magnitudes, and causes. Annu Rev Environ Resour, 2012; 34:179-204.
25. Nordt LC, Wilding LP, Drees LR. in Global Climate Change and Pedogenic Carbonates, Lal R, Kimble JM, Eswaran H, Stewart BA, Eds. (CRC/Lewis, Boca Raton, FL, 2001, 43-63.
26. Nziguheba G, Vargas R, Bationo A, Black H *et al.*, Soil Carbon, a critical natural resource: Wide scale goals, urgent actions In: Banwart S, Noellemeyer E, Milne E. (Eds). Soil Carbon – Science, management and policy for multiple benefits. SCOPE series 71. Wallingford, United Kingdom, 2014. CABI.org.

27. Petchawee S, W Chaitep W. In Organic Matter Management in Upland Systems in Thailand (Australian Center for International Agricultural Research, Canberra, Australia), 1995, 21–26.

28. Pieri C. Fertilisation des cultures vivrières et fertilité des sols en agriculture paysanne subsaharienne. Agron Trop. 1986; 41:1–20.

29. Rosegrant MW, Cline SA. Global food security: challenges and policies. Science, 2003; 302:1917-1919.

30. Roudier P, Sultan B, Quirion P, Beys A. The impact of future climate change on West African crop yields. What does the recent literature say? Global Environ Change- Human and Policy Dimensions. 2011; 20:1073-1083.

31. Ruddiman W. The anthropogenic greenhouse era began thousands of years ago. Clim. Change. 2003; 61:261-293.

32. Sa JCM, Cerri CC, Lal R, Dick WA *et al.*, Organic matter dynamics and sequestration rates for a tillage chronosequence in a Brazilian Oxisol. Soil Sci. Soc. Am. J, 2001; 65:1486-1499.

33. Sanchez PA. Ecology, Soil fertility and hunger in Africa. Science. 2002; 295:2019-20.

34. Sauerbeck D.R. CO_2 emissions and C sequestration by agriculture-perspectives and limitations. Nutr. Cycling Agroecosyst. 2001; 60(1-3):253-266.

35. Schlenker W, Lobell DB. Robust negatie impacts of claimte change on African agriculture. Environ Res Lett 5, 2010. doi:10.1088/1748-9326/1/014010.

36. Seck PA, Diagne A, Mohanty S, Wopereis MCS. Crops that feed the world 7: rice. Food Security. 2012; 4:7-24.

37. Shiferaw B, Prasanna BM, Hellin J, Banziger M. Crops that feed the world 6. Past successes and future challenges to the role played by maize in global food security. Food Security. 2011; 3:307-327.

38. Silver WL, Ostertage R, Lugo AE. The potential for carbon sequestration through reforestation of abandoned tropical agricultural and pasture lands. Restor. Ecol. 2000; 8 (4):394-407.

39. Six J, Feller C, Denef K, Ogle S *et al.*, Soil organic matter, biota and aggregation in temperate and tropical soils-Effects of no-tillage. Agronomie, EDP Sci. 2002; 22(7-8):755-775.

40. Smith P, Goulding KW, Smith KA, Powlson DS, Smith Jo U, Falloon P *et al*. Enhancing the carbon sink in European agricultural soils: including trace gas fluxes in estimates of carbon mitigation potential. Nutr. Cycl. Agroecosyst. 2001; 60(1-3):237-252.

41. Smith SV, Renwick WH, Buddemeier RW, Crossland CJ. Budgets of soil erosion and deposition for sediments and sedimentary organic carbon across the conterminous United States. Global Biogeochem. Cycles. 2001; 15(3):697-707.

42. Song YL, Zhao YX. Effects of drought on winter wheat yields in North America during 2012-2100. Acta Meteorlogica Sinica. 2012; 26:516-528.

43. Venkateswarlu B, Shanker AK. Dryland Agriculture: bringing resilience to crop production under changing climate. In: Venkatewarlu *et al*., (Eds) Crop stress and its management: perspectives and strategies, 2012. Doi: 10.1007/978-94-007-2220-0.

44. West TO, Post WM. Soil Organic Carbon Sequestration Rates by Tillage and Crop Rotation: A Global Data Analysis. Soil Sci. Soc. Am. J, 2002; 66:930-1946.

45. Yamoah CF, Andre Bationo A., Shapiro B, Koala S. Trend and stability analyses of millet yields treated with fertilizer and crop residues in the Sahel. Field Crops Res. 2002; 75:53-62.

Chapter - 14
Climate Change and Animal Health

Authors

Dr. R.C. Rathi
K.V.K., Muzaffarnagar, Sardar Vallabhbhai Patel University of Agriculture & Technology, Meerut, Uttar Pradesh, India

Dr. Satyaveer Singh
Department of Agronomy, Bihar Agricultural University Sabour, Bihar, India

Chapter - 14

Climate Change and Animal Health

Dr. R.C. Rathi and Dr. Satyaveer Singh

Climate change and global warming now being accepted facts have affected all the ecosystems and will do so if left uncontrolled. Impacts on some components have gained more attention while others have been neglected. Animals belong to latter. Even among the aspects relating to impacts of climate change on animals, production related impacts have gained attention when the impacts on health in general and on infectious diseases in particular are neglected. Despite not being studied in depth, it can be assumed that as in the case of humans, climate change, in particular global warming, is likely to greatly affect the health of animals, both directly and indirectly. Direct effects include temperature-related illness and death, and the morbidity of animals during extreme weather events. Indirect impacts follow more intricate pathways and include those deriving from the attempt of animals to adapt to thermal environment or from the influence of climate on microbial populations, distribution of vector-borne diseases, host resistance to infectious agents, feed and water shortages, or food-borne diseases. Infectious diseases especially vector-borne diseases and diseases having a stage in the environment are likely to be more affected by climate change than other diseases, but may vary depending on different factors.

There are in general many factors operating, and considerably more work is needed on disease dynamics and how these may adapt to a changing climate. These things make impact assessment of animal diseases (in developing countries) particularly challenging. By assessing the risk of potential disease through the identification of potential pathogens and routes of disease spread, can generate forecasts of disease risk areas and implement surveillance efforts to monitor spread and plan intervention efforts in order to reduce the risk of animal disease. From the years 1906 to 2005, global average temperature has warmed by 0.74°C, and since 1961, sea level has risen on average by approximately 2 mm per year. Arctic sea ice extent has declined by 7.4% per decade and snow cover and glaciers have diminished in both hemispheres. The rate of change in climate is faster now than in any period in the last 1000 years. According to the United Nations Intergovernmental Panel on Climate Change, in 90 years, average global temperatures will increase between 1.8°C and 4.0°C and sea level will rise between 18 and 59 cm. Extremes of the hydrologic cycle (e.g. floods and droughts) are also expected to accompany global warming trends.

Human population is expected to increase from 7.2 to 9.6 billion by 2050 (UN, 2013). This represents a population increase of 33%, but as the global standard of living increases, demand for agricultural products will increase by about 70% in the same period (FAO, 2009a). Meanwhile, total global cultivated land area has not changed since 1991 (O'Mara, 2012), reflecting increased productivity and intensification efforts. Livestock products are an important agricultural commodity for global food security because they provide 17% of global kilocalorie consumption and 33% of global protein consumption (Rosegrant *et al.*, 2009). The livestock sector contributes to the livelihoods of one billion of the poorest population in the world and employs close to 1.1 billion people (Hurst *et al.*, 2005). There is a growing demand for livestock products, and its rapid growth in developing countries has been deemed the "livestock revolution" (Thornton, 2010; Wright *et al.*, 2012). Worldwide milk production is expected to increase from 664 million tonnes (in 2006) to 1077 million tonnes (by 2050), and meat production will double from 258 to 455 million tonnes (Alexandratos and Bruinsma, 2012). Livestock production is likely to be adversely affected by climate change, competition for land and water, and food security at a time when it is most needed (Thornton, 2010).

Global climate change is primarily caused by greenhouse gas (GHG) emissions that result in warming of the atmosphere (IPCC, 2013). The livestock sector contributes 14.5% of global GHG emissions (Gerber *et al.*, 2013), and thus may increase land degradation, air and water pollution, and declines in biodiversity (Bellarby *et al.*, 2013; Thornton and Gerber, 2010). At the same time, climate change will affect livestock production through competition for natural resources, quantity and quality of feeds, livestock diseases, heat stress and biodiversity loss while the demand for livestock products is expected to increase by 100% by mid of the 21st century (Garnett, 2009). Therefore, the challenge is to maintain a balance between productivity, household food security, and environmental preservation (Wright *et al.*, 2012). There is growing interest in understanding the interaction of climate change and agricultural production and it is motivating a significant amount of research (Aydinalp and Cresser, 2008). There is still limited research regarding the impacts of climate change on livestock production (IPCC, 2014). The impacts of climate change on livestock production and food security, and the livestock sector's contribution to climate change (Fig 14.1b). The objectives are to: (1) address the impacts of climate change on livestock production; (2) describe the impacts of the livestock sector on climate change; and (3) summarize climate change adaptation and mitigation strategies.

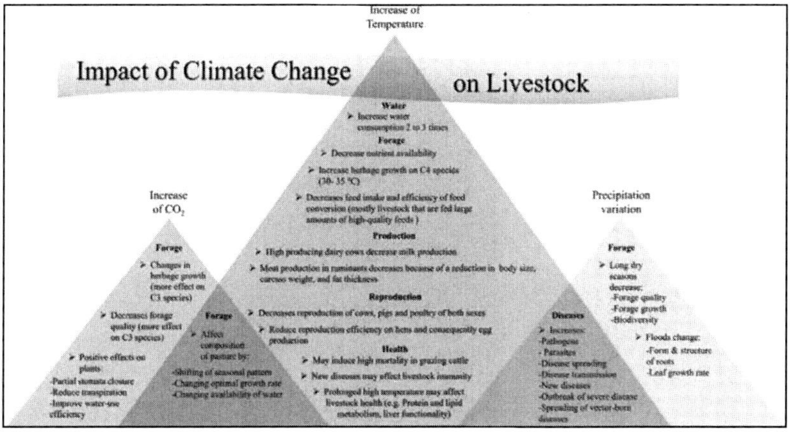

Fig 14.1a: Impacts of Climate Change on Livestock

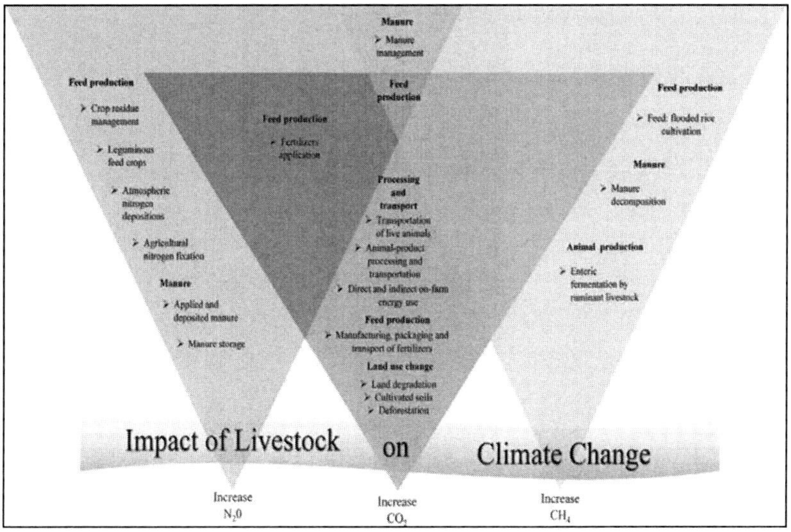

Fig 14.1b: Impact of livestock on Climate Change

Despite uncertainties in climate variability, the IPCC Fifth Assessment Report identified the "likely range" of increase in global average surface temperature by 2100, which is between 0.3^{0}C and 4.8^{0}C (IPCC, 2013). The potential impacts on livestock include changes in production and quality of feed crop and forage (Chapman *et al.*, 2012; Polley *et al.*, 2013), water availability (Henry *et al.*, 2012; Nardone *et al.*, 2010), animal growth and milk production (Henry *et al.*, 2012; Nardone *et al.*, 2010), diseases (Nardone *et al.*, 2010), reproduction (Nardone *et al.*, 2010), and biodiversity (Reynolds *et al.*, 2010). These impacts are primarily due to an increase in temperature and atmospheric carbon dioxide (CO_2) concentration, precipitation variation, and a combination of these factors (Henry *et al.*, 2012; IFAD, 2010; Nardone *et al.*, 2010; Polley *et al.*, 2013). The impacts of climate change on livestock production factors are presented in (Fig 14.1a). Temperature affects most of the critical factors for livestock production, such as water availability, animal production, reproduction and health. Forage quantity and quality are affected by a combination of increases in temperature, CO_2 and precipitation variation. Livestock diseases are mainly affected by an increase in temperature and precipitation variation (Fig 14.2).

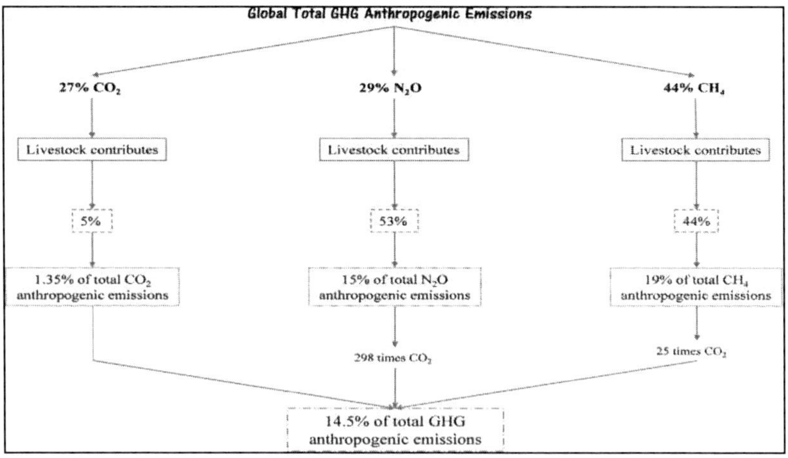

Fig 14.2: Contribution of livestock to the total GHG anthropogenic emissions

Quantity and Quality of Feeds

Quantity and quality of feed will be affected mainly due to an increase in atmospheric CO_2 levels and temperature (Chapman *et al.*, 2012). The effects of climate change on quantity and quality of feeds are dependent on location, livestock system, and species (IFAD, 2010). Some of the impacts on feed crops and forage are:

- Increase of CO_2 concentration will result in herbage growth changes, with greater effect on C_3 species and less on grain yields (Chapman *et al.*, 2012; Thornton *et al.*, 2009, 2015). The effects of

CO_2 will be positive due to inducing partial closure of stomata, reducing transpiration, and. improving some plants' water-use efficiency (Wand *et al.*, 1999).

- C_4 species (which account for less than 1% of plants on Earth) are found in warm environments, and have higher water use efficiency than C_3 plants. Temperature increases to 30–35^{0}C could increase herbage growth, with larger effects on C_4 species. However, the effects may vary depending on the location, production system used, and plant species (Hatfield and Prueger, 2011; IFAD, 2010).
- Changes in temperature and CO_2 levels will affect the composition of pastures by altering the species competition dynamics due to changes in optimal growth rates (Thornton *et al.*, 2009, 2015). Plant competition is influenced by seasonal shifts in water availability (Polley *et al.*, 2013). Primary productivity in pastures may be increased due to changes in species composition if temperature, precipitation, and concurrent nitrogen deposition increase (IPCC, 2007).
- Quality of feed crops and forage may be affected by increased temperatures and dry conditions due to variations in concentrations of water-soluble carbohydrates and nitrogen. Temperature increases may increase lignin and cell wall components in plants (Polley *et al.*, 2013; Sanz-Saez *et al.*, 2012), which reduce digestibility and degradation rates (Polley *et al.*, 2013), leading to a decrease in nutrient availability for livestock (Thornton *et al.*, 2009). However, as CO_2 concentration rises forage quality will improve more in C_3 plants than C_4 plants. C_3 plants also have greater crude protein content and digestibility than C_4 plants (Polley *et al.*, 2013).
- Extreme climate events such as flood, may affect form and structure of roots, change leaf growth rate, and decrease total yield (Baruch and Mérida, 1995).

Impacts on forage quantity and quality depend on the region and length of growing season (Polley *et al.*, 2013; Thornton *et al.*, 2009). An increase of 2^{0}C will produce negative impacts on pasture and livestock production in arid and semiarid regions and positive impacts in humid temperate regions. The length of growing season is also an important factor for forage quality and quantity because it determines the duration and periods of available forage. A decrease in forage quality can increase methane emissions per unit of gross energy consumed

(Benchaar *et al.*, 2001). Therefore, if forage quality declines, it may need to be offset by decreasing forage intake and replacing it with grain to prevent elevated methane emissions by livestock (Polley *et al.*, 2013).

Water

Global agriculture uses 70% of fresh water resources, making it the world's largest consumer (Thornton *et al.*, 2009). However, global water demand is moving towards increased competition due to water scarcity and depletion, where 64% of the world's population may live under water-stressful conditions by 2025 (Rosegrant *et al.*, 2002).

Water availability issues will influence the livestock sector, which uses water for animal drinking, feed crops, and product processes (Thornton *et al.*, 2009). The livestock sector accounts for about 8% of global human water use and an increase in temperature may increase animal water consumption by a factor of two to three (Nardone *et al.*, 2010). To address this issue, there is a need to produce crops and raise animals in livestock systems that demand less water (Nardone *et al.*, 2010) or in locations with water abundance.

As sea level rises, more saltwater will be introduced into coastal freshwater aquifers (Karl *et al.*, 2009). Salination adds to chemical and biological contaminants and high concentrations of heavy metals already found in waterbodies worldwide and may influence livestock production (Nardone *et al.*, 2010). Water salination could affect animal metabolism, fertility, and digestion. Chemical contaminants and heavy metals could impair cardiovascular, excretory, skeletal, nervous and respiratory systems, and impair hygienic quality of production (Nardone *et al.*, 2010).

There is a lack of research related to implications of reduced water availability for land-based livestock systems due to climate change (Thornton *et al.*, 2009). Therefore, it is important to consider water availability and appropriate mitigation strategies in the context of sustainable livestock production.

Livestock Diseases

The effects of climate change on livestock diseases depend on the geographical region, land use type, disease characteristics, and animal susceptibility (Thornton *et al.*, 2009). Animal health can be affected directly or indirectly by climate change, especially rising temperatures (Nardone *et al.*, 2010). The direct effects are related to the increase of temperature, which increases the potential for morbidity and death. The indirect effects are related to the impacts of climate change on microbial communities (pathogens or parasites), spreading of vector-borne diseases, food-borne diseases, host resistance, and feed and water scarcity (Nardone *et al.*, 2010; Thornton *et al.*, 2009).

Temperature increases could accelerate the growth of pathogens and/or parasites that live part of their life cycle outside of their host, which negatively affects livestock (Karl *et al.*, 2009). Climate change may induce shifts in disease spreading, outbreaks of severe disease, or even introduce new diseases, which may affect livestock that are not usually exposed to these types of diseases (Thornton *et al.*, 2009). Evaluating disease dynamics and livestock adaptation will be important to maintain their resilience. Global warming and changes in precipitation affect the quantity and spread of vector-borne pests such as flies, ticks, and mosquitoes (Thornton *et al.*, 2009). In addition, disease transmission between hosts will be more likely to happen in warmer conditions (Thornton *et al.*, 2009). For example, White *et al.*, (2003) simulated the impacts of climate change on Australian livestock, finding that livestock lost about 18% of their weight due to increased tick infestations. Wittmann *et al.*, (2001) also used a model to simulate the response of Culicoides imicola in Iberia, which is the main vector of the bluetongue virus that affects mainly sheep and sometimes cattle, goat, and deer. They reported that the vector would spread extensively with a 2^0C increase in global mean temperature. However, these predicted spreads may be prevented by disease surveillance and technologies, such as DNA fingerprinting, genome sequencing, tests for understanding resistance, antiviral medications, cross-breeding, and more (Perry and Sones, 2009; Thornton, 2010). Meanwhile, there is high probability that emergence of new diseases may act as a mixing vessel between human and livestock, facilitating combination of new genetic material and their transmissibility. This makes it difficult to estimate actual disease risk because of the dependence of diseases on animal exposure and interactions factors, (Randolph, 2008)

Heat Stress

All animals have a thermal comfort zone, which is a range of ambient environmental temperatures that are beneficial to physiological functions (FAO, 1986). During the day, livestock keep a body temperature within a range of $\pm 0.5^0$ C (Henry *et al.*, 2012). When temperature increases more than the upper critical temperature of the range (varies by species type), the animals begin to suffer heat stress (FAO, 1986). Animals have developed a phenotypic response to a single source of stress such as heat called acclimation (Fregley, 1996). Acclimation results in reduced feed intake, increased water intake, and altered physiological functions such as reproductive and productive efficiency and a change in respiration rate (Nardone *et al.*, 2010).

Heat stress on livestock is dependent on temperature, humidity, species, genetic potential, life stage, and nutritional status.

Livestock in higher latitudes will be more affected by the increase of temperatures than livestock located in lower latitudes, because livestock in lower latitudes are usually better adapted to high temperatures and droughts (Thornton *et al.*, 2009). Confined livestock production systems that have more control over climate exposure will be less affected by climate change (Rotter and van de Geijn, 1999).

Heat stress decreases forage intake, milk production, the efficiency of feed conversion, and performance (Haun, 1997). Warm and humid conditions cause heat stress, which affects behavior and metabolic variations on livestock or even mortality. Heat stress impacts on livestock can be categorized into feed nutrient utilization, feed intake, animal production, reproduction, health, and mortality. The following presents these in more detail.

Feed Nutrient Utilization and Feed Intake

Livestock have several nutrient requirements including energy, protein, minerals, and vitamins, which are dependent on the region and type of animal (Thornton *et al.*, 2009). Failure to meet the dietary needs of cattle during heat stress affects metabolic and digestive functions (Mader, 2003). Sodium and potassium deficiencies under heat stress may induce metabolic alkalosis in dairy cattle, increasing respiration rates (Chase, 2012). Most of the research concerning feed intake in livestock animals has been focused on cattle. Thermal livestock stress decreases feed intake (Thornton *et al.*, 2009) and efficiency of feed conversion (McDowell, 1968), especially for livestock that are fed large amounts of high quality feeds (Haun, 1997). In the case of cattle, feed intake reduction leads to a negative energy balance and reduced weight gain (Lacetera *et al.*, 2003). Reduction of water intake may also decrease sweating and feed intake (Henry *et al.*, 2012).

Animal Production.

One of the major causes of decreased production in the dairy and beef industry is heat stress (Nardone *et al.*, 2010) and significant economic losses have been related to this. The United States livestock industry has an annual economic loss between 1.69 and 2.36 billion US dollars due to heat stress, of which 50% occurs in the dairy industry (St-Pierre *et al.*, 2003). High-producing dairy cows generate more metabolic heat than low-producing dairy cows. Therefore, high-producing dairy cows are more sensitive to heat stress. Consequently, when metabolic heat production increases in conjunction with heat stress, milk production declines (Kadzere *et al.*, 2002). Heat stress also affects ewe, goat, and buffalo milk production (Nardone *et al.*, 2010). In general, ewes are more sensitive to the combined temperature and relative humidity affect (the temperature humidity index) than actual temperature or relative humidity. However, the index values that trigger heat stress on ewes varies by breed type (Finocchiaro *et al.*, 2005). Heat stress also impacts goat milk composition and amount. For example, in lactating goats, a water loss reduction mechanism is activated during hot seasons. This mechanism reduces water loss in urine in favor of milk production during seasons with limited water resources (Olsson and Dahlborn, 1989). Buffalo exposure to high temperatures also reduces milk production because it affects the animal physiological functions, such as pulse, respiration rate, and rectal temperature (Seerapu *et al.*, 2015). However, less attention has been given to these animals because of their adaptability to warm conditions and lower demand for their milk (Nardone *et al.*, 2010).

In the case of meat production, beef cattle with high weights, thick coats, and darker colors are more vulnerable to warming (Nardone *et al.*, 2010). Global warming may reduce body size, carcass weight, and fat thickness in ruminants (Mitloehner *et al.*, 2001). The same is true in pig production, where larger pigs will have more reduction in growth, carcass weight, and feed intake (Nardone *et al.*, 2010). Piglets' survival may be reduced because of a reduction of sows feed intake during suckling periods with temperatures greater than 25^{0}C, which reduces the milk yield of the sow (Lucas *et al.*, 2000).

The poultry industry may also be compromised by low production at temperatures higher than 30^{0}C (Esminger *et al.*, 1990). Heat stress on birds will reduce body weight gain, feed intake and carcass weight, and protein and muscle calorie content (Tankson *et al.*, 2001). Heat stress on hens will reduce reproduction efficiency and consequently egg production because of reduced feed intake and interruption of ovulation (Nardone *et al.*, 2010). Egg quality, such as egg weight and shell weight and thickness may also be negatively affected under hotter conditions (Mashaly *et al.*, 2004).

Reproduction

Reproduction efficiency of both livestock sexes may be affected by heat stress. In cows and pigs, it affects oocyte growth and quality (Barati *et al.*, 2008), impairment of embryo development, and pregnancy rate (Hansen, 2007; Nardone *et al.*, 2010). Cow fertility may be compromised by increased energy deficits and heat stress (King *et al.*, 2006). Heat stress has also been associated with lower sperm concentration and quality in bulls, pigs, and poultry (Kunavongkrita *et al.*, 2005).

Health

Nardone *et al.*, (2010) presented several livestock health problems related to climate change. Prolonged high temperature may affect metabolic rate (Webster, 1991), endocrine status (Johnson, 1980), oxidative status (Bernabucci *et al.*, 2002), glucose, protein and lipid metabolism, liver functionality (reduced cholesterol and albumin) (Bernabucci *et al.*, 2006), non-esterified fatty acids (NEFA) (Ronchi *et al.*, 1999), saliva production, and salivary HCO_3 - content. In addition, greater energy deficits affect cow fitness and longevity (King *et al.*, 2006).

Mortality

Warm and humid conditions that cause heat stress can affect livestock mortality. Howden *et al.*, (2008) reported that increases in temperature between 1 and 5^0C might induce high mortality in grazing cattle. As a mitigation measure, they recommend sprinklers, shade, or similar management practices to cool the animals.More information is needed concerning the nature and extent of how heat stress affects feed nutrient utilization and feed intake, animal production, reproduction, and health. With greater knowledge related to nutritional and metabolism processes of livestock, management practices could be adapted to increase animal performance.

Biodiversity

Biodiversity refers to a variety of genes, organisms, and ecosystems found within a specific environment (Swingland, 2001) and contributes to human well-being (MEA, 2005). Populations that are decreasing in genetic biodiversity are at risk, and one of the direct drivers of this biodiversity loss is climate change (UNEP, 2012). Climate change may eliminate 15% to 37% of all species in the world (Thomas *et al.*, 2004). Temperature increases have affected species reproduction, migration, mortality, and distribution (Steinfeld *et al.*, 2006). The Intergovernmental Panel on Climate Change Fifth Assessment Report states that an increase of 2 to 3 ^{0} C above pre-industrial levels may result in 20 to 30% of biodiversity loss of plants and ani-mals (IPCC, 2014). By 2000, 16% of livestock breeds (ass, water buffalo, cattle, goat, pig, sheep, and horse) were lost (Thornton *et al.*, 2009). In addition, the FAO (2007) has stated that from 7,616 livestock breeds reported, 20% were at risk, and almost one breed per month was being extinguished. Cattle had the highest number of extinct breeds (N = 209) of all species evaluated.

The livestock species that had the highest percentages of risk of breed elimination were chicken (33% of breeds), pigs (18% of breeds), and cattle (16% of breeds). However, the breeds at risk depend on the region. Developing regions had between 7% and 10% of mammalian species at risk (not restricted to livestock), but between 60% and 70% of mammalian species are classified as of unknown risk. Conversely, in developed regions, where the livestock industry is very specialized and based on a small number of breeds, the mammalian species at risk were between 20% and 28% (FAO, 2007). Thornton *et al.*, (2009) states that this biodiversity loss is mainly because of the practices used in livestock production that emphasize yield and economic returns and marginalization of traditional production systems where other considerations are also important (such as ability to withstand extremes). Livestock and plants will be highly affected by climate change and biodiversity loss. These breeds and species cannot be replaced naturally; therefore, future work that studies the inherent genetic capabilities of different breeds and identifies those that can better adapt to climate conditions is vital.

Agro-Ecological Zones

Agricultural practices such as livestock production and management varies around the world. In order to describe this variability's, the Food and Agricultural Organization of the United Nations and the International Institute for Applied Systems Analysis established the agro-ecological zones (AEZs) for assessing agricultural resources (FAO, 2017). These zones are defined based on climate, landform, soils, land cover, and land use (FAO, 1996). The AEZs are broadly classified into five categories including tropics, subtropics, temperate, boreal, and artic (FAO, 1996). Meanwhile, climate change variabilities and impacts on agricultural productions are not necessarily aligned with the AEZs or even can be significantly different within an AEZ, which makes it difficult to generalize the impacts of climate change on AEZs.

In general, climate change impacts within an AEZ can be both negative and positive. For example, climate change impacts in a form of increase in CO_2 level can improve the photosynthetic activity and the water use efficiency, which consequently increases crop productivity. Meanwhile, rise in average temperature can trigger plant diseases while increasing water stress, which consequently decreases crop productivity (Fischer *et al.*, 2002). These fluctuations in crop yields are indirectly affecting livestock productivity especially in rainfed systems. At the same time, higher temperatures will directly affect livestock production by changing the animals' migration pattern and reproduction time. Therefore, livestock species with restricted habitat, small population, limited mobility, and low breeding rates will be the most vulnerable (Steinfeld *et al.*, 2006).

The majority of worldwide ruminants are found in tropics and subtropics AEZs, especially in arid or semi-arid regions, where climate conditions limit animal productivity and yield (Herrero *et al.*, 2012). Among climate factors, heat stress is common due to high temperature. For example, within the tropics and subtropics AEZs of Europe, North America, Africa, and Australia, heat stress resulted in increased livestock morbidity and mortality (Renaudeau *et al.*, 2012).

Even though the largestnumbers of livestock are found in the tropics and subtropics AEZs, the largest productivity is found in the temperate AEZs (Herrero *et al.*, 2012). In these regions, the impacts of climate change ruminants' performance can be minimal (except extreme events) while crop production is expected to increase due increase in average temperature (Rowlinson, 2008; Seguin, 2008). Meanwhile, warmer winters may negatively impact animal productivity due to increase in incidences of vertebrate pests (Rowlinson, 2008).

In the boreal AEZs, the impacts of climate change on crop and livestock productions are generally positive (Bajzˇelj and Richards, 2014). Climate change impacts in forms of increased temperature, winter rainfall, and the length of the growing season will improve crop and grass productions and the opportunity for planting new crops (Iglesias *et al.*, 2007). In addition, livestock housing costs might be reduced due to the longer growing seasons (Iglesias *et al.*, 2007). Meanwhile, these changes may result in improving the yields in low fertile soils while reducing the yields in high fertile soils (Iglesias *et al.*, 2007).

Food Security

About 842 million people (one in eight people worldwide) went hungry between 2011 and 2013, not receiving enough food to maintain an active and healthy life (FAO *et al.*, 2013). Livestock contributes greatly to food security because: (1) they are suppliers of global calories, proteins, and essential micronutrients, (2) they are produced in areas that have difficulty growing crops, (3) most of the feed for livestock is not appropriate for human consumption, and (4) they provide manure for crop production (FAO, 2011). However, there are also concerns that livestock production is detrimental to food security.

First, the use of grains as feed in livestock production is a worldwide concern because they are produced for animal feed and not for human consumption. For example, in 2002, one-third of the global cereal harvest was used as livestock feed (Steinfeld *et al.*, 2006).The bulk of the livestock feed comes from grasses and legume forage that grows on land not suitable to agriculture (O'Mara, 2012), and in many countries livestock do not receive cereal supplements. In such areas, livestock are a positive contributor to food security. The debate occurs in areas where cattle are pastured in areas perfectly suitable for agriculture, or where they are fed substantial cereal supplements. Second, climate change, mostly via an increase in temperature, may decrease intake of digestible nutrients. Therefore, livestock production may decrease through declining forage quality and quantity and/or by reducing animal feed intake. These two factors affect livestock production because animals will usethe available nutrients to first maintain their physiological needs, then for growth or milk production, and finally for reproduction (Hatfield *et al.*, 2008). Third, climate change also affects nutritional content of livestock products because of potential increases in pathogens and diseases in their food and effects on the animals themselves (Harvell *et al.*, 2002). As new pathogens and diseases emerge and spread, pesticide and veterinary medicine use will change, consequently changing the principal transfer process of environmental contaminants to food (Lake *et al.*, 2012; van der Spiegel *et al.*, 2012).

Sustainable livestock production needs more research, extension, and demonstration. Livestock are an important contributor to food security, but it is important to maintain an efficient conversion of natural resources to human food to sustain a neutral food balance (FAO, 2011). This can be accomplished through efficient production of protein from livestock (FAO, 2013). However, climate change will influence this conversion by affecting the nutritional content of livestock products (Karl *et al.*, 2009) and reducing livestock production (Hatfield *et al.*, 2008). Currently, the livestock sector's best approach to contribute to food security is by addressing the primacy of food balance (FAO, 2013).

Impact of Livestock on Climate Change

Livestock contribute 14.5% of the total annual anthropogenic GHG emissions globally (Gerber *et al.*, 2013). Livestock influence climate through land use change, feed production, animal production, manure, and processing and transport (Fig 2). Feed production and manure emit CO_2, nitrous oxide (N_2O), and methane (CH_4), which consequently affects climate change. Animal production increases CH4 emissions. Processing and transport of animal products and land use change contributes to the increase of CO_2 emissions (Fig 2).

The livestock sector is often associated with negative environmental impacts such as land degradation, air and water pollution, and biodiversity destruction (Bellarby *et al.*, 2013; Thornton and Gerber, 2010). Increases in livestock production are expected to originate from a declining natural resource base, which will cause further environmental damage without proper natural resources management (Thornton and Herrero, 2010a,).

GHG Emissions

The primary livestock GHG emissions are CO_2, CH_4, and N_2O. CH_4 contributes the most to anthropogenic GHG emissions (44%), followed by N_2O (29%) and CO_2 (27%) (Gerber *et al.*, 2013). Globally livestock contribute 44% of anthropogenic CH_4, 53% of anthropogenic N_2O and 5% of anthropogenic CO_2 emissions (Fig 3). Higher concentrations of these gases can be explained by lower efficiency and productivity of livestock system due to excess loss of nutrients, energy, and organic matter (Gerber *et al.*, 2013).

The global warming potential (GWP) of CH_4 and N_2O over a 100 year time horizon differs between the IPCC (2013), and the United Nations Framework Convention on Climate Change (UNFCC, 2014). Current assessments have estimated higher GWP than what was previously thought. IPCC (2006) is most commonly used in the literature for estimating livestock GHG emissions, and reported that the warming potential of CH_4 is 25 CO_2-eq, while N_2O is 298 CO_2-eq. However, the latest IPCC (2013) reported a warming potential for CH_4 of 34 CO_2-eq, while N_2O is 310 CO_2-eq. The UNFCCC (2014) stated that CH_4 emissions have a warming potential of 21 CO_2-eq, and N_2O is 310 CO_2-eq over a 100 year time horizon.

The world's transportation sector emits around 5656 Tg CO_2-eq yr^{-1} and the livestock sector emits 7100 Tg CO_2-eq yr^{-1} (DSI MSU, 2015; Gerber *et al.*, 2013). Emissions from livestock production contribute more GHG to the atmosphere than the entire global transportation sector. The livestock sector contributes directly and indirectly to GHG emissions, including through animal physiology, animal housing, manure storage, manure treatments, land application, and chemical fertilizers (Casey *et al.*, 2006). Direct emissions from animal sources include enteric fermentation, respiration, and excretions (Jungbluth *et al.*, 2001). Indirect emissions refers to emissions derived from feed crops, manure application, farm operations, livestock products processing, transportation, and land use allocation for livestock production (e.g. deforestation, desertification, carbon released from cultivated soils) (Mosier *et al.*, 1998). In the livestock sector, indirect emissions play a greater role in the release of carbon to the atmosphere than direct emissions (Steinfeld *et al.*, 2006).

The livestock sector's contribution of 14.5% of total anthropogenic GHG emissions was evaluated by Gerber *et al.*, (2013) using a global livestock environmental assessment model (GLEAM). GLEAM performs an analysis of the emissions of global livestock production along supply chains. The main elements of the livestock supply chains that are analyzed by GLEAM are:

Herd, feed, manure, animals' energy requirement, feed intake, production, and emissions, allocation of the total emissions at the farmgate (physical farm boundaries) to co-products and services (emission per kg of product), and post-farmgate emissions (transport and processing). GLEAM also considers land use change as the conversion of forest to pasture or arable land for crops.

The contributors of the 14.5% of livestock GHG emissions are presented in Fig 4. Enteric fermentation is the largest contributor of the sector's emissions with 39.1%, followed by manure management, application, and direct deposit with 25.9%, feed production with 21.1%, land use change with 9.2%, post-farmgate with 2.9%, and direct and indirect energy with 1.8% (Gerber *et al.*, 2013). However, contribution to GHG emissions varies depending on the type of farming system and region.

For example, intensification of any of the three major livestock production systems and expansion of industrialized (or landless) systems will increase CO_2 emissions due to greater fossil fuel use and less solar energy utilized by photosynthesis (Steinfeld *et al.*, 2006). Gerber *et al.*, (2013) also estimated livestock GHG emissions by region, finding that, Asia produce the highest, followed by Latin America and the Caribbean, Europe, North America, Africa, and Oceania.

Land Use

Forests and natural habitats have been steadily converted to pasture and cropland since the 1850s (Goldewijk and Battjes, 1997). Agriculture lands cover about 38.5% of global total land area, which consists of 28.4% arable land and 68.4% permanent meadows and pasture (DSI MSU, 2015). Pasturelands have expanded by a factor of six since 1800, now covering 35 million km2 (White *et al.*, 2000). Agricultural land use change is related to two concepts: profit per unit of land and opportunity cost (Steinfeld *et al.*, 2006). Profit per unit of land refers to the willingness of farmers to manage a specific landuse. Profit will vary depending on several factors, such as the land's biophysical characteristics and price, access to markets, inputs, and services. The opportunity cost concept compares the social and economic cost of different ways to use the same land area. Opportunity costs include private production costs and ecosystem service costs. Therefore, when non-marketable ecosystem services have no associated cost, land use decisions are based on private profit per unit of land (Steinfeld *et al.*, 2006).

The increasing demand for livestock products has significantly changed the natural landscape. Land degradation is the deterioration of physical, chemical, and biological properties of soil. Land degradation has been recognized as one of the drivers of land conversion from forest to croplands and pastures because producers exhaust their soil resources and thus search for more suitable land (Steinfeld *et al.*, 2006). However, Asner *et al.*, (2004) stated that due to climate and soil characteristics, the expansion of pasture into marginal areas is limited; therefore, they could only expand into areas with agro-ecological potential.

Land use change affects the natural carbon cycle, which consequently releases high amounts of carbon into the atmosphere, increasing GHG emissions. Natural habitats, mainly forests, sequester more carbon in soil and vegetation than croplands and pasturelands. Soil and terrestrial vegetation sequester up to 40% of global anthropogenic CO_2 emissions (The Royal Society, 2001). Furthermore, pasturelands contain more carbon than croplands. Croplands sequester 6% of global carbon, while tropical savannas and temperate pasturelands together sequester 27% (IPCC, 2000). However, soils sequester the most carbon in the terrestrial carbon cycle, and double that of vegetation (Steinfeld *et al.*, 2006). Sundquist (1993) estimated that 1,100 to 1,600 billion tonnes of carbon is stored in soils. However, soil carbon can be lost through burning, volatilization, erosion, land use change, and agricultural management practices (Bolin *et al.*, 1982). Therefore, when a forest is converted to cropland and pasture by logging or burning, high amounts of carbon are released into the atmosphere (Steinfeld *et al.*, 2006).

Feed Production

The use of manure and synthetic fertilizers for forage and feed crop production, processing of feed, and transport of feed are the most important contributors of GHG emissions related to the livestock sector (IFAD, 2010).

These make up 45% of global livestock anthropogenic GHG emissions, consisting primarily as CO_2, N_2O and NH_4 (Gerber *et al.*, 2013). By considering the amount of fertilizer use, packaging, transport, and application in the livestock sector, the manufacturing process of fertilizers contributes more than 40 million tonnes of CO_2 annually (Steinfeld *et al.*, 2006). Forty percent of all the nitrogen up-take by crops comes from synthetic fertilizers (Smil, 2001). Ammonia volatilization loss from synthetic nitrogen fertilizer is an indirect contributor to GHG emissions. Between 4 to 5 million tonnes of mineral fertilizer is used for livestock feed production.

The average loss due to ammonia volatilization from mineral fertilizer is about 14%. Therefore, it was estimated that the livestock sector contributes 3.1 million tonnes of global ammonia volatilization from mineral fertilizers per year (Steinfeld *et al.*, 2006). N_2O is another contributor to GHG emissions. Fertilizer use, agricultural nitrogen fixation, and atmospheric nitrogen deposition generally increase N_2O emissions (Bouwman, 1996). By following the same assumptions as ammonia, with a 1% average of N_2O-N (tonnes of nitrogen in nitrous oxide form) loss rate from mineral fertilizer, the livestock sector contributes 0.2 million tonnes N_2O-N of global N_2O emission from mineral fertilizer per year (Steinfeld *et al.*, 2006). In addition, leguminous feed crops for livestock account for additional N_2O emissions. Steinfeld *et al.*, (2006) estimated their contribution by considering the global area of soybeans used for livestock feed and doubling it to include alfalfa and clover, because there are no estimates of worldwide alfalfa and clover production. Therefore, the contribution of leguminous feed crops is more than 0.5 million tonnes per year of N_2O-N emissions. By adding both contributors (mineral fertilizer and leguminous feed crops), the total N_2O-N emissions are 0.7 million tonnes per year (Steinfeld *et al.*, 2006). As growth in fertilizer and manure use continues, a 35–60% increase of N_2O emissions (0.9 to 1.1 million tonnes per year of total N_2O-N emissions) is expected by 2030 (Bruinsma, 2003).

On-farm fossil fuel use in livestock production produces 50% more CO_2 emissions than manufacturing N fertilizers for feed. The livestock sector includes direct and indirect (e.g. electricity) on-farm fossil fuel use, which is used for machinery operations, irrigation, heating, cooling, ventilation, production of herbicides and pesticides, and more. More than half of fossil-fuel use is attributed to feed production. By assuming CO_2 emissions from on-farm fossil fuel use are double that of manufacturing N fertilizers, and adding emissions related to livestock rearing, on-farm fossil fuels account for 90 million tones of CO_2 per year (Steinfeld *et al.*, 2006).

Animal Production

In general, livestock respiration is not counted as a net source of CO_2 emissions because they are part of the global biological system cycle. The vegetation consumed by the animal originates from the conversion of atmospheric CO_2 to organic compounds or biomass. Therefore, under the Kyoto Protocol (2005) it is assumed that the consumed amounts of CO_2 in vegetative form are equivalent to those emitted by the livestock. Conversely, the animal is a carbon sink because a fraction of the carbon consumed is absorbed in the live tissue of the animal (UNFCCC, 1998) and products such as milk.

Livestock contributes 44% of the world's anthropogenic CH_4 emissions through their normal digestive processes (enteric fermentation) and manure management (Gerber *et al.*, 2013). Enteric fermentation and manure account for 80% of 52 agricultural emission sources (Steinfeld *et al.*, 2006). During the animals' digestive process, enteric fermentation converts the feed consumed into digestible feed. Enteric fermentation releases a CH_4 by-product through exhalation (Beauchemin *et al.*, 2009). Therefore, this by-product is considered an energy loss (Gerber *et al.*, 2013). Feed composition and feed intake can vary enteric fermentation and hence methane emissions. Increasing the concentrate (high energy feeds containing cereal grains and oil meals) proportion in the animal diet can reduce methane emissions from the animal (Dourmad *et al.*, 2008).

Methane emissions vary depending on production systems and regional characteristics (e.g. climate and landscape) (Gerber *et al.*, 2013). The enteric fermentation produced by ruminant livestock (e.g. cattle, sheep, and goat) emits globally between 87 and 94 Tg of methane annually (IPCC, 2013). Mixed crop-livestock systems account for 64% of global enteric fermentation methane emissions; grazing systems account for 35%, and industrial 1% (Steinfeld *et al.*, 2006). The high percentage from mixed crop-livestock systems reflects that two-thirds of total livestock animals are present in those systems (Steinfeld *et al.*, 2006). The countries that contribute the most methane emissions related to livestock production are India, China, Brazil, and the United States (IPCC, 2013; Olivier and Janssens-Maenhout, 2012). India, with the largest livestock population in the world, emitted 11.8 Tg of CH_4 in 2003, from which 91% derives from enteric fermentation and 9% from manure management (Chhabra *et al.*, 2013).

In Africa, methane emissions are expected to increase due to increases in livestock populations. Herrero *et al.*, (2008) estimated that African cattle, goats, and sheep, which produced about 7.8 million tonnes of methane in 2000, are likely to increase to 11.1 million tonnes by 2030. If this linear relationship between methane emissions and livestock population continues, global methane emissions from livestock production may increase 60% by 2030 (Bruinsma, 2003). However, changing feeding practices and manure management could moderate methane emissions (Thornton and Herrero, 2010). Animals that contribute the most to livestock GHG emissions are beef and dairy cattle, accounting for 65% of the total livestock GHG emissions (Gerber *et al.*, 2013). Pigs, poultry, buffaloes, and small ruminants contribute about 7 to 10%. If GHG emissions are estimated based on commodities, beef cattle contribute the most with **41% of the sector's emission, follow**ed by dairy cattle (20%), swine (9%), buffalo (8%), poultry (8%), and small ruminant (6%) (Gerber *et al.*, 2013).

Enteric fermentation is the largest source of GHG emissions from cattle, buffalo, and small ruminants, comprising between 43% and 63% of the livestock sector emissions (Fig 5). However, for pigs and chickens the largest source of emissions is due to feed production (between 25% and 27%), which includes fertilizer production, machinery use, and feed transportation. Enteric fermentation from pigs is much lower than in ruminants because their digestive process does not produce as much methane as a by-product (Gerber *et al.*, 2013).

Manure

Livestock manure releases CH_4 and N_2O gas. The decomposition of the organic materials found in manure under anaerobic conditions releases methane (EPA, 1999). Liquid manure found in lagoons or holding tanks releases more methane than dry manure (Burke, 2001). Manure methane emissions are a function of air temperature, moisture, pH, storage time, and animal diet (IFAD, 2010). Steinfeld *et al.*, (2006) estimated global methane emissions from manure decomposition of 17.5 million tonnes of CH4 per year. Pig manure comprises almost half of global manure-related methane emissions. At the country level, China has the highest global methane manure-related emissions, primarily due to pig manure (Steinfeld *et al.*, 2006).

N_2O emissions from manure storage are dependent on environmental conditions, handling systems, and duration of waste management. Manure must be handled aerobically and then anaerobically to release N_2O emissions, which is more likely to occur in dry waste-handling systems. Steinfeld *et al.*, (2006) reported that N_2O emissions from stored manure are equivalent to 10 million tonnes N per year. Nitrous oxide soil emissions from manure application are the largest source of global N_2O emissions (Steinfeld *et al.*, 2006). Nitrogen emissions from applied or deposited manure are dependent on soil infiltration, organic carbon amount, pH, soil temperature, precipitation, and the plant/crop uptake rate (Mosier *et al.*, 2004). Steinfeld *et al.*, (2006) estimated that 1.7 million tonnes of manure soil N_2O are released per year. N_2O emissions from applied manure are 40% higher in mixed crop-livestock systems than the N_2O emissions from excreted manure deposited on pasture systems. Industrial production systems have 90% less N_2O emissions than mixed crop-livestock systems (Steinfeld *et al.*, 2006).

Summarizing Adaptation and Mitigation Practices

There is several climate change adaptation and mitigation recommendations that can be made based on the above discussion. Adaptation strategies can improve the resilience of crop and livestock productivity to climate change (USDA, 2013). Mitigation measures could significantly reduce the impact of livestock on climate change (Dickie *et al.*, 2014). Adaptation and mitigation can make significant impacts if they become part of national and regional policies (FAO, 2009b).

Adaptation Measures

Adaptation measures involve production and management system modifications, breeding strategies, institutional and policy changes, science and technology advances, and changing farmers' perception and adaptive capacity (IFAD, 2010; USDA, 2013). Research is needed on assessments for implementing these adaptation measures and tailoring them based on location and livestock system. This could be accomplished with GIS and remote sensing technologies applicable at broad and local scales (Thornton *et al.*, 2008).

Livestock Production and Management Systems: An adaptation such as the modification of production and management systems involves diversification of livestock animals and crops, integration of livestock systems with forestry and crop production, and changing the timing and locations of farm operations (IFAD, 2010). Diversification of livestock and crop varieties can increase drought and heat wave tolerance, and may increase livestock production when animals are exposed to temperature and precipitation stresses. In addition, this diversity of crops and livestock animals is effective in fighting against climate change-related diseases and pest outbreaks (Batima *et al.*, 2005; IFAD, 2010).

Agroforestry (establishing trees alongside crops and pastures in a mix) as a land management approach can help maintain the balance between agricultural production, environmental protection and carbon sequestration to offset emissions from the sector. Agroforestry may increase productivity and improve quality of air, soil, and water, biodiversity, pests and diseases, and improves nutrient cycling (Jose, 2009; Smith *et al.*, 2012).

Changes in mixed crop-livestock systems are an adaptation measure that could improve food security (Herrero *et al.*, 2010; Wani *et al.*, 2009). This type of agricultural system is already in practice in two-thirds of world, producing more than half of the milk, meat, and crops such as cereal, rice and sorghum (Herrero *et al.*, 2012). Changes in mixed crop-livestock systems can improve efficiency by producing more food on less land using fewer resources, such as water (Herrero *et al.*, 2012; Steinfeld *et al.*, 2006).

Improving feeding practices as an adaptation measure could indirectly improve the efficiency of livestock production (Havlík *et al.*, 2013). Some of the suggested feeding practices include, modification of diets composition, changing feeding time and/or frequency (Renaudeau *et al.*, 2012), incorporating agroforestry species in the animal diet (Thornton and Herrero, 2010a,), and training producers in production and conservation of feed for different agro-ecological zones (IFAD, 2010). These practices can reduce the risk from climate change by promoting higher intake or compensating low feed consumption, reducing excessive heat load (Renaudeau *et al.*, 2012), decreasing the feed insecurity during dry seasons (Thornton and Herrero, 2010a,b), and reducing animal malnutrition and mortality (IFAD, 2010), respectively.

Shifting locations of livestock and crop production could reduce soil erosion and improve moisture and nutrient retention (Kurukulasuriya and Rosenthal, 2003). Another adaptive measure could be adjusting crop rotations and changing timing of management operations (e.g. grazing, planting, spraying, irrigating). This measure can be adapted to changes in duration of growing seasons, heat waves and precipitation variability (IFAD, 2010).

Breeding Strategies: Changes in breeding strategies can help animals increase their tolerance to heat stress and diseases and improve their reproduction and growth development (Henry *et al.*, 2012). Therefore, the challenge is in increasing livestock production while maintaining the valuable adaptations offered by breeding strategies, all of which will require additional research (Thornton *et al.*, 2008). In addition, policy measures that improve adaptive capacity by facilitating implementation of adaptation strategies will be crucial (USDA, 2013). For example, developing international gene banks could improve breeding programs and serve as an insurance policy, such as has been done for plants with the In-Trust plant collections in the CGIAR gene banks (Thornton *et al.*, 2008). This would be a major breakthrough that requires significant investment and international collaboration to succeed.

Farmers' Perception and Adaptive Capacity

One of the limiting factors for these changes to succeed is the disposition and capability of farmers to recognize the problem and adopt climate change adaptation and mitigation measures (Jones *et al.*, 2013). Because of this, it is important to collect information about farmers' perceptions to mitigation and adaptation measures. One approach for collecting information about farmers' perceptions that has been used for mitigation and adaptation research is qualitative; using open-ended survey questions or group discussion at workshops to understand individual and group opinions (Barnes *et al.*, 2008). By understanding farmers' perceptions and including them in rural policy development, there is a greater chance of accomplishing food security and environmental conservation objectives (Barnes, 2013; Oliver *et al.*, 2012).

Risk perception within farmer decision-making can be increased through education, family farm succession, and social interaction among farmers and farming communities. Barnes (2013) applied a latent class clustering approach (which uses statistical methodology to construct results) to evaluate the heterogeneity of dairy farmers' risk perception of climate change. Their results show that the drivers of risk perception due to climate change are family members and the influence of succession planning. They recommended increasing the social capital of farming communities to promote acceptance of communication strategies for climate change adaptation and mitigation measures.

Mitigation Measures

There is potential to reduce livestock sector GHG emissions through the implementation of different technologies and practices. However, they are not widely used (Gerber *et al.*, 2013). Some of the technical options for mitigating the impact of livestock on climate change are carbon sequestration, improving diets to reduce enteric fermentation, improving manure management, and more efficient use of fertilizers (Thornton and Gerber, 2010). Mitigation measures need public policy support to be effective (Dickie *et al.*, 2014).

Carbon Sequestration: Carbon sequestration can be achieved through decreasing deforestation rates, reversing of deforestation by replanting (Carvalho *et al.*, 2004), targeting for higher-yielding crops with better climate change adapted varieties, and improvement of land and water management (Steinfeld *et al.*, 2006). A beef sector study performed in Brazil estimated a reduction of up to 25% of GHG emissions related to grazing land use and land use change, accomplished by improving animal and herd efficiency (Gerber *et al.*, 2013).

Soil organic carbon can be restored in cultivated soils through conservation tillage, erosion reduction, soil acidity management, double-cropping, crop rotations, higher crop residues, mulching and more (Steinfeld *et al.*, 2006). Improving pasture management can also lead to carbon sequestration by incorporating trees, improving plant species, legume interseeding, introducing earthworms, and fertilization (Conant *et al.*, 2001). In addition, grass productivity and soil carbon sequestration could be improved by increasing grazing pressure in grasslands that have a lower amount of grazing animals than the livestock carrying capacity (Holland *et al.*, 1992). Improving grazing land management could sequester around 0.15 gigatonnes CO_2-eq yr^{-1} globally (Henderson *et al.*, 2015).

Enteric Fermentation: Enteric fermentation is a source of methane emissions that can be reduced through practices such as improvement of animal nutrition and genetics (US-EPA, 1999). Examples of practices for mitigating enteric fermentation are: increasing dietary fat content (Martin *et al.*, 2010), providing higher quality forage (Hristov *et al.*, 2013), increasing protein content (ICF International, 2013), providing supplements (e.g. bovine somatotropin, feed antibiotics) (Boadi *et al.*, 2004), and the use of antimethanogens (vaccines to suppress methane emissions) (EPA, 2013). However, there is high uncertainty in the efficacy of these practices because various studies have demonstrated that the initial reductions of enteric fermentation achieved are only temporary (ICF International, 2013).

A one percent increase of dietary fat can decrease enteric methane emissions between 4 to 5% (Martin *et al.*, 2010). However, rumens need to limit fat content to 8% of dry matter to avoid a decrease in livestock performance (Hales *et al.*, 2012). Providing higher quality forage also results in a reduction of methane emissions because it increases digestibility (Hristov *et al.*, 2013). An increase of protein content of feed can also improve digestibility and reduce overall methane emissions per unit of product (ICF International, 2013).

Providing supplements, such as feed antibiotics, which tend to increase weight gain and reduce feed intake per metric ton of meat produced, can reduce enteric fermentation (Boadi *et al.*, 2004). In the case of milk, bovine somatotropin (a bovine growth hormone) increases production. An increase in milk production leads to less animals needed to produce the same amount of milk and less emissions produced (EPA, 2013). Antimethanogen vaccines are another practice that directly reduces methane emissions in the rumen. However, this is a new technology with limited research on emission reduction efficiency and animal health (EPA, 2013).

Manure Management: Most methane emissions from manure management are related to storage and anaerobic treatment. Although manure deposited on pasture can produce nitrous oxide emissions, the mitigation measures are often difficult to apply because of the manure dispersion on pasture (Dickie *et al.*, 2014). Therefore, most mitigation practices involve shortening storage duration, improving timing and application of manure, used of anaerobic digesters, covering the storage, using a solids separator, and changing the animal diets (ICF International, 2013).

Anaerobic digestion can reduce methane emissions while producing biogas (Gerber *et al.*, 2008). Anaerobic digesters are lagoons or tanks that maintain manure under anaerobic conditions to capture biogas and combust it for producing energy or flaring. This process reduces the potential of GHG emissions by converting methane into CO2 (ICF International, 2013). Unfortunately, anaerobic digesters are costly for producers; the best approach for implementing digesters is through policies that create enough incentive for adaptation (Dickie *et al.*, 2014). Similar to digesters, the covering of ponds, tanks or lagoons reduces emissions by capturing and destroying methane (ICF International, 2013).

Other storage and handling practices can also reduce GHG emissions. Such practices include reducing storage time, improving housing and waste management systems to handle manure, and removing bedding from manure by using a solids separator (Dickie *et al.*, 2014). The solids separator is mostly used in confinement systems to remove solids from manure streams that are entering the treatment or storage systems. By removing the solids from manure streams methane emissions are reduced, the time between storage system cleaning is increased, and crust formation is prevented (ICF International, 2013). These practices, compared to anaerobic digesters, are usually low-cost and low-tech. However, they require more time and effort from the producer (Dickie *et al.*, 2014).

Adjusting animal diets can also be used as a mitigation measure, by changing the volume and composition of manure. GHG emissions can be reduced by balancing dietary proteins and feed supplements. If protein intake is reduced, the nitrogen excreted by animals can also be reduced. Supplements such as tannins are also known to have the potential to reduce emissions. Tannins are able to displace the nitrogen excretion from urine to feces to produce an overall reduction in emissions (Dickie *et al.*, 2014; Hess *et al.*, 2006).

Fertilizer Management: Fertilizer application on animal feed crops increases nitrous oxide emissions (Bouwman, 1996). Therefore, mitigation measures such as increasing nitrogen use efficiency, plant breeding and genetic modifications (Dickie *et al.*, 2014), using organic fertilizers (Denef *et al.*, 2011), regular soil testing, using technologically advanced fertilizers, and combining legumes with grasses in pasture areas may decrease GHG emissions in feed production (Dickie *et al.*, 2014).

Nitrogen use efficiency can be improved by applying the required amount that the crop will absorb and when it needs the nutrients, and placing it where the plant can easily reach it. Regular soil testing can be a part of a nutrient management plan depending on the region and crop, and improve efficiency of nitrogen use (Dickie *et al.*, 2014). Plant breeding and genetic modifications can reduce the use of fertilizers by increasing a crop's nitrogen uptake (Dickie *et al.*, 2014). Increasing the use of organic fertilizers would also decrease emissions because organic fertilizers do not produce as much nitrogen oxide as synthetic fertilizers (Denef *et al.*, 2011). Furthermore, fertilizer technology has improved through regulating the release of nutrients from the fertilizer and inhibiting nitrification to slow the degradation of the fertilizer and maintain the nutrients available for the plant. However, these technologically advanced fertilizers are more costly than the other practices mentioned above (Dickie *et al.*, 2014). In the case of pasturelands, the use of synthetic nitrogen can be reduced by combining legumes with grasses. Legumes fix nitrogen through Rhizobium bacteria; therefore, the need for supplementary nitrogen is reduced (USDA-NRCS, 2007).

Shifting Human Dietary Trends: Most studies are focused on reducing GHG emissions on the Supply-side of the livestock production system. However, less research has focused on the demand section related to consumption of livestock products.Because beef accounts for a large portion of GHG emissions from the livestock sector and it is the least resource-efficient animal protein producer (Stehfest *et al.*, 2009), the mitigation potential is high for the beef component of the livestock sector. Research to understand why populations feel compelled to increase animal protein consumption when they rise above the poverty line is needed, as well as why those at the top of the economic ladder are compelled to improve their diets by reducing meat consumption and returning to a more vegetarian diet. This conundrum stands at the center of the challenge faced by the state of knowledge and policies surrounding the livestock sector.

References

1. Alexandratos N, Bruinsma J. World agriculture towards 2030/2050: the revision. ESA Working paper No. 12-03. FAO, Rome, 2012.
2. Asner GP, Elmore AJ, Olander LP, Martin RE, Harris AT. Grazing systems, ecosystems responses, and global change. Annu. Rev. Environ. Resour. 2004; 29:261–299.
3. Aydinalp C, Cresser MS. The effects of climate change on agriculture. Agric. Environ. Sci. 5, 2008, 672–676.
4. Barati F, Agung B, Wongsrikeao P, Taniguchi M *et al.*, Meiotic competence and DNA damage of porcine oocytes exposed to an elevated temperature. Theriogenology. 2008; 69:767–772.
5. Bajzˇelj B, Richards K. The positive feedback loop between the impacts of climate change and agricultural expansion and relocation. Land. 2014; 3:898–916.
6. Barnes A, Beechener S, Cao Y, Elliot J *et al.*, Market Segmentation in the Agriculture Sector: Climate Change, DEFRA Project FF0201. ADAS, UK, 2008.
7. Barnes AP. Heterogeneity in climate change risk perception amongst dairy farmers: a latent class clustering analysis. Appl. Geogr. 2013; 41:105–115.
8. Beauchemin KA, Eriksen L, Norgaard P, Rode LM. Nutritional management for enteric methane abatement: a review. Aust. J Exp. Agric. 2008; 48:21–27.
9. Beauchemin KA, McAllister TA, McGinn SM. Dietary mitigation of enteric methane from cattle. CAB reviews: perspectives in agriculture, veterinary science. Nutr. Nat. Resour. 2009; 4(35):1–8.
10. Bellarby J, Tirado R, Leip A, Weiss F *et al.*, Livestock greenhouse gas emissions and mitigation potential in Europe. Glob. Change Biol. 2013; 19:3–18.
11. Berman AJ. Estimates of heat stress relief needs for Holstein dairy cows. J Anim. Sci. 2005; 83:1377–1384.

12. Bernabucci U, Lacetera N, Basirico L, Ronchi B *et al.*, Hot season and BCS affect leptin secretion of periparturient dairy cows. Dairy Sci. 2006; 89:348–349.

13. Campbell I, Durant D, Hunter K, Hyatt K. Agriculture and climate change: a priarie perspective. In: Warren F, Lemmen D. (Eds.), Canada in a Changing Climate: Sector Perspectives on Impacts and Adaptation. Government of Canada, Ottawa, ON, 2004, 99–134.

14. Casey KD, Bicudo JR, Schmidt DR, Singh A *et al.*, Air quality and emissions from livestock and poultry production/waste management systems. In: Rice JM, Caldwell DF, Humenik FJ. (Eds.), Animal Agriculture and the Environment: National Center for Manure and Waste Management White Papers. American Society of Agricultural and Biological Engineers, St. Joseph, Mich, 2006, 40.

15. Chapman SC, Chakraborty S, Dreccer MF, Howden SM. Plant adaptation to climate change: opportunities and priorities in breeding. Crop Pasture Sci. 2012; 63:251–268.

16. Chase LE. Climate change impacts on dairy cattle. Climate change and agriculture: Promoting practical and profitable responses, 2012.

17. Chhabra A, Manjunath KR, Panigrahy S, Parihar JS. Greenhouse gas emissions from Indian livestock. Climatic Change. 2013; 117:329–344.

18. De Rensis F, Scaramuzzi RJ. Heat stress and seasonal effects on reproduction in the dairy cow: a review. Theriogenology. 2003; 60:1139–1151.

19. Denef K, Archibeque S, Paustian K. Greenhouse gas emissions from U.S. agriculture and forestry: A review of emission sources, controlling factors, and mitigation potential, 2011.

20. Dickie A, Streck C, Roe S, Zurek M *et al.*, Strategies for mitigating climate change in agriculture: Abridged report. Climate focus and California environmental associates, prifadred with the support of the climate and land use Alliance, 2014.

21. Dourmad J, Rigolot C, Van Der Werf H. Emission of Greenhouse Gas: developing management and animal farming systems to assist mitigation. In: Rowlinson P, Steel M, Nefzaoui A. (Eds.), Livestock and Global Climate Change Conference Proceeding. Cambridge University Press, Tunisia, 2008, 36–39.

22. Esminger ME, Oldfield JE, Heinemann WW. Feeds and Nutrition:

Formerly Feeds & Nutrition, Complete. Ensminger Publishing Company, Clovis, CA, 1990.

23. FAO (Food and Agriculture Organization of the United Nations). Agro-Ecological Zoning Guidelines. Rome, 1996.
24. FAO (Food and Agriculture Organization of the United Nations). The state of the world's animal genetic resources for food and agriculture: in brief, edited by Barbara Rischkowsky & Dafydd Pilling, Rome, 2007.
25. FAO (Food and Agriculture Organization of the United Nations). Global agriculture towards 2050. High Level Expert Forum Issues Paper. FAO, Rome, 2009a.
26. FAO (Food and Agriculture Organization of the United Nations). The state of food and agriculture: Livestock in the balance. Food and Agriculture Organization of the United Nations, 2009b.
27. FAO (Food and Agriculture Organization of the United Nations). World livestock: Livestock in food security. FAO, Rome, 2011.
28. FAO (Food and Agriculture Organization of the United Nations). Climate-smart agriculture: Sourcebook. FAO, Rome, 2013.
29. FAO (Food and Agriculture Organization of the United Nations), IFAD (International Fund for Agricultural Development), WFP (World Food Programme). The State of Food Insecurity in the World: The Multiple Dimensions of Food Security. FAO, Rome, 2013.
30. FAO (Food and Agriculture Organization of the United Nations). GAEZ - Global Agro-Ecological Zones, 2017.
31. Finocchiaro R, Van Kaam J, Portolano B, Misztal I. Effect of heat stress on production of dairy sheep. J Dairy Sci. 2005; 88:1855–1864.
32. Fischer G, Shah M, Van H. Climate Change and Agricultural Vulnerability. World Summit on Sustainable Development, Vienna, 2002.
33. Fregley MJ. Adaptations: some general characteristics. In: Fregley MJ, Blatteis CM. (Eds.), Handbook of physiology, Section 4: Environmental physiology. Oxford University Press, 1996, 3–15.
34. Garnett T. Livestock-related greenhouse gas emissions: impacts and options for policymakers. Environ. Sci. Policy. 2009; 12:491–503.
35. Gerber PJ, Carsjens GJ, Pak-uthai T, Robinson TP. Decision support for spatially targeted livestock policies: diverse examples from Uganda and Thailand. Agric. Syst. 2008; 96:37–51.

36. Gerber PJ, Steinfeld H, Henderson B, Mottet A *et al.*, Tackling Climate Change through Livestock: A Global Assessment of Emissions and Mitigation Opportunities. FAO, Rome, 2013.

37. Hales KE, Cole NA, MacDonald JC. Effects of corn processing method and dietary inclusion of wet distillers grain with solubles on energy metabolism and enteric methane emissions of finishing cattle. J Anim. Sci. 2012; 90(9):3174–3185.

38. Hansen PJ. Exploitation of genetic and physiological determinants of embryonic resistance to elevated temperature to improve embryonic survival in dairy cattle during heat stress. Theriogenology. 2007; 68(4):S242–S249.

39. Harvell CD, Mitchell CE, Ward JR, Altizer S *et al.*, Climate warming and disease risks for terrestrial and marine biota. Science. 2002; 296:2158–2162.

40. Hatfield JL, Boote K, Fay P, Hahn L *et al.*, Agriculture, in: Walsh M, The effects of climate change on agriculture, land resources, water resources, and biodiversity in the United States. Climate Change Science Program and the Subcommittee on Global Change Research. Washington, DC., USA, 2008, 362.

41. Hatfield JL, Prueger JH. Agroecology: implications for plant response to climate change. In: Yadav SS, Redden RJ, Hatfield JL, Lotze-Campen H, Hall AE. (Eds.), Crop Adaptation to Climate Change. Wiley-Blackwell, Chichester, UK, 2011, 27–43.

42. Havlík P, Valin H, Mosnier A, Obersteiner M *et al.*, Crop productivity and the global livestock sector: Implications for land use change and greenhouse gas emissions. Am. J Agric. Econ. 2013; 95:442–448.

43. Henderson BB, Gerber PJ, Hilinski TE, Falcucci A *et al.*, Greenhouse gas mitigation potential of the world's grazing lands: modeling soil carbon and nitrogen fluxes of mitigation practices. Agric. Ecosyst. Environ. 2015; 207:91–100.

44. Henry B, Charmley E, Eckard R, Gaughan JB, Hegarty R. Livestock production in a changing climate: adaptation and mitigation research in Australia. Crop Pasture Sci. 2012; 63:191–202.

45. Herrero M, Thornton PK, Kruska R, Reid RS. Systems dynamics and the spatial distribution of methane emissions from African domestic ruminants to 2030. Agric. Ecosyst. Environ. 2008; 126:122–137.

46. Herrero M, Thornton PK, Notenbaert A, Msangi S *et al.*, Drivers of Change in Crop–Livestock Systems and Their Potential Impacts on Agro-Ecosystems Services and Human Wellbeing to 2030: A Study Commissioned by the CGIAR Systemwide Livestock Programme. International Livestock Research Institute, Nairobi, Kenya 160 M.M. Rojas-Downing *et al.*, Climate Risk Management. 2012, 2017: 16:145-163.

47. Herrero M, Thornton PK, Notenbaert AM, Wood S *et al.*, Smart investments in sustainable food production: revisiting mixed crop-livestock systems. Science. 2010; 327:822–825.

48. Hess HD, Tiemann TT, Noto F, Carulla JE, Kreuzer M. Strategic use of tannins as means to limit methane emission from ruminant livestock. International Conference on Greenhouse Gases and Animal Agriculture, Elsevier International Congress Series, Zurich, Switzerland. 2006; 1293:164–167.

49. Howden SM, Crimp SJ, Stokes CJ. Climate change and Australian livestock systems: impacts, research and policy issues. Aust. J Exp. Agric. 2008; 48:780–788.

50. Hristov AN, Oh J, Firkins JL, Dijksta J, Kebreab E *et al.*, Mitigation of Methane and Nitrous Oxide Emissions from Animal Operations: I A review of enteric methane mitigation options. Am Soc Ani Sci, 2013, 91.

51. Hurst P, Termine P, Karl M. Agricultural workers and their contribution to sustainable agriculture and rural development. FAO, Rome, 2005.

52. ICF International. Greenhouse gas mitigation options and costs for agricultural land and animal production within the United States, 2013.

53. IFAD (International Fund for Agricultural Development). Livestock and climate change, 2010.

54. Iglesias A, Avis K, Benzie M, Fisher P *et al.*, Adaptation to climate change in the agricultural sector. AEA Energy & Environment and Universidad de Politécnica de Madrid, 2007.

55. IPCC (Intergovermental Panel on Climate Change). IPCC/OECD/IEA Programme on national greenhouse gas inventories. Intergovernmental panel on climate change, 1997.

56. IPCC (Intergovermental Panel on Climate Change). IPCC Guidelines for national greenhouse gas inventories. In: Eggleston HS, Buendia L, Miwa K, Ngara T, Tanabe K. (Eds.), Prepared by the National Greenhouse Gas Inventories Programme. IGES, Japan, 2006.

57. IPCC (Intergovermental Panel on Climate Change). Climate Change: Synthesis Report. In: Pachauri RK, Reisinger A. (Eds.), Contribution of Working Groups I, II and III to the Fourth assessment report of the Intergovernmental Panel on Climate Change. IPCC, Geneva, Switzerland, 2007, 104.

58. IPCC (Intergovermental Panel on Climate Change). Climate change: The physical science basis. In: Stocker TF, Qin D, Plattner GK, Tignor M, Allen SK, Boschung J, Nauels A, Xia Y, Bex V, Midgley PM. (Eds.), Contribution of Working Group I to the Fifth Assessment Report of the Intergovernmental Panel on Climate Change. Cambridge University Press, Cambridge, United Kingdom and New York, NY, USA, 2013, 1535.

59. IPCC (Intergovermental Panel on Climate Change). Climate Change: impacts, adaptation, and vulnerability. Part A: global and sectoral aspects. In: Field CB, Barros VR, Dokken DJ, Mach KJ, Mastrandrea MD, Bilir TE, Chatterjee M, Ebi KL, Estrada YO, Genova RC, Girma B, Kissel ES, Levy AN, MacCracken S, Mastrandrea PR, White LL. (Eds.), Contribution of Working Group II to the Fifth Assessment Report of the Intergovernmental Panel on Climate Change. Cambridge University Press, Cambridge, United Kingdom and New York, NY, USA, 2014, 1132.

60. Jones AK, Jones DL, Edwards-Jones G, Cross P. Informing decision making in agricultural greenhouse gas mitigation policy: a best-worst scaling survey of expert and farmer opinion in the sheep industry. Environ. Sci. Policy. 2013; 29:46–56.

61. Jose S. Agroforestry for ecosystem services and environmental benefits: an overview. Agrofor. Syst. 2009; 76:1–10.

62. Karl TR, Melillo JM, Peterson TC. Global Climate Change Impacts in the United States. U.S. Global Change Research Programme. Cambridge University Press, 2009.

63. King JM, Parsons DJ, Turnpenny JR, Nyangaga J *et al.*, Modellig energy metabolism of Friesians in Kenya smallholdings shows how heat stress and energy deficit constrain milk yield and cow replacement rate. Anim. Sci. 2006; 82:705–716.

64. Kunavongkrita A, Suriyasomboonb A, Lundeheimc N, Learda TW, Einarsson S. Management and sperm production of boars under differing environmental conditions. Theriogenology. 2005; 63:657–667.

65. Lake IR, Hooper L, Abdelhamid A, Bentham G *et al.*, Climate change and food security: health impacts in developed countries. Environ. Health Perspect. LEAD (Livestock Environment and Development). Livestock role in deforestation. Food and Agriculture Organization of the United Nations. 2012, 2014; 120:1520-1526.

66. Martin C, Morgavi DP, Doreau M. Methane mitigation in ruminants: from microbe to the farm scale. Animal. 2010; 4(3):351–365.

67. Mosier A, Kroeze C, Nevison C, Oenema O *et al.*, Closing the global N_2 budget: Nitrous oxide emissions through the agricultural nitrogen cycle: OEDC/IPCC/IEA phase II development of IPCC guideline for national greenhouse gas methodology. Nutr. Cycl. Agroecosyst. 1998; 52:225–248.

68. Mosier A, Wassmann R, Verchot L, King J, Palm C. Methane and nitrogen oxide fluxes in tropical agricultural soils: sources, sinks and mechanisms. Environ. Dev. Sustainability. 2004; 6:11–49.

69. Nardone A, Ronchi B, Lacetera N, Ranieri MS, Bernabucci U. Effects of climate change on animal production and sustainability of livestock systems. Livest. Sci. 2010; 130:57–69.

70. Oliver DM, Fish RD, Winter M, Hodgson CJ *et al.*, Valuing local knowledge as a source of expert data: farmer engagement and the design of decision support systems. Environ. Modell. Software. 2012; 36:76–85.

71. Olivier JG, Janssens-Maenhout G. Part III: Greenhouse gas emissions, in: International Energy Agency (IEA), CO_2 Emissions from Fuel Combustion, Ed, Paris, France, III. 1–III. 51. 2012.

72. O'Mara FP. The role of grasslands in food security and climate change. Ann. Bot-London. 2012; 110:1263–1270.

73. Perry B, Sones K. Global Livestock Disease Dynamics over the Last Quarter Century: Drivers, Impacts and Implications. FAO, Rome, 2009.

74. Polley HW, Briske DD, Morgan JA, Wolter K *et al.*, Climate change and North American rangelands: trends, projections, and implications. Rangeland Ecol. Manage. 2013; 66:493–511.

75. Randolph SE. Dynamics of tick-borne disease systems: minor role of recent climate change. Rev. Sci. Technol. Oie. 2008; 27:367–381.

76. Renaudeau D, Collin A, Yahav S, De Basilio V *et al.*, Adaptation to hot climate and strategies to alleviate heat stress in livestock production. Animal. 2012; 6(05):707–728.

77. Reynolds C, Crompton L, Mills J. Livestock and climate change impacts in the developing world. Outlook Agric. 2010; 39:245–248.

78. Rosegrant MW, Fernandez M, Sinha A. Looking into the future for agriculture and AKST. In: McIntyre BD, Herren HR, Wakhungu J, Watson RT. (Eds.), International Assessment of Agricultural Knowledge, Science and Technology for Development (IAASTD). Agriculture at a crossroads, Island Press, Washington, DC, 2009, 307-76.

79. Rowlinson P. Adapting livestock production systems to climate change: temperate zones. In: Rowlinson P, Steel M, Nefzaoui A. (Eds.), Livestock and Global Climate Change Conference Proceeding. Cambridge University Press, Tunisia, 2008, 61–63.

80. Rowlinson P, Steele M, Nefzaoui A. Livestock and global climate change: Adaptation I and II. In: Rowlinson, P., Steel, M., Nefzaoui, A. (Eds.), Livestock and Global Climate Change Conference Proceeding. Cambridge University Press, Tunisia, 2008, 56-85.

81. Sanz-Saez A, Erice G, Aguirreolea J, Muñoz F *et al.*, Alfalfa forage digestibility, quality and yield under future climate change scenarios vary with Sinorhizobium meliloti strain. Plant Physiol. 2012; 169:782–788.

82. Seerapu SR, Kancharana AR, Chappidi VS, Bandi ER. Effect of microclimate alteration on milk production and composition in Murrah buffaloes. Vet. World. 2015; 8:1444–1452.

83. Seguin B. The consequences of global warming for agriculture and food production. In: Rowlinson P, Steele M, Nefzaoui A. (Eds.), Livestock and Global Climate Change. Cambridge University Press, Hammamet, Tunisia, 2008, 9–11.

84. Smith JP, Pearce BD, Wolfe MS. Reconciling productivity with protection of the environment: is temperate agroforestry the answer? Renewable Agric. Food Syst. 2012; 28(1):80–92.

85. Stehfest E, Bouwman L, Van Vuuren DP, Den Elzen M *et al.*, Climate benefits of changing diet. Climatic Change. 2009; 95(1–2):83–102.

86. Steinfeld H, Gerber P, Wassenaar T, Castel V *et al.*, Livestock's Long Shadow: Environmental Issues and Options. FAO, Rome, 2006.

87. Tennigkeit T, Wilkies A. Carbon finance in rangelands. World Agroforestry Centre (ICRAF). The Royal Society. The role of land

carbon sinks in mitigating global climate change. Policy document 10/01. The Royal Society, London, 2001, 2008.

88. Thomas CD, Cameron A, Green RE, Bakkenes M *et al.*, Extinction risk from climate change. Nature. 2004; 427:145-148.

89. Thornton PK. Livestock production: recent trends, future prospects. Philos. Trans. Roy. Soc. B. 2010; 365:2853-2867.

90. Thornton PK, Boone RB, Ramirez-Villegas J. Climate change impacts on livestock. CGIAR Resrach program on Climate Change, Agriculture and Food Security (CCAFS), Working, 2015, 120.

91. Thornton PK, Gerber PJ. Climate change and the growth of the livestock sector in developing countries. Mitig. Adapt. Strategies Global Change. 2010; 15:169–184.

92. Thornton PK, Herrero M. Potential for reduced methane and carbon dioxide emissions from livestock and pasture management in the tropics. Proc. Natl. Acad. Sci. U.S.A. 2010a; 107(46):19667–19672.

93. Thornton PK, Herrero M. The Inter-linkages between rapid growth in livestock production, climate change, and the impacts on water resources, land use, and deforestation. World Bank Policy Research Working Paper, WPS 5178. World Bank, Washington, DC

94. Thornton PK, Herrero M, Freeman A, Mwai O, Rege E, Jones P, *et al.* Vulnerability, climate change and livestock: Research opportunities and challenges for poverty alleviation. International Livestock Research Institute (ILRI), Kenya. 2008, 2010.

95. Thornton PK, Van De Steeg J, Notenbaert A, Herrrero M. The impacts of climate change on livestock and livestock systems in developing countries: A review of what we know and what we need to know. Agric. Syst. 2009; 101:113–127.

96. Tubiello F, Schmidhuber J, Howden M, Neofotis PG *et al.*, Climate Change Response Strategies for Agriculture: Challenges and Opportunities for the 21st Century. The World Bank, Washington, DC, 2008.

97. UN (United Nations). World population projected to reach 9.6 billion by 2050. United Nations Department of Economic and Social Affairs. 162 M.M. Rojas-Downing *et al.*, Climate Risk Management. 2013, 2017; 16:145–163.

98. UNEP (United Nations Environment Programme). Global environment outlook 5: Chapter, 2012.

99. UNFCCC (United Nations Framework Convention on Climate Change). Challenges and opportunities for mitigation in the agricultural sector: technical paper, 2008.

100. UNFCCC (United Nations Framework Convention on Climate Change). Global Warming Potentials, 2014.

101. USDA (United States Department of Agriculture). Climate Change and Agriculture in the United States: Effects and Adaptation. USDA technical bulletin, Washington, DC, 2013.

102. USDA-NRCS (United States Department of Agriculture- Natural Resources Conservation Service). Conservation practices that save: Energy conservation in confined animal operations, 2006.

103. Van Der Spiegel M, Van Der Fels-Klerx HJ, Marvin HP. Effects of climate change on food safety hazards in the dairy production chain. Food Res. Int. 2012; 46:201–208.

104. Wani SP, Rockstrom J, Oweis T. Rainfed Agriculture: Unlocking the Potential. CAB International, Wallingford, UK, 2009.

105. Wassenaar T, Gerber P, Verburg PH, Rosales M *et al.*, Projecting land use changes in the neotropics: the geography of pasture expansion into forest. Global Environ. Change. 2007; 17:86–104.

106. Wright IA, Tarawali S, Blummel M, Gerard B *et al.*, Integrating crops and livestock in subtropical agricultural systems. J Sci. Food Agric. 2012; 92:1010–1015.

Chapter - 15
Soil Health Indicators under Climate Change

Authors

Dr. Vineet Kumar
Indian Institute of Farming System Research Modipuram, Meerut, Uttar Pradesh, India

K.S. Krishna Prasad
Sardar Vallabhbhai Patel University of Agriculture & Technology, Meerut, Uttar Pradesh, India

Chapter - 15

Soil Health Indicators under Climate Change

Dr. Vineet Kumar and K.S. Krishna Prasad

To manage and maintain soil in a sustainable fashion, the definitionof soil health must be broad enough to encompass the many functions of soil, e.g. environmental filter, plant growth and water regulation (Doran *et al.*, 1997). Definitions of air and water quality standards have existed for a long time, while a similar definition does not exist for soil. There is, however, little if any parallel between air or water quality and soil health (Sojka *et al.*, 1999). Air and water quality standards are usually based on maximum allowable concentration of materials hazardous to human health. A definition of soil health based on this concept would encompass only a small fraction of the many roles soil play (Singer *et al.*, 2000). Soil health is the net result of on-going conservation and degradation processes, depending highly on the biological component of the soil ecosystem, and influences plant health, environmental health, food safety and quality (Halvorson *et al.*, 1997; Parr *et al.*, 1992).

Definition of Soil Health

Several definitions of soil health have been proposed during the last decades. Historically, the term soil quality described the status of soil as related to agricultural productivity or fertility (Singer *et al.*, 2000).

In the 1990s, it was proposed that soil quality was not limited to soil productivity but instead expanded to encompass interactions with the surrounding environment, including the implications for human and animal health. In this regard, several examples of definitions of soil quality have been suggested (Doran *et al.*, 1994). In the mid-1990s, the term soil health was introduced. For example, a programme to assess and monitor soil health in Canada used the terms quality and health synonymously to describe the ability of soil to support crop growth without becoming degraded or otherwise harming the environment (Acton *et al.*, 1995). Others broadened the definition of soil health to capture the ecological attributes of soil, and went beyond its capacity to simply produce particular crops. These attributes are chiefly associated with biodiversity, food web structure, and functional measures (Pankhurst *et al.*, 1997). In 1997, Doran & Safley (Doran *et al.*, 1997) proposed the following definition of soil health:

- The continued capacity of soil to function as a vital living system, within ecosystem and land-use boundaries, to sustain biological productivity, promote the quality of air and water environments, and maintain plant, animal and human health

The definition encompasses a time component, reflecting the importance of continuous functions over time and the dynamic nature of soil. Soil health thus focuses on the continued capacity of a soil to sustain plant growth and maintain its functions regardless of the fitness for any certain purposes (Pankhurst *et al.*, 1997). Examples of dynamic soil properties could be organic matter content, the number or diversity of organisms, and microbial constituents or products (Singer *et al.*, 2000). We have adopted the definition by Doran & Safley in the present report.

Soil Health and Human Health

Soil is a finite and non-renewable resource because regeneration of soil through chemical and biological weathering of underlying rock requires geological time (Huber *et al.*, 2001). Deterioration of soil, and thereby soil health, is of concern for human, animal and plant health because air, groundwater and surface water consumed by humans can be adversely affected by mismanaged and contaminated soil (Singer *et al.*, 2000). As such, deteriorating soil health and the benefits of soil management has become of political concern. A healthy soil functions to buffer nutrients as well as contaminants and other solutes via sorption to or incorporation into clay particles and organic materials. The soil itself thus serves as an environmental filter for removing undesirable solid and gaseous constituents from air and water (Parr *et al.*, 1992). The extent to which a soil immobilises or chemically alters substances that are toxic, thus effectively detoxifying them, reflects the degree of soil health in the sense that humans or other biological components of the system are protected from harm (Singer *et al.*, 2000).

Physical, Chemical and Biological Components of Soil

Soil is dominated by a solid phase consisting of particles of different size surrounded by water and gases, the amount and composition of which fluctuate markedly in time and space. Water is normally discontinuous, except when the soil is water saturated. The pore space without water is filled with air and other gases and volatiles (Stotzky 1997). There is continual interchange of molecules and ions between solid, liquid and gaseous phases which are mediated by physical, chemical and biological processes (Doran *et al.*, 1994). These processes represent a unique balance between physical, chemical and biological components (Doran *et al.*, 1994). Maintaining this balance is of great importance to soil health.

Microorganisms have Key Functions in Soil

The biological activity in soil is largely concentrated in the topsoil, the depth of which may vary from a few to 30 cm. In topsoil, the biological components occupy a tiny fraction (<0.5%) of the total soil volume and make up less than 10% of the total organic matter in soil. These biological components consist mainly of soil organisms, especially microorganisms. Despite their small volume in soil, microorganisms are key players in the cycling of nitrogen, sulphur, and phosphorus, and the decomposition of organic residues. Thereby they affect nutrient and carbon cycling on a global scale (Pankhurst *et al.*, 1997). That is, the energy input into the soil ecosystems is derived from the microbial decomposition of dead plant and animal organic matter. The organic residues are, in this way, converted to biomass or mineralized to CO_2, H_2O, mineral nitrogen, phosphorus, and other nutrients (Bloem *et al.*, 1997). Mineral nutrients immobilised in microbial biomass are subsequently released when microbes are grazed by microbivores such as protozoa and nematodes (Bloem *et al.*, 1997). Microorganisms are further associated with the transformation and degradation of waste materials and synthetic organic compounds (Torstensson *et al.*, 1998).

In addition to the effect on nutrient cycling, microorganisms also affect the physical properties of soil. Production of extra-cellular polysaccharides and other cellular debris by microorganisms help in maintaining soil structure, as these materials function as cementing agents that stabilise soil aggregates. Thereby, they also affect water holding capacity, infiltration rate, crusting, erodibility, and susceptibility to compaction (Elliott *et al.*, 1996).

Microorganisms as Indicators of Soil Health

Microorganisms possess the ability to give an integrated measure of soil health, an aspect that cannot be obtained with physical/chemical measures and/or analyses of diversity of higher organisms. Microorganisms respond quickly to changes, hence they rapidly adapt to environmental conditions. The microorganisms that are best adapted will be the ones that flourish. This adaptation potentially allows microbial analyses to be discriminating in soil health assessment, and changes in microbial populations and activities may therefore function as an excellent indicator of change in soil health (Kennedy *et al.*, 1995; Pankhurst *et al.*, 1995).

Microorganisms also respond quickly to environmental stress compared to higher organisms, as they have intimate relations with their surroundings due to their high surface to volume ratio. In some instances, changes in microbial populations or activity can precede detectable changes in soil physical and chemical properties, thereby providing an early sign of soil improvement or an early warning of soil degradation (Pankhurst *et al*., 1995). An example is the turnover rate of the microbial biomass. This is much faster, e.g. 1-5 years, than the turnover of total soil organic matter (Carter *et al*., 1999).

The bioavailability of chemicals, e.g. heavy metals or pesticides, is also an important issue of soil health because of its connection with microbial activities. The impact of such chemicals on soil health is dependent on microbial activities. For example, the concentration of heavy metals in soil will not change over small time periods, but their bioavailability may. It has thus been shown that the bioavailability of poly-aromatic hydrocarbons was lower in autumn compared to early spring due to a higher microbial activity after the growing season (H. Harms, pers. comm. 2001). Therefore, the total content of chemicals in soil is not a reliable indicator of its bioavailability (Logan 2000) and thereby soil health. Instead, bioavailability has to be measured in relation to bioassays and specific microbial processes. In context of this, microbial responses also integrate the effect of chemical mixtures, information not obtained by studying the chemical mixtures themselves.

Definition of Microbial Indicator

- A microbial parameter that represents properties of the environment or impacts to the environment, which can be interpreted beyond the information that the measured or observed parameter represents by itself.

Indicators of soil health have further been defined as measurable surrogates for environmental processes that collectively tell us whether the soil is functioning normally (Pankhurst *et al*., 1997). In the context of microbial indicators, these measurements will cover soil microbial processes and related parameters. Indicators may be used as an indirect measure of soil function, serving to assess soil quality or health and its direction of change with time, by linking functional relationships among measurable attributes and monitoring for sustainable land management, including environmental impacts (Dalal *et al*., 2003a,).

Kinyangi (2007) proposes that soil health assessment involves an evaluation process consisting of a series of actions: (1) selection of soil health indicators, (2) determination of a minimum data set, (3) development of an interpretation scheme of indices, and (4) on-farm assessment and validation. While the use of integrative soil health tests is increasing, limited information exists to evaluate the applicability of soil health indicators for monitoring soil functions within the context of climate change. A range of frameworks identifying soil health indicators to measure change and implement strategies to adapt to climate change exist (Kibblewhite *et al.*, 2008; Nuttall 2007).

Evaluating the Soil Health

Evaluation of soil health should be considered relative to the many different land uses, e.g. agriculture, forestry, urbanisation, recreation, and preservation. The objective for evaluating soil health in an e.g. agricultural ecosystem may, consequently, be different from objectives used for assessing urban or natural ecosystems (Singer *et al.*, 2000). Thus, in agriculture, soil may be managed to maximise production without adverse environmental effects, whereas in a natural ecosystem, soil may be managed by a set of baseline values against which future changes in the system may be compared (Karlen *et al.*, 2000).

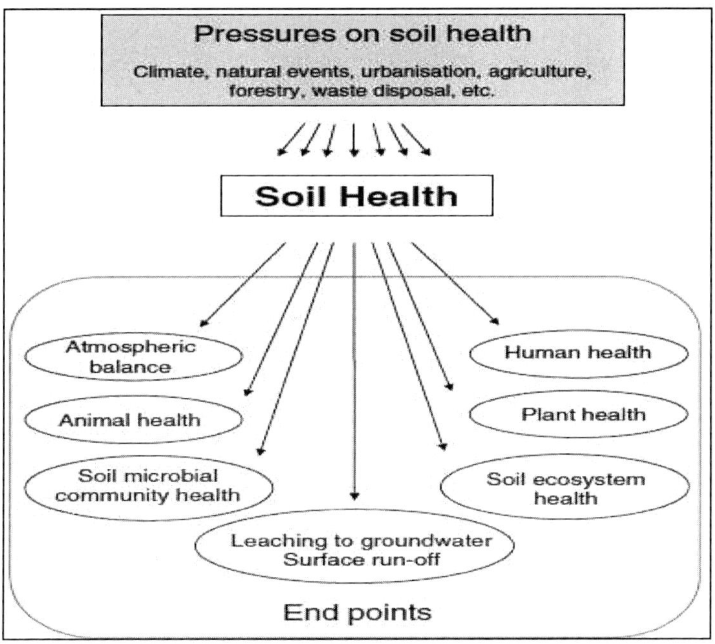

Fig 15.1a: Policy-relevant end points of soil health monitoring. Several examples of pressures on soil health are presented (grey box) and this may impact several end points of soil health (elliptical boxes).

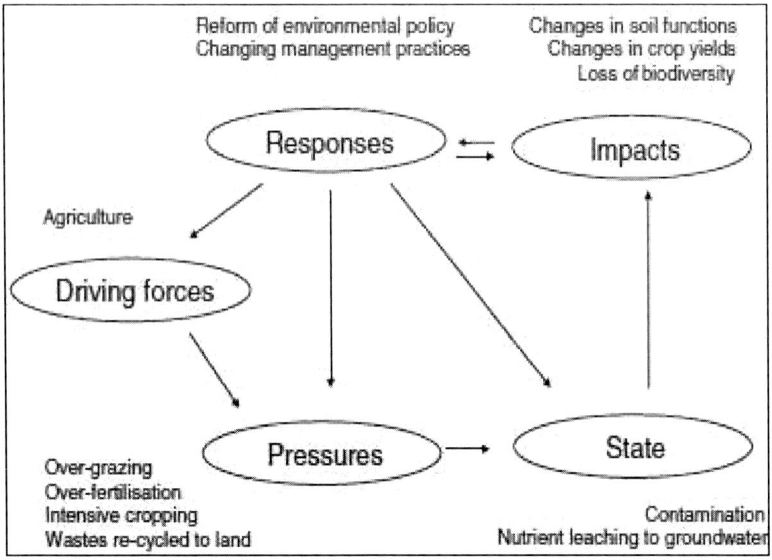

15.1b: The DPSIR assessment framework applied to soil. Modified from Huber *et al.*, (2001).

A framework for soil health evaluation is critical for the development of a useful monitoring programme covering the different functions and land-uses and it must identify priorities and relevant indicators relating to policy-relevant end points (Huber *et al.*, 2001).

Policy-Relevant end Points of Soil Health

Definition of policy-relevant end points is very important as monitoring programmes are developed. End points should be pragmatic in the sense of providing logical categories for regulatory decisions and they should be integrated for indicators that are ecologically related. After reviewing environmental monitoring programmes, it is clear that end points for soil health need to be clearly specified and then used as guidance in the identification of indicators. As a consequence, we suggest an end point matrix (Fig 15.1a) that, when integrated together, provide a comprehensive and effective assessment of soil health.

Integrated Environmental Assessment

Relevant indicators of specific end points can be identified using the Integrated Environmental Assessment method, which is based on the Driving force-Pressure-State-Impact-Response (DPSIR) assessment framework, that has been developed primarily for environmental issues (Holten-Andersen *et al.*, 1995). The DPSIR framework analyses the complex relationships between the environment and the impact of economic activities and societal behaviour. The driving force (D) lead to pressures (P) on the environment affecting the state (S) and leading to impacts (I), which finally results in response (R) by the society.The DPSIR framework has recently been adopted for soil issues (Fig 15.1b) and is recommended for the terrestrial monitoring programme (Iversen *et al.*, 2001). A prerequisite for the use of the DPSIR framework is a clear definition of the problems and a scientific understanding of the causal mechanisms (Christensen *et al.*, 2001). Further, the development of indicators for each of the PSI-elements is necessary (Huber *et al.*, 2001). These indicators should relate to the policy-relevant end points of soil health.

Soil Physical Properties

Soil physical properties provide information related to water and air movement through soil, as well as conditions affecting germination, root growth and erosion processes. Many soil physical properties thus form the foundation of other chemical and biological processes, which may be further governed by climate, landscape position and land use. A range of soil physical properties are highlighted as potential soil health indicators, and key soil physical indicators in relation to climate change include soil structure, water infiltration, bulk density, rooting depth, and soil surface cover etc.

Soil Structure (Aggregate Stability, Porosity)

Aggregate stability, the resistance of soil aggregates to external energy such as high intensity rainfall and cultivation, is determined by soil structure, as well as a rangeof chemical and biological properties and management practices (Moebius *et al.*, 2007). It is considered a useful soil health indicator since it is involved in maintaining important ecosystem functions in soil including organic carbon (C) accumulation, infiltration capacity, movement and storage of water, and root and microbial community activity; it can also be used to measure soil resistance to erosion and management changes (Moebius *et al.*, 2007; Rimal and Lal 2009). Because of its association with the storage of soil organic carbon (SOC) and water, its measurement can be useful to guide climate adaptation strategies, especially in areas that are likely to experience high and intense rainfall and consequently increased erosion events. Since aggregate stability is measured in many different ways, standardised procedures are required within a soil health monitoring framework under climate change scenarios (Salvador Sanchis *et al.*, 2008).

Porosity, a measure of the void spaces in a material as a fraction (volume of voids to that of the total volume), and pore size distribution provide a direct, quantitative estimate of the ability of a soil to store root-zone water and air necessary for plant growth (Reynolds *et al.*, 2002). Pore characteristics are strongly linked to soil physical quality; bulk density and macroporosity are functions of pore volume, while soil porosity and water release characteristics directly influence a range of soil physical indices including soil aeration capacity, plant available water capacity and relative field capacity (Reynolds *et al.*, 2009). Recent studies to model soil water balance and ecosystem conditions under present-day and projected climatic scenarios use porosity as a model parameter (Porporato *et al.*, 2005). Since root development and soil enzyme activities are closely related to soil porosity and pore size distribution (Piglai and De Nobili 1993) and because future climate change scenarios (e.g. elevated CO_2 and temperature, and variable and extreme rainfall events) may alter root development and soil biological activities, soil porosity and pore size distribution and consequently soil functions are likely to be affected in unexpected directions; Moreover, pore size distribution and aeration status, besides other factors, are the key factors in governing methane (CH_4) fluxes, both CH_4 emission and uptake (Dalal *et al.*, 2008), and nitrous oxide (N_2O) emissions from soil.

Infiltration, Soil Available Water and Distribution

Soil water infiltration, the rate at which water enters the soil surface and moves through soil depth, is gaining increasing interest within soil water modelling, as field-based measurements continue to improve (Dalal and Moloney 2000). Since infiltration rate may change significantly with soil use, management and time, it has been included as an indicator of soil health for assessments of land use change impacts (O'Farrell *et al.*, 2010).

The availability of water for plant growth and important soil processes is governed by a range of soil properties including porosity, field capacity, lower limit of plant available water (thus excluding osmotic potential) and hence plant available water capacity, macropore flow and texture (Jarvis 2007). Plant available water capacity has been used as part of integrative soil health tests to assess management impacts, although variation in methodology may complicate the interpretation of results (Idowu *et al.*, 2009). Furthermore, the soil available water and distribution may respond rapidly to climate change, especially to variable and high intensity rainfall or drought events, and thus management strategies, such as the planting of cover crops, conservation tillage and incorporation of organic matter, that maintain or even enhance water infiltration and available water in soil may help in mitigating the impacts of severe rainfall and drought events or severe erosion events (Salvador Sanchis *et al.*, 2008).

Bulk Density

Bulk density is routinely assessed in agricultural systems to characterise the state of soil compactness in response to land use and management (Hakansson and Lipiec 2000). It is considered as a useful indicator for the assessment of soil health with respect to soil functions such as aeration and infiltration (Reynolds *et al.*, 2009). Since bulk density is in general negatively correlated with soil organic matter (SOM) or SOC content, loss of organic C from increased decomposition due to elevated temperatures may lead to increase in bulk density and hence making soil more prone to compaction via land management activities and climate change stresses, for example, from variable and high intensity rainfall and drought events (Birka´s *et al.*, 2009).

Rooting Depth

Rooting depth is considered an important indicator of soil health, since changes in this property is likely to affect plant available water capacity, subsoil salinity, SOC content or other properties to indicate physicochemical constraints in the soil profile (Birka´s *et al.*, 2009). Under prolonged drought, the impact of subsoil constraints such as salinity and high chloride concentrations (Rengasamy 2010) is likely to be greater on plant available water and hence plant productivity. Also, Birka´s *et al.*, (2008) included rooting depth as a soil health parameter for monitoring of soil condition and plant growth under extreme drought and variable rainfall events to indicate the potential for adaptability and mitigation of climate stresses through alteration of rooting depth.

Soil Surface Cover

Soil surface cover (e.g. a layer of crop residues or biological soil crust) provides a range of important ecological functions including protection of soil surface by dissipating raindrop impact energy, soil stabilisation, and reduction in erodible surface area, water and nutrient retention, C fixation and, in some instances, N fixation and support of native seed germination. Dalal and Moloney (2000) note that soil surface cover provides an integrated indicator of soil physical management, organic matter input and effects associated with erosion and runoff, although correct timing of monitoring of soil surface cover change is required to evaluate impacts of management and climate change on soil health.

Soil structural conditions such as soil crust and soil seal formation, primarily related to sodicity, are also indicators that may be used to characterise soil health under climate change. The formation of soil crusts and seals can affect a range of soil processes, including water infiltration, oxygen diffusion, runoff, surface water evaporation and wind erosion. A range of methods exist to measure their thickness and strength, although research effort is needed to relate these properties with soil processes (Assouline 2004) affecting ecosystem functions and plant productivity, as well as to evaluate their role in mitigating adverse climate change impacts, thereby assisting in climate change adaptation.

Soil Chemical Properties

pH

Soil pH, a function of parent material, time of weathering, vegetation and climate, is considered as one of the dominant chemical indicators of soil health, identifying trends in change for a range of soil biological and chemical functions including acidification, salinisation, crop performance, and nutrient availability and cycling and biological activity (Dalal and Moloney 2000). Soil pH has thus been included in integrative soil health tests to assess impacts of land use change and agricultural practices (Gil *et al.*, 2009). While Brinkman and Sombroek (1999) suggested that most soils would not be subjected to rapid pH changes resulting fromdrivers of climate change such as elevated temperatures, CO_2 fertilisation, variable precipitation and atmospheric N deposition (McCarthy *et al.*, 2001), it is likely, however, that these drivers of climate change will affect organic matter status, C and nutrient cycling, plant available water and hence plant productivity, which in turn will affect soil pH (Reth *et al.*, 2005).

Electrical Conductivity

Soil electrical conductivity (EC), a measure of salt concentration, is considered an easily measured, reliable indicator of soil quality/health (Arnold *et al.*, 2005). It can inform trends in salinity, crop performance, nutrient cycling (particularly nitrate) and biological activity and, along with pH, can act as a surrogate measure of soil structural decline especially in sodic soils (Arnold *et al.*, 2005). Electrical conductivity has been used as a chemical indicator to inform soil biological quality in response to crop management practices (Gil *et al.*, 2009). Using elevation gradient as a surrogate for increasing temperatures and decreasing precipitation under climate change scenarios, Smith *et al.*, (2002) found that EC decreased and pH increased in a semi-arid environment. Pariente (2001) examined the dynamics of soluble salts concentration in soils from four climatic regions (Mediterranean, semi-arid, mildly arid and arid) and found a non-linear relationship between the soluble salts content and rainfall, with sites that received <200 mm rainfall contained significantly high soluble contents and vice versa. Clearly, there is a need for comprehensive assessment of the influence of drivers of climate change on soil EC as an important soil health indicator in different ecosystems.

Sorption Capacity and Cation Exchange Capacity

Sorption capacity and cation exchange capacity (CEC) are considered important determinants of soil chemical quality, particularly the retention of major nutrient cations Ca, Mg and K, and immobilisation of potentially toxic cations Al and Mn; these properties can thus be useful indicators of soil health, informing of a soil's capacity to absorb nutrients, as well as pesticides and chemicals (Ross *et al.*, 2008). This may qualify CEC as a useful soil health indicator that is also required to be monitored in response to climate change. Since CEC of coarse-textured soils and low-activity clay soils is attributed to that of SOM (Weil and Magdoff 2004), the increasing decomposition and loss of SOM due to elevated temperatures (Davidson and Janssens 2006) may lead to the loss of CEC of these soils. Low CEC of soil may result in increased leaching of base cations in response to high and intense rainfall events, thus transporting alkalinity from soil to waterways.

Plant Available Nutrients

In their identification of basic soil properties to meet requirements of indicators for screening soil quality/health, Doran *et al.*, (1999) list extractable nutrients N, phosphorus, potassium, since "they provide information on plant available nutrients and potential loss from soil providing indication of productivity and environment quality". Measurement of extractable nutrients may provide indication of a soil's capacity to support plant growth; conversely, it may identify critical or threshold values for environmental hazard assessment (Dalal and Moloney 2000). Nutrient cycling, especially N, is intimately linked with soil organic C cycling (Weil and Magdoff 2004), and hence drivers of climate change such as elevated temperatures, variable precipitation and atmospheric N deposition are likely to impact on N cycling and possibly the cycling of other plant available nutrients such as phosphorus and sulphur, although direction and exact magnitude of change in plant available nutrients need to be investigated.

Soil Biological Properties

"While the chemistry (and physics) of the soil system provides the context... It is the soil biota which is adaptive to changes in environmental circumstances" (Kibblewhite *et al.*, 2008). Under conditions of climate change, biological indicators form an integral component in soil health assessment, since, by virtue, they involve complex adaptive systems (i.e. the biota) by integrating key soil processes in ways that other indicators do not (Ritz *et al.*, 2009). Ritz *et al.*, (2009) note that published information on potential biological indicators has shown an almost exponential increase since the 1970s; however, many of these publications cover a range of processes, for example, microbial, invertebrate or ecological processes. Consequently, a comprehensive explanation of the factors including formal criteria identifying a collective set of biological indicators is limited.

Soil Organic Matter

SOM comprises an extensive range of living and non-living components; it has been widely acknowledged that SOM is one of the most complex and heterogeneous components of soils, which vary in their properties, functions and turnover rates (Weil and Magdoff 2004). It is considered an important attribute of soil quality (or soil health) due to the many functions it provide and/or support, including the contribution to the charge characteristics of soils, a sink for and source of C and N, and to a variable extent regulates phosphorus and sulphur cycling, possesses an ability to complex with multivalent ions and organic compounds, provides microbial and faunal habitat and substrates, as well as affecting aggregate stability, trafficability, water retention and hydraulic properties (Haynes 2008). Main indicators for evaluating SOM status include: SOC, since it comprises about 50% of SOM; organic N, since it is closely associated with organic C and is the most important nutrient for plant productivity; and readily mineralisable C and N (Haynes 2008).

As SOM drives the majority of soil functions, decreases in SOM can lead to a decrease in fertility and biodiversity, as well as a loss of soil structure, resulting in reduced water holding capacity, increased risk of erosion and increased bulk density and hence soil compaction (Weil and Magdoff 2004). Land use and management practices that lead to build up of SOM will help in absorbing CO_2 from the atmosphere, thus mitigating global warming. By increasing water storage, SOM can play an important role in the mitigation of flooding impacts following extreme rainfall events, while storing water in the event of droughts thus increasing soil resilience.

SOM indicators have been used in long-term soil experiments for the assessment of climate change (Rinnan *et al.*, 2007), although the response of SOM to elevated temperature remains scientifically controversial and without consensus (Agren and Wetterstedt 2007). In general, increases in temperature have been reported to enhance decomposition of SOM, but rising temperature and precipitation, CO_2 fertilisation and atmospheric N deposition may support high plant productivity and organic matter input to soil and consequently increase SOM. According to Kuzyakov and Gavrichkova (2010), it is the accessibility and availability of SOM to microorganisms that govern SOM losses rather than the rate-modifying climate factor (i.e. temperature).

Soil Carbon

Soil C can be used as an indicator of change for a number of reasons: it is familiar, direct, linked to ecosystem performance and it has "memory", that is, changes across time; however, it may not be reliable as a solitary indicator of soil quality/health, as it may not encompass all ecosystem traits (Janzen 2005). While soil contains C in diverse forms and residence times, considerable research attention has focused on the SOC form, since it (1) has been largely modified through human activities, and (2) is predicted to decline with increase in mean global temperatures, which would have adverse effects on important soil functions and processes and soil quality/health (Lal *et al.*, 2007). However, in his review on the role of soil C in the changing global C cycle, Janzen (2005) proposed that "if soil C is to be used as an indicator of ecosystem performance, we may need to rethink how it is measured and interpreted. since, although we have been measuring soil C for a long time, our methods were not developed to measure changing C stocks". Since then substantial progress has been made, especially in geospatial and geostatistical methodologies to measure soil C across the landscape and over time (Allen *et al.*, 2010; Worsham *et al.*, 2010).

Light Fraction and Macro-Organic Matter (Labile Organic Matter)

Light (or low-density) fraction and macro-organic components of SOM consist mainly of mineral-free particulate plant and animal residues, which serve as readily decomposable substrate for soil micro-organisms, as well as a labile nutrient reservoir (Gregorich *et al.*, 1994;Wagai *et al.*, 2009). Since light fraction and macro-organic matter are responsive to management practices, they may act as early indicators to measure the effectiveness of changing management practice in adaptive response to climate change (Gregorich *et al.*, 1994). For example, the labile soil organic C is rapidly depleted as the temperature rises (Davidson and Janssens 2006). In addition, elevated CO_2 in the future may reduce sequestration of root-derived soil C, a major source of labile, light fraction C (Heath *et al.*, 2005).

Potentially Mineralisable C and N

The amount of mineralisable organic matter in soil is an indicator of organic matter quality, acting as the interface between autotrophic and heterotrophic organisms during the nutrient cycling process (Gregorich *et al*., 1994). While potentially mineralisable C and N may be measured in both the field and laboratory, it is generally treated as a relative rather than an absolute value due to inconsistencies in methods (Haynes 2008). However, mineralisable organic matter may be a useful indicator to assess soil health under climate change, since it affects nutrient dynamics within single growing seasons, and may be used to compare management regimes and C sequestration over extended periods of time (Gregorich *et al*., 1994).

Soil Respiration

Soil respiration is often used as a biological indicator for soil health since it is positively correlated with SOM content and often with microb.

References

1. Agren GI, Wetterstedt M. What determines the temperature response of soil organic matter decomposition? Soil Biol Biochem. 2007; 39:1794-1798.
2. Acton DF, Gregorich EG. Executive summary. In: Acton DF, Gregorich EG (Eds) the health of our soils. Towards sustainable agriculture in Canada. Center for Land and Biological Resources, Research Branch Agriculture and Agri-food Canada, ON, Canada, 1995.
3. Allen DE, Pringle MJ, Page KL, Dalal RC. A review of sampling designs for the measurement of soil organic carbon in Australian grazing lands. Rangel J, 2010; 32:227–246.
4. Arnold SL, Doran JW, Schepers J, Wienhold B. Portable probes to measure electrical conductivity and soil quality in the field. Commun Soil Sci Plant Anal. 2005; 36:2271–2287.
5. Assouline S. Rainfall-induced soil surface sealing: a critical review of observations, conceptual models, and solutions. Vadose Zone J, 2004; 3:570–591.
6. Bloem J, De Ruiter P, Bouwman LA. Food webs and nutrient cycling in agro-ecosystems. In: Van Elsas JD, Trevors JT, Wellington EMH (eds) Modern soil microbiology. Marcel Dekker, New York, 1997, 245–278.
7. Birka SM, Dexter A, Szemo KA. Tillage-induced soil compaction, as a climate threat increasing stressor. Cereal Res Commun. 2009; 37:379–382.
8. Brinkman R, Sombroek W. The effects of global change on soil

conditions in relation to plant growth and food production. In: Bazzaz F, Sombroek W (eds) Global climate change and agricultural production. Food and Agriculture Organization of the United Nations and John Wiley and Sons Ltd., Rome, Italy, 1999.

9. Dalal RC, Moloney D. Sustainability indicators of soil health and biodiversity. In: Hale P, Petrie A, Moloney D, Sattler P (eds) Management for sustainable ecosystems. Centre for Conservation Biology, Brisbane, 2000, 101–108.

10. Dalal RC, Eberhard R, Grantham T, Mayer DG. Application of sustainability indicators, soil organic matter and electrical conductivity, to resource management in the northern grains region. Aust J Exp Agric. 2003a; 43:253–259.

11. Dalal RC, Wang WJ, Robertson GP, Parton WJ. Nitrous oxide emission from Australian agricultural lands and mitigation options: a review. Aust J Soil Res. 2003b; 41:165–195.

12. Dalal RC, Allen DE, Livesley SJ, Richards G. Magnitude and biophysical regulators of methane emission and consumption in the Australian agricultural, forest, and submerged landscapes: a review. Plant Soil. 2008; 309:43–76.

13. Davidson EA, Janssens IA. Temperature sensitivity of soil carbon decomposition and feedbacks to climate change. Nature. 2006; 440:165–173.

14. Doran JW, Parkin TB. Defining and assessing soil quality. In: Doran JW, Coleman DC, Bezdicek DF, Stewart BA (eds) Defining soil quality for a sustainable environment. Soil Science Society of America, Madison, 1994, 3–21.

15. Doran JW, Safley M. Defining and assessing soil health and sustainable productivity. In: Pankhurst CE, Doube BM, Gupta VVSR (eds) Biological indicators of soil health. CABI, Wellingford, 1997, 1–28.

16. Doran JW, Jones AJ, Arshad MA, Gilley JE. Determinants of soil quality and health. Pankhurst CE, Doube BM, Gupta VVSR (eds) Biological indicators of soil health. CAB International, Oxon, 1999.

17. Elliott LF, Lynch JM, Papendick RI. The microbial component of soil quality. In: Stotzky G, Bollag JM (eds) Soil biochemistry. Marcel Dekker, New York, 1996, 1-21.

18. Gil SV, Meriles J, Conforto C, Figoni G *et al.*, Field assessment of soil

biological and chemical quality in response to crop management practices. World J Microbiol Biotechnol. 2009; 25:439–448.

19. Gregorich EG, Carter MR, Angers DA, Monreal CM, Ellert BH. Towards a minimum data set to assess soil organic matter quality in agricultural soils. Can J Soil Sci. 1994; 74:367–385.

20. Hakansson I, Lipiec J. A review of the usefulness of relative bulk density values in studies of soil structure and compaction. Soil Tillage Res. 2000; 53:71–85.

21. Haynes RJ. Soil organic matter quality and the size and activity of the microbial biomass: their significance to the quality of agricultural soils. In: Huang Q, Huang PM, Violante A (eds) Soil mineral-microbe-organic interactions: theories and applications.Springer, Berlin, 2008, 201–230.

22. Heath J, Ayres E, Possell M, Bardgett RD *et al.*, Rising atmospheric CO_2 reduces sequestration of root-derived soil carbon. Science. 2005; 309:1711–1713.

23. Huber S, Syed B, Freudenschuss A, Ernsten V, Loveland P. Proposal for a European soil monitoring and assessment framework-Technical report no.61, Europe Environmental Agency, Copenhagen, Denmark, 2001.

24. Idowu OJ, Van Es HM, Abawi GS, Wolfe DW *et al.*, Use of an integrative soil health test for evaluation of soil management impacts. Renew Agric Food Syst. 2009; 24:214 224.

25. Janzen HH. Soil carbon: a measure of ecosystem response in a changing world? Can J Soil Sci. 2005; 85:467–480.

26. Jarvis NJ. A review of non-equilibrium water flow and solute transport in soil macropores: principles, controlling factors and consequences for water quality. Eur J Soil Sci. 2007; 58:523–546.

27. Karlen DL, Doran JW, Andrews SS, Wienhold BJ. Soil Quality-Humankinds' foundation for survival. J Soil Water Conserv. 2003; 58:171–179.

28. Kibblewhite MG, Ritz K, Swift MJ. Soil health in agricultural systems. Philos Transact R Soc. 2008; 363:685–701.

29. Kinyangi J. Soil health and soil quality: a review. Draft publication, 2007.

30. Kennedy AC, Papendick RI. Microbial characteristics of soil quality. J soil water conserv. May–June, 1995, 243–248.

31. Kibblewhite MG. Soil quality assessment and management. In Grassland: a global resource McGilloway DA. Eds. Wageningen, the Netherlands: Wageningen Academic Publishers, 2005, 219-226.

32. Kuzyakov Y, Gavrichkova O. Time lag between photosynthesis and carbon dioxide efflux from soil: a review of mechanisms and controls. Global Change Biology. 2010; 16:3386–3406.

33. Lal R, Follett RF, Stewart BA, Kimble JM. Soil carbon sequestration to mitigate climate change and advance food security. Soil Sci. 2007; 172:943–956.

34. McCarthy JJ, Canziani OF, Leary NA, Dokken DJ, White KS. Climate Change: impacts, adaptation, and vulnerability. Contribution of Working Group II to the Third Assessment Report of the Intergovernmental Panel on Climate Change. Cambridge University Press, Cambridge, 2001.

35. Moebius BN, Van Es HM, Schindelbeck RR, Idowu OJ *et al.*, Evaluation of laboratory-measured soil properties as indicators of soil physical quality. Soil Sci. 2007; 172:895–912.

36. Nuttall JG. Climate changes-identifying the impacts on soil and soil health. Department of Primary Industries, Future Farming Systems Research Division, Victori, 2007.

37. O'Farrell PJO, Donaldson JS, Hoffman MT. Vegetation transformaion, functional compensation, and soil health in a semi-arid environment. Arid Land Res Manage. 2010; 24:12–30.

38. Parr JF, Papendick RI, Hornick SB, Meyer RE. Soil quality: attributes and relationship to alternative and sustainable agriculture. Am J Altern Agric. 1992; 7:5–11.

39. Pankhurst CE, Hawke BG, McDonald HJ, Kirkby CA *et al.*, Evaluation of soil biological properties as potential bioindicators of soil health. Aust J Exp Agric. 1995; 35:1015–1028.

40. Pankhurst CE, Doube BM, Gupta VVSR. Biological indicators of soil health: synthesis. In: Pankhurst CE, Doube BM, Gupta VVSR (eds) Biological indicators of soil health. CABI, Wallingford, Oxfordshire, 1997, 419–435.

41. Pariente S. Soluble salts dynamics in the soil under different climatic conditions. Catena. 2001; 43:307–321.

42. Porporato A, Daly E, Rodriguez-Iturbe I. Soil water balance and ecosystem response to climate change. Am Nat. 2005; 164:625–632.

43. Rinnan R, Michelsen A, Ba˚a˚th E, Jonasson S. Fifteen years of climate change manipulations alter soil microbial communities in a subarctic heath ecosystem. Glob Change Biol. 2007; 13:28–39.

44. Rengasamy P. Soil processes affecting crop production in salt-affected soils. Funct Plant Biol. 2010; 37:613–620.

45. Reth S, Reichstein M, Falge E. The effect of soil water content, soil temperature, soil pHvalue and root mass on soil CO_2 efflux. Plant Soil. 2005; 268:21–33.

46. Reynolds WD, Bowman BT, Drury CF, Tan CS, Lu X. Indicators of good soil physical quality: density and storage parameters. Geoderma. 2002; 110:131–146.

47. Reynolds WD, Drury CF, Tan CS, Fox CA, Yang XM. Use of indicators and pore volumefunction characteristics to quantify soil physical quality. Geoderma. 2009; 152:252–263.

48. Rimal BK, Lal R. Soil and carbon losses from five different land management areas under simulated rainfall. Soil Tillage Res. 2009; 106:62–70.

49. Ritz K, Young IM. Interactions between soil structure and fungi. Mycologist. 2004; 18:52–59.

50. Ross DS, Matschonat G, Skyllberg U. Cation exchange in forest soils: the need for a new perspective. Eur J Soil Sci. 2008; 59:1141–1159.

51. Salvador Sanchis MP, Torri D, Borselli L, Poesen J. Climate effects on soil erodibility. Earth Surf Process Landf. 2008; 33:1082–1097.

52. Singer MJ, Ewing S. Soil quality. In: Sumner ME (Ed) Handbook of soil science. CRC, Boca Raton, FL, 2000, 271–298.

53. Wagai R, Mayer LM, Kitayama K. Nature of the “occluded” low-density fraction in soil organic matter studies: a critical review. Soil Sci Plant Nutr. 2009; 55:13–25.

54. Weil RR, Magdoff F. Significance of soil organic matter quality and health. In: Significance of Soil Organic Matter to Soil Quality and Health. (Eds F. Magdoff and R.R.Weil.) (CRCPress: Boca Raton, FL.). 2004, 1-43.

55. Worsham L, Markewitz D, Nibblelink N. Incorporating spatial dependence into estimates of soil carbon contents under different land covers. Soil Sci Soc Am J, 2010; 74:635–64.

Chapter - 16
Smart Strategies for Enhanced Agricultural Resilience and Food Security under a Changing Climate

Authors

Dr. S.S. Tomar
Rajmata Vijayaraje Scindia Krishi Vishwavidyalaya, Gwalior Zara-a, b. Road Morena, Madhya Pradesh, India

Dr. Sandeep Chaudhary
Sardar Vallabhbhai Patel University of Agriculture & Technology, Meerut, Uttar Pradesh, India

Chapter - 16

Smart Strategies for Enhanced Agricultural Resilience and Food Security under a Changing Climate

Dr. S.S. Tomar and Dr. Sandeep Chaudhary

Enhancing the resilience of Indian agriculture to cope with climate variability and climate change is imperative to the livelihood security of millions of small and marginal farmers in the country. Devising appropriate adaptation strategies will enable farmers to cope with various climate risks, promote efficient use of natural resources to bring sustainability to farm production and stability to their incomes. By 2050, the world population will reach nearly 9.5 billion, which effectively means that we will have to produce 70% more food for over two billion additional mouths. Hence, the food and agriculture systems need to adapt fast to the changing climate and become more resilient, productive and sustainable. This would require judicious use of natural resources and minimised post-harvest losses coupled with improved harvesting, storage, packaging, transportation and marketing practices as well as appropriate infrastructural facilities.

Climate change impacts on agriculture are being witnessed all over the world, but countries like India are more vulnerable in view of the high population depending on agriculture and excessive pressure on natural resources. The warming trend in India over 0 the past 100 years (1901 to 2007) was observed to be $0.51^{0}C$ with accelerated warming of $0.21^{0}C$ per every10 years since 1970 (Krishna Kumar, 2009). The projected impacts are likely to further aggravate yield fluctuations of many crops with impact on food security and prices. Climate change impacts are likely to vary in different parts of the country. Parts of western Rajasthan, Southern Gujarat, Madhya Pradesh, Maharashtra, Northern Karnataka, Northern Andhra Pradesh, and Southern Bihar are likely to be more vulnerable in terms of extreme events. The rainfall analysis of India showed that significant negative trends of rainfall were observed in the eastern parts of Madhya Pradesh, Chhattisgarh and parts of Jharkhand, U.P., and northeast India. Lalet.al. (2001) reported that annual mean area-averaged surface warming over the Indian sub-continent 0is to likely to range between 3.5 and 5.5 ^{0}C by 2080s.These projections showed more warming in winter season over summer.The spatial distribution of surface warming 0suggests a mean annual rise in surface temperatures in North India by 3^{0}C or more by 2050. The study also indicated that during winter, the surface mean air temperature could 00rise by 3 C in Northern and Central parts, while it would rise by 2^{0}C in Southern parts by the same period. In case of rainfall, a marginal increase of 7 to10% in annual rainfall is projected over the sub-continent by 2080.Rainfall is the key variable influencing crop productivity in rainfed farming. Intermittent and prolonged droughts are a major cause of yield reduction in most crops. Long term data for India indicates that rainfed areas witness 3-4 drought years in every 10-year period. However, no definite trend is seen on the frequency of droughts as a result of climate change so far.

Since the beginning of 21st Century, India has experienced droughts in quick succession, of which the 2009 was the most recent one that significantly affected kharif crop production. It was the second largest all India monsoon rainfall deficit since 1972 (23% below normal). Incidentally, 2009 also achieved the distinction of being the warmest year in past several centuries across the world. However, after 2009, 2010 thproved to the warmest year on record since 1850. 2011 is now the 11 warmest years on record since 1850. Apart from that, 1998 was one of the warmest years; 2003 experienced unprecedented heat and cold waves across the globe; occurrence of high temperature in March 2004 adversely affected crops like wheat, apple, potato etc. across northern India; 2005 witnessed destructive hurricanes/cyclones across the globe and again, 2007 was as warm as 1998 in the entire northern hemisphere and unusual summer rains and floods were experienced in many parts of India. Besides that, the amount and distribution of rainfall is becoming more and more erratic which is causing greater incidences of droughts and floods globally. The increase in frequency of heavy rainfall events in last 50 years over Central India points towards a significant change in climate pattern in India (Goswami, 2006).The data clearly indicates increase in heavy (>10 cm) and very heavy (>15cm) rainfall events and decrease in light to moderate rainfall events.

Agriculture is the source of livelihood for nearly two-thirds of the population in India. The sector currently accounts for 14.2% of the GDPand employs 55% of country's total workforce. The sharp fluctuations in agricultural growth are mainly attributed to the vagaries of monsoon. Some part of the country or the other experiences monsoon failure almost every year but most states encounter drought once in 2 to 4 years. The south west monsoon accounts for nearly 75% of the natural precipitation received in the country and therefore exert a strong influence on khariffood grain production and the economy in terms of agricultural output, farmers' income and price stability. Both the amount of rainfall and its distribution are crucial factors influencing performance of agriculture. The probability of monsoon rains being erratic is 40% of the time which implies that in 4 out of 10 years there would be an adverse impact on crop production in the absence of appropriate strategy to deal with such eventualities.Monsoon failures result in drought which has serious implications for small and marginal farmers and livelihoods of the rural poor. The impact of climate change on agriculture will be one of the major deciding factors influencing future food security of the world including India. Amajor part of the agriculture in India is rainfed (~80 million ha out of 142 million ha net cultivated area and will remain so for at least for a foreseeable future. The crop losses due to climate variability will vary from region to region depending on regional climate, crop and cropping systems, soils and management practices. Rainfed crops are likely to be worst hit by climate change because of the limited options for coping with variability of rainfall and temperature. The major crops like wheat and rice are expected to undergo all the weather aberrations and their sustainability will be determined by crop's capacity of natural adaptability as well as appropriate mitigation measures adopted. Assessment of climate change impact over Indian agriculture (Aggarwal, 2009).

Weather Variability and Impacts on Agriculture

- Rainfall drives water availability and determines sowing time (rainfed crops).
- Temperature drives crop growth and duration.
- Temperature and Relative Humidity influence pest and diseases incidence on crops.
- Radiation influences the photosynthetic productivity.
- Wet and dry spells cause significant impact on standing crops, physiology, loss of economic products (eg. fruit drop).
- Extreme events (eg. high rainfall/floods/heat wave/cold wave/cyclone /hail/frost) cause enormous losses of standing crop.

Food security refers to an ability to access/utilize sufficient quantities of safe and nutritious food; however, the related challenges are afflicting the urban/rural populations in wealthy/poor nations alike. FAO estimates nearly 194.6 million Indians (15.2%) were undernourished during 2014-16.

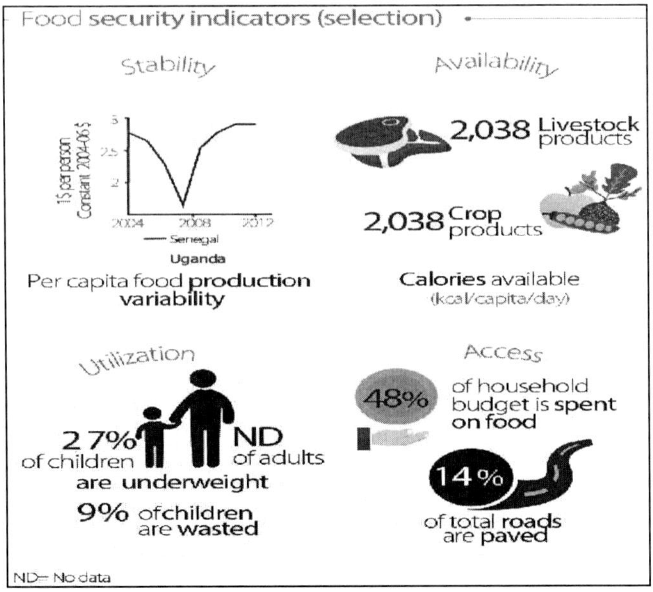
Food security indicators (selection)
Stability
Availability
$ per person Constant 2004-06 $
2004 2008 2012
Senegal
Uganda
Per capita food production variability
2,038 Livestock products
2,038 Crop products
Calories available (kcal/capita/day)
Utilization
Access
48% of household budget is spent on food
27% of children
ND of adults
are underweight
9% of children are wasted
14% of total roads are paved
ND= No data

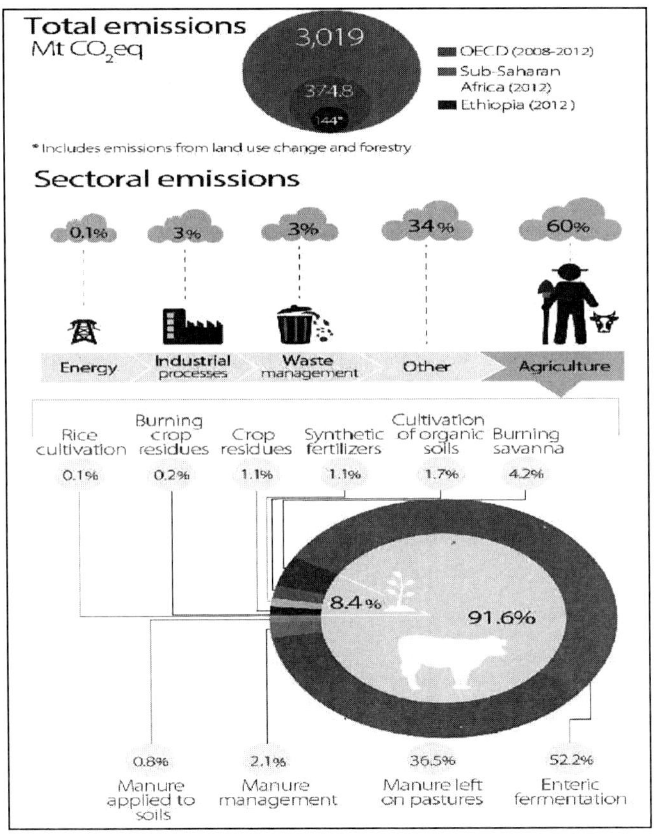
Total emissions Mt CO2eq
3,019
374.8
144*
OECD (2008-2012)
Sub-Saharan Africa (2012)
Ethiopia (2012)
* Includes emissions from land use change and forestry
Sectoral emissions
0.1%
3%
3%
34%
60%
Energy
Industrial processes
Waste management
Other
Agriculture
Rice cultivation
Burning crop residues
Crop residues
Synthetic fertilizers
Cultivation of organic soils
Burning savanna
0.1%
0.2%
1.1%
1.1%
1.7%
4.2%
8.4%
91.6%
0.8%
2.1%
36.5%
52.2%
Manure applied to soils
Manure management
Manure left on pastures
Enteric fermentation

Climate Change – A Catalyst of Crisis and Food/Nutrition Insecurity

By the end of 21st century, global temperature is predicted to rise by nearly 1.4-5.8°C leading to a substantial reduction in food production. As per ISRO, the Himalayan glaciers already on retreat (shrinkage during the last 15 years: 3.75 km) may disappear by 2035. Ill effects of climate change include growing deserts and escalation in extreme weather events like droughts, cyclones, floods and droughts. Such situations often pose worst effects on the poorest of the poor (many being farmers) and are, thus, a serious threat to our goal – ending hunger by 2030! Hence, concerted action on climate change is crucial for sustainable development. Ironically, agriculture is also considered amongst the big contributors to climate change. On 2 October 2016, India has ratified the Paris Agreement which aims to combat climate change and limit global temperature rise to well below 2°C.

To quote, our Prime Minister Shri Narendra Modi Ji "The world is today worried about climate change, global warming, and natural disasters. Pandit Deen Dayal Upadhayay had understood the need for striking the fine balance between human development and the need to preserve natural resources... to is vigilant about the exploitation of natural resources. Human race has only now realised the disastrous impact of our material development on the nature".

Since a consistent increase in greenhouse gases is the major cause for climate change, it is imperative to ensure the wellbeing of ecosystems by reducing their emissions. In the context of Indian agriculture, key issues of climate change include – vastness of the nation with diverse climatic conditions; varied cropping/farming systems; excessive monsoon dependency; climate-change hampering water availability; small land holdings; lack of coping mechanisms; poor penetration of risk management strategies; extreme rainfall events (droughts/floods – esp. in coastal regions); high incidence of pests/diseases; speedy oxidation of carbon-print affecting soil fertility and extinction of biodiversity. Though, India has been successful in achieving self-sufficiency in grain production, it has not been able to address chronic household food insecurity. It is likely that climate change will exacerbate food insecurity, particularly in areas vulnerable to hunger/under-nutrition.

For our country, where a large chunk of our population is poor and nearly half the children are malnourished, ensuring food security is of utmost importance. While access to food is directly/indirectly affected via collateral effects on household/individual incomes, food utilization gets impaired due to poor access to drinking water and its adverse health effects. India is likely to be hit harder by global warming – affecting more than 1.2 billion, particularly those residing in flood/cyclone/drought prone areas. Climate change is a significant '**hungerrisk multiplier**' **which can affect all the** dimensions of food/nutrition security – Food availability, accessibility, utilization and stability.

Attaining and sustaining food security is one of the biggest challenges worldwide. Food security plans must emphasise on effective handling of threats, efficient storage/distribution of food along with suitable monitoring/surveillance according priority to corrective actions. Adaptive measures such as modified cropping patterns, innovative technologies and water conservation become rather important, particularly in arid/semi-arid areas. Therefore, necessary efforts should be directed towards carbon sequestration and mitigation of green-house gases. In this regard, there is a dire need for awareness generation and efficient involvement of the public at every step.

1. Improved community mobilization to collectively plan and undertake climate change adaptation
2. Building resilience through increased water availability and efficient water use in the region
3. Adoption of climate smart agriculture technologies and farm diversification options for climate resilient livelihoods
4. Improved potential of livestock resources as an option for livelihood stabilization in the areas
5. Knowledge generation based on field actions and wider dissemination to enhance awareness of communities and stakeholders as well as for better policy inputs

Smart Strategies for Enhanced Agriculture Resilience & Food Security under a Changing Climate in India

i. Crop Management Strategies

1. Primary and Secondary Tillage Practices

Minimum, reduced or conservation tillage technologies, in combination with planting of cover crops and green manure crops, offer substantial possibilities to reverse existing soil organic matter, soil moisture, soil erosion, and nutrient loss to combat further losses due to climate change.

2. Choice of Crops and Cultivars

Selecting the crops to cope with the most frequent weather aberrations in the region such as drought, flood and extreme events like heat wave, cold wave, frost, hail storm and untimely excess rainfall event. Based on the length of growing season and tolerance to weather aberrations including delayed onset of monsoon, the varieties/hybrids/composites have to be selected. Longer- season cultivars can be sown to provide a steadier yield under more variable conditions.

3. Resilient Cropping System

The benefits of efficient cropping systems include (a) maintaining and enhancing soil fertility as some crops are soil exhausting while others help to restore soil-fertility, (b) enhanced crop growth as crops may provide mutual benefits to each other, for example, reducing lodging, improving winter survival, or even acting as windbreaks to improve growth, (c) minimize spread of diseases, the more diverse the species of plants and the longer period before the soil is reseeded with the same crop, the more likely disease problems will be avoided, (d) control weeds, as crops planted at different times of the year have different weed species associated with them.

The research on intercropping systems (ICSs) in All India Coordinated Research Project for Dryland Agriculture (AICRPDA) revealed that additive series was the most successful under various seasonal drought situations than replacement series with multiple benefits of higher output and returns, spread labour peaks, maintenance of soil fertility (with inclusion of legume) and stability in production. The performance of ICSs were strongly correlated with the amount of seasonal rainfall, when it was the above normal, optimum productivity was achieved; under normal rainfall conditions, fairly high values of land equivalent ratios were achieved; and under low rainfall conditions, one of the two crops reasonably yielded providing an insurance against the risk due to weather aberrations (Ravindra Chary *et al.*, 2012).

4. Seed Rates and Spacing

Since the plant population influence crop-water requirements and soil-moisture extraction pattern, the seed rates and spacing, particularly between rows have to be optimized to cope with situations like delayed onset of monsoon and early season drought wherein the re-sowing is inevitable. Plant populations may also have to be adjusted after conforming the crop water requirements and or potential water-use efficiency based on observed early season rainfall.

5. Sowing/Planting Time

Normal sowing window is the period which commensurate with normal onset of monsoon and most ideal for taking up sowing/ planting operations of a particular crop. However, under this situations like delayed onset of monsoon and resowing during early season drought, the sowing operations have to be done under ideal soil moisture conditions or with protective irrigation, if available. Here, precision sowing with improved implements like ferti-cum-seed drills and other till drills is important under dryland condition to cover large area in short moisture-availability period. Shifts on sowing date of spring crops will allow more effective use of the soil-moisture content formed by snow melting. In flood-prone areas, adjusting the sowing time of crops is another strategy to avoid waster-logging conditions during critical stages. Late-maturing varieties and alteration of time of sowing to take advantage of the longer growing seasons needs to be adopted.

6. Contingency Crop Planning

Contingency crop planning refers to implementing a plan for making alternate crop or cultivar choices in tune with the actual rainfall situation and soil moisture status in a given location (Venkateswarlu *et al.*, 2011). The crops experience agricultural drought at various stages due to prolonged dry spells after sowing at normal onset of monsoon. A seasonal drought during rainy season from July to September leads to complete crop failure. Early season drought generally occurs either due to delayed onset of monsoon or due to prolonged dry spell soon after the onset of rainy season. Mid season drought occurs due to inadequate soil moisture availability between 2 successive rainfall events during the crop-growing period. Its effect varies with the crop-growth stage, duration and intensity of the drought spell. Late season or terminal droughts occur as a result of early cessation of monsoon rains and it can be anticipated to occur with greater certainty during the years with late commencement or weak monsoon activity. Effect of an early season drought in July, mid-season drought in August and the terminal drought in September due to early withdrawal of monsoon has a varying effect on the plant growth and production. Experience reveals that the early season drought is of little consequence due to less growth and leaf-area index, mid-season drought reduces the potential yield, while the terminal drought is devastating leading to heavy reduction in productivity and even crop failure. Real-time contingency planning is "any contingency measure, either technology related (land, soil, water, crop) or institutional and policy based, which is implemented based on real time weather pattern (including extreme events) in any crop growing season". The agronomic practices as contingency measures to be adopted during various weather aberrations on real time basis are given below.

a. **Delayed Onset of Monsoon (South-West):** In rainfed areas, as a general rule early sowing of crops with the onset of monsoon is the best-bet practice that gives higher realizable yield. Major crops affected due to monsoon delays are those crops that have a narrow sowing window and therefore cannot be taken up if the delay is beyond this cut-off date. Crops with wider sowing windows can still taken up till the cut-off date without major yield loss and only the change warranted could be the choice of short duration cultivars. Beyond the sowing window, choice of alternate crops or cultivars depends on the farming situation, soil, rainfall and cropping pattern in the location and extent of delay in the onset of monsoon.

b. **Early Season Drought:** Early season drought may at times result in seedling mortality needing re-sowing or may result in poor crop stand and seedling growth. Further, the duration of water availability for crop growth gets reduced due to the delayed start, and the crops suffer from an acute shortage of water during reproductive stage due to early withdrawal of monsoon. The effect of early season drought is less on the crop, because during this period sowing is carried out. Various operations carried out are primary tillage, sowing, fertilizer application and intercultural operations. Other agronomic measures include re-sowing within a week of 10 days with subsequent rains for better plant stand when germination is less than 30%, thinning in small-seeded crops, inter-culture to break soil crust and remove weeds and create soil mulch for conserving soil moisture, avoiding top dressing of fertilizers till favourable soil moisture, opening conservation furrows at 10-15m intervals, ridge and furrow across the slope for effective moisture conservation as well in as rainwater in wide spaced crops (greater than 30cms), pot watering may be taken up along with gap filling when the crop stand is less than 75% in crops like cotton, foliar spray of 2% urea during prolonged dry spells wherever ground/ surface water is available.

c. **Mid Season Drought:** Stunted growth takes place if mid-season drought occurs at vegetative phase. If it occurs at flowering or early reproductive stage, it will have an adverse effect on the ultimate crop yield. In-situ soil-moisture conservation is a vital component of dryland crop-management practices. During mid season drought plant protection, top-dressing of fertilizer, inter-cultural and supplemental irrigation are the usual practices. In case of long dry spells, crop-based production system (location) related specific contingency plans are needed. Other agronomic measures include repeated inter-culture to remove weeds and create soil mulch to conserve soil moisture, thinning, avoiding top-dressing of fertilizers until receipt of rains, opening conservation furrows for moisture conservation, foliar spray of 2% KNO_3or 2% urea solution or 1% water soluble fertilizers like 19-19-19, 20-20-20, 21-21-21 to supplement nutrition during dry spells, open alternate furrows, surface mulching with crop residues, providing supplementary irrigation (10mm depth), if available.

d. **Terminal Drought:** If there is a terminal drought, crop-management Strategies like plant protection, soil and water conservation, inter-culture, supplemental irrigation and harvesting are to be adopted. Terminal droughts are more critical as the grain yield is strongly related to water availability during the reproductive stage. Further these conditions are often associated with an increase in ambient temperatures leading to forced maturity. The agronomic measure include providing life- saving or supplemental irrigation, if available, from harvested pond water or other sources, harvesting crop at physiological maturity with some realizable yield or harvest for fodder and prepare for winter (rabi) sowing in double- cropped areas. Ratoon maize or pearl-millet or adopt relay crops as chickpea, safflower, rabisorghum and sunflower with minimum tillage after soybean in medium to deep black soils in Maharashtra or take up contingency crops (horse-gram/cowpea) or dual-purpose forage crops on receipt of showers under receding soil moisture conditions.

e. **High Intense Rainfall Events:** Suggested contingency measures include re-sowing, providing surface drainage, application of hormones/nutrient sprays to prevent flower drop or promote quick flowering/fruiting and plant-protection measures against pest/ disease outbreaks with need-based prophylactic/curative interventions. At crop maturity stage suggested measures include prevention of seed germination and harvesting of produce. If untimely rains occur at vegetative stage, the contingency measures include: draining out the excess water as early as possible, application of 20kg N+ 10kg K/acre (0.4 ha) after draining excess water, application of 50kg urea+50kg mutriate of potash (MOP)/acre (0.4ha) after draining excess water, gap filling either with available nursery or by splitting the tillers from the surviving hills in rice, weed control, suitable plant protection measures in anticipation of pest and disease out breaks, foliar spray with 1% KNO_3 or water-soluble fertilizers like 19-19-19, 20-20-20, 21-21-21 at 1% to support nutrition, need-based fungicidal spray with COC 0.3% or carbendazim 0.1% or mancozeb 0.25% 2 to 3 times by rotating the chemicals, inter-culture at optimum soil-moisture condition to loosen and aerate the soil and to control weeds, earthing up the crop for anchorage etc.

f. **Floods:** Crop/field management depends on nature of material (sand or silt) deposited during floods. In sand-deposited crop fields/fallows, ameliorative measures include early removal of ploughing in of sand (depending on the extent of deposit) for facilitating Rabi crop or next kharif. In silt-deposited indo-gangetic plains, early Rabi crop plan are suggested in currect cropped areas and current fallow lands. Other methods include draining out of stagnant water and strengthening of field bunds etc. In diara (flood prone) land areas, measures include alternate crop plans for receding situations. Usually rice cropped areas are flood prone causing loss of nurseries, delayed transplanting or damage to the already transplanted fields etc. Suggested measures include community nursery raising, scheduling bushenings, re-transplanting in damaged fields and transplanting new areas or direct seeding depending on seed availability so that the season is not lost.Other steps include prevention of pre-mature germinationof submerged crop at maturity or of harvested produce by spray of salt solution.

7. Nutrient Management

In arid and semi-arid regions where frequent dry spells and droughts are common, relying totally on basal soil application of nutrients can be risky. It is better to try for about 50-60% of the nutrients to be supplied through basal application and the balance through foliar spray. In conditions of lacking soil moisture, plants are unable to get optimal amount of nutrients which has negativeeffects on the overall condition of plants, especially their growth and grain quality. Therefore, it is beneficial to supply selected plant nutrients with or without plant growth substances for fast recovery of stressedplants. For example, pulses are raised under rainfed and rice fallow conditions without fertilizers and irrigation, their yields are poor. So to augment their production, foliar application of nutrients can be done. Also the time gap between the maximum vegetative and reproductive stage in pulses is less. To avert all these problems, foliar nutrition is a better tool in nutrient management of pulses. Foliar spray of 2% di-ammonium phosphate (DAP) in rainfed and rice fallow pulses.

8. Weed Management

Timely weed management is essential during early and mid-season droughts with appropriate bullock- or tractor drawn implements or power weeders. Preventive measures include cultural practices like stale seedbed method, manipulation of time of sowing in such a way ecological conditions for the germination of weed seeds are not met, mulching or covering with dead or live mulch, selection of competitive crop varieties, crop diversification, crop substitution and crop cultivation of weed-smothering crops, land levelling, conservation tillage, chemical weed control and integrated approach with effective, dependable and workable and economical weed-management practices.

9. Water Management Strategies

The water and agriculture sectors are likely to be most sensitive and hence vulnerable to climate change induced impacts in arid and semi-arid tropical regions. Since climate change will result in increased frequencies of extreme events (droughts, cyclones, floods), and higher rainfall variability in terms of time, space and amounts, a potential adaption measure would be to secure water availability for crop and livestock production. One way of doing this is to harvest rainwater and runoff. Irrigation requirement in arid and semi-arid regions is estimated to increase by 10% by every 1° C rise in temperature. About 74%of annual rainfall in India is received during south-west monsoon (June to September). Any aberrations during south-west monsoon such as delay in onset, in season drought and early withdrawal, strongly influence the crop productivity. Conservation of water used for irrigated agriculture therefore should be given priority attention. With increased evapo-transpiration, any adoption strategy in agriculture should be oriented toward a shift from conventional crops to types of agriculture that are not vulnerable to evapo-transpiration. Increasing efforts should be directed towards rainwater harvesting and other water- conserving practices to slow the decline in water levels in aquifers. Recycling of waste water should be encouraged in drought-prone countries in tropical regions. There is necessity to adapt to changing hydrological regimes by changing crops. Improvements in runoff management and irrigation technology (e.g. river runoff control by reservoirs, water transfers, and land-conservation practices) will be crucial as it is dependent upon the availability of water supplies.

Under dryland condition, water is the key factor for successful crop production. Therefore, efforts have to be made to harvest rainwater both in-situ and ex- situ, as much as possible, store and to be used efficiently to cope with agricultural drought.

In-situ moisture-conservation practices vary with rainfall situation, soil type, physiography and planting patterns, i.e., spacing/row ratios. These practices include off-season tillage, sowing across the slope, field bunding, broadbed furrow system, raised bed and sunken system, ridge and furrow system, compartmental bunding, inter-plot rainwater harvesting etc.

Rainwater harvesting for crop production is an old-farm-management technology that is being re-examined due to its potential to address climate change impacts through stabilizing on-farm water supply. Through harvesting rainwater, the destructive effects of runoff are reduced thereby contributing to environmental protection. Reduced soil erosion means that the silt-load of rivers are reduced, thereby enhancing river flow, protecting river fauna and helping to keep the costs of water purification for urban consumers low. Because of the harvested water, the range of crops that can be grown is expanded, thereby contributing to the diversity of the cropping system, and by extension to resilience. In the absence of water or under conditions of increased rainfall variability due to climate change, irrigation of crops from rivers, lakes and shallow groundwater offer opportunities for adoption to climate change. Runoff water harvesting in structures like farm ponds, percolation tanks, check dams etc need to be emphasized. The design and size of the structures vary with the catchment- command area- rainfall-runoff relationship and soil type.

Contributionto Enhancing Food Security, Resilience and Productivity in a Sustainable Manner

- **Productivity/food Security:** Crop productivity can be increased through the breeding of higher yielding crop varieties, though crop and crop nutrient managementand through the choice of crop species that have higher yield potentials under given environmental conditions.
- **Resilience/Adaptation through Short-Term Risk Management:** Crop breeding for greater drought tolerance or the choice of earlier maturing varieties for 'terminal drought escape' provide substantially reduced risk of yield reduction or crop failure. Similarly, many outbreaks of crop pest and diseases are 'triggered' by single or a combination of weather events. Breeding fo resistance to such pests and diseases provides an important source

of climate risk reduction. Plant breeding for drought, pest and disease resistance will increase in importance since the risk of drought and heat is projected to increase in many regions and the distribution and severity of pest and disease outbreaks will also change as climates change (FAO 2008). Other strategies include making incremental changes to cropping systems which help cope with impact such as water stress.

- **Resilience/Adaptation Through Longer-Term Risk Management:** Longer-term adaptation by farmers will become necessary through the planting of more heat tolerant crop varieties or by changing the crop species they grow to those which can tolerate higher temperatures and the greater risk of drought. For example, dryland cereals like millets and sorghum are the hardiest, resilient and climate adaptable crops for harsh, hot and dry environments (ICRISAT 2014). Farmers who currently rely on maize may well have to switch to these alternative cereals in the future (ICRISAT 2015). Farmers growing water intensive crops such as rice and wheat may shift to maize. Another adaptation strategy is the substitution of potentially vulnerable annual crops with more hardy perennials. Furthermore, in regions which are already marginal for annual crop production, farmers may well have to adapt more radically by abandoning cropping altogether for livestock production (Jones and Thornton 2008; Thierfelder *et al*., 2014). Another strategy, which is likely to be the most successful, is the integrative diversification and intensification of farming systems, moving them from mono-cropping or not integrated cropping systems into a diverse, highly integrated, both in time and space, mixed crop-livestockagroforestry system. This strategy is synergistically complementary to the use of better-adapted varieties or novel crop species.
- **Mitigation Co-Benefits:** The mitigation potential of crop production largely stems from the soil and/or water management or agroforestry system under which they are grown. However, perennial crops with life spans of 10 to 20 years are able to sequester greater amounts of carbon below ground in their root systems than annual crops. Inevitably, fertilizer inputs will have to increase in regions in order to address current and future food security needs in a changing climate (Sanchez 2014). A key challenge will be to identify practices that will narrow the yield gap and also have a low GHG emissions intensity (Bellarby *et al*., 2014).

Climate-Smart Agricultural Production Systems: Relevant Practices

The production, processing and marketing of agricultural goods are central to food security and economic growth. Products derived from plants and animals include foods, fibers, fuels and raw materials and inputs for other industries. Production has been achieved through a number of production systems which range from smallholders mixed cropping and livestock systems to intensive farming practices such as large monocultures and intensive livestock rearing. The overall efficiency, resilience, adaptive capacity and mitigation potential of the production systems can be enhanced through improving its various components, some of the key ones are highlighted here to illustrate the feasibility and constraints of developing CSA. Two important manifestations of such interventions could be reduced deforestation and lesser encroachment of land systems for agriculture purposes. Other key issues, such as access to markets, inputs, knowledge, finances and issues related to land tenure are also fundamental for ensuring food security. These issues are also highlighted here:

Soil and Nutrient Management

The availability of nitrogen and other nutrients is essential to increase yields. This can be done through composting manure and crop residues, more precise matching of nutrients with plant needs, controlled release and deep placement technologies or using legumes for natural nitrogen fixation. Using methods and practices that increase organic nutrients inputs, retention and use are therefore fundamental and reduce the need for synthetic fertilizers which, due to cost and access, are often unavailable to smallholders and, through their production and transport, contribute to GHG emissions.

Water Harvesting and use

Improved water harvesting and retention and water-use efficiency are fundamental for increasing production and addressing increasing irregularity of rainfall patterns. Today, irrigation is practiced on 20 % of the agricultural land in developing countries but can generate 130 % more yields than rain-fed systems. The expansion of efficient management technologies and methods, especially those relevant to smallholders, is fundamental.

Pest and Disease Control

There is evidence that climate change is altering the distribution, incidence and intensity of animal and plant pests and diseases as well as invasive alien species. The recent emergence in several regions of multi-virulent, aggressive strains of wheat yellow rust adapted to high temperatures is a good indication of the risks associated with pathogen adaptation to climate change. These new aggressive strains have spread at unprecedented speed in five continents resulting in epidemics in new cropping areas, previously not favorable for yellow rust and where well-adapted, resistant varieties are not available.

Resilient Ecosystems

Improving ecosystem management and biodiversity can provide a number of ecosystem services, which can lead to more resilient, productive and sustainable systems that may also contribute to reducing or removing GHG. Services include, control of pests and disease, regulation of microclimate, decomposition of wastes, regulating nutrients cycles and crop pollination. Enabling and enhancing the provision of such services can be achieved through the adoption of different natural resource management and production practices.

Genetic Resources

Genetic make-up determines a plants and animals tolerance to shocks such as temperature extremes, salts, drought, flooding and pests and diseases. It also regulates the length of growing season/production cycle and the response to inputs such as fertilizer, water and feed. The preservation of genetic resources (establishing gene banks, genetic engineering) of crops and breeds and their wild relatives is therefore fundamental in developing resilience to shocks, improving the efficient use of resources, shortening production cycles and generating higher yields per area of land. Generating varieties and breeds, which are tailored to ecosystems and the needs of farmers, is crucial.

Harvesting, Processing and Supply Chains

Efficient harvesting and early transformation of agricultural produce can reduce post-harvest losses and preserve food quantity, quality and nutritional value of the product. It also ensures better use of co-products and by-products, either as feed for livestock, to produce renewable energy in integrated systems or to improve soil fertility. As supply chains become longer and more complex, it becomes ever more important to increase the operational efficiency of processing, packaging, storage, transport, etc. to ensure increased shelf life, retain quality and reduce carbon footprints. Food processing allows surplus to be stored for low production years and allows a staggered sale. This ensures availability of food and income throughout the season and in years and low production. Food processing creates job and income opportunities, especially for women.

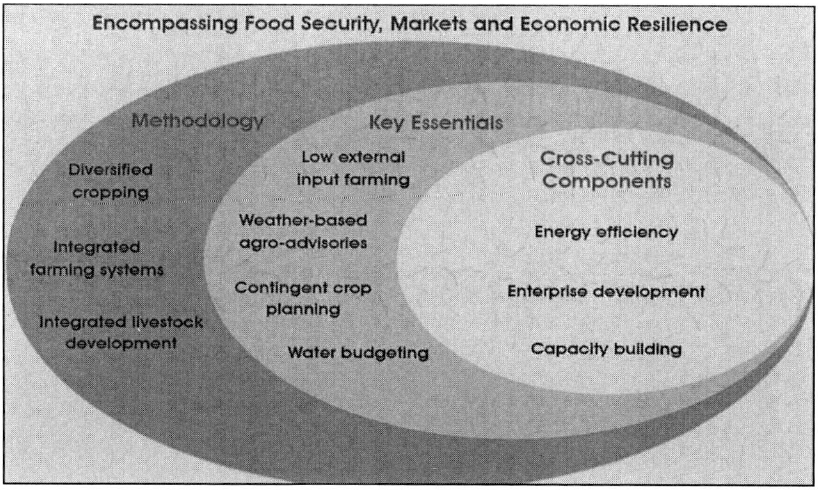

Climate-Resilient Agriculture: An Approach

Climate-Smart Adaptive Sustainable Agriculture is the way forward

- Stabilisation and management of the natural resource base (land, water, and biodiversity) is important for all, particularly the semi-arid and rain-fed, regions. An ecosystems-based approach to Participatory Watershed Management as a central point of activity is essential for building the adaptive capacities to climate change.
- Assessing Vulnerability of a cluster of villages/sub-region to climate change is essential for developing a road map for building locale-specific resilience of the people and their land to varying weather extremes.
- It is imperative that we integrate a package of climate-smart agriculture practices into ongoing programmes which includes weather-based localespecific agro-advisories, contingent crop planning, promotion of low-externalinput technology, water budgeting, livelihood diversification, and promotion of local agro-biodiversity. These, together, would build the resilience of the farming community, while simultaneously improving the quality of the resource base.
- Reduction of wastage of agriculture produce from farm to plate, especially during the post-harvest phase, will greatly contribute to conservation of precious resources (water, energy, soil health, etc.), while decreasing the carbon footprint of agriculture. Decentralised good quality storage facilities would greatly reduce losses.

Simultaneously, the promotion of village/household level storage facilities and value-addition would protect the income of small-holder producers.

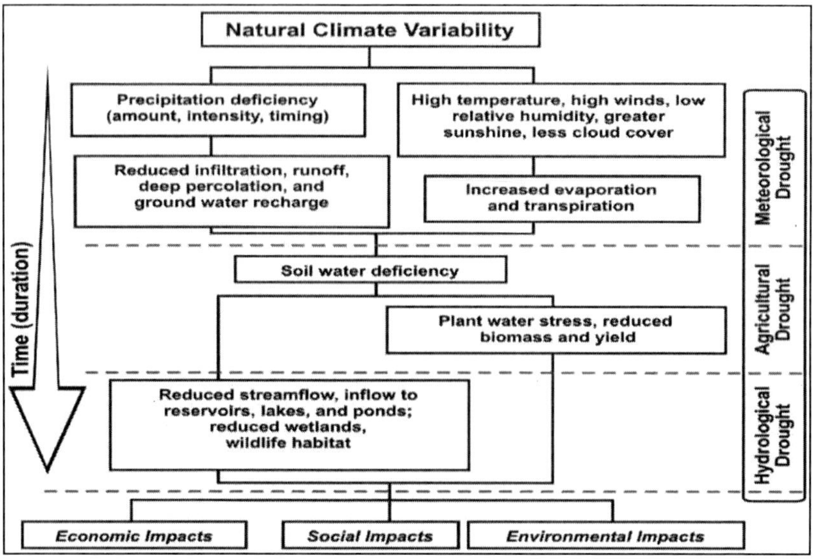

Types of droughts and related impacts over time

References

1. Aggarwal PK. Global Climate Change and Indian Agriculture: Case studies from the ICAR Network Project. ICAR, New Delhi, 2009, 148.
2. Bellarby J, Stirling C, Vetter SH, Kassie M, *et al.*, Identifying secure and low carbon food production practices: A case study in Kenya and Ethiopia. Agri Ecosyst Environ. 2014; 197:137-146.
3. FAO. Climate related transboundary pests and diseases. Technical Background Document for the Expert Consultation. 25 to 27 Feb. Rome: FAO, 2008.
4. FAO. Scaling-up Conservation Agriculture in Africa: Strategy and Approaches. Addis Ababa, Ethiopia: FAO Subregional Office for Eastern Africa, 2009.
5. FAO. Climate-Smart Agriculture Sourcebook. Rome: FAO, 2013a.
6. FAO. Climate change guidelines for forest managers. FAO Forestry, 2013b, 172.
7. FAO. State of the World's Forests. Rome: FAO, 2014, 2014a.

8. Goswami BN. Increasing trend of extreme rain events and possibility of extremes of seasonal mean Indian monsoon in a warming world, 2006. (http://saarcsdmc.nic.in/ pdf/ workshops/ Kathmandu /pres 16.pdf).

9. ICRISAT. Millets and sorghum: Climate-smart grains for a warmer world. CGIAR Development Dialogues. Montpellier, France: CGIAR, 2014.

10. ICRISAT. Go for sorghum, say climate smart Kenyan farmers. ICRISAT Happenings In-house Newsletter no. 1660. Telangana, India: ICRISAT, 2015.

11. Jones PG, Thornton PK. Croppers to livestock keepers: livelihood transitions to 2050 in Africa due to climate change. Environ Sci Policy. 2008; 12(4):427-437.

12. Krishna Kumar. Impact of Climate Change on India's monsoon climate and development of high resolution climate change scenarios for India. Presented at MoEF, New Delhi on October 14, 2009.

13. Lal M, Nozawa T, Emori S, Harasawa H *et al.*, Future Climate Change: Implications for Indian summermonsoon and its variability. Curr Sci. 2001; 81:1196-1207.

14. Mohamed Behnassi, Mohamed Boussaid, R Gopichandran. Achieving Food Security in a Changing Climate: The Potential of Climate-Smart Agriculture, 2014, 27-42.

15. Sanchez PA. En route to plentiful food production in Africa. Nature Plants. 2014; 1:1-2.

16. Thierfelder C, Mutenje M, Mujeyi A, Mupangwa W. Where is the limit? Lessons learned from long-term conservation agriculture research in Zimuto Communal Area, Zimbabwe. Food Security. 2014; 7:15–31.